High-Tech Concrete Materials

Advanced Chemical Products and Materials

High-Tech Concrete Materials: Design, Preparation, and Applications, First Edition, 2026. Fazhou Wang and Shuguang Hu

High-Tech Concrete Materials

Design, Preparation, and Applications

Fazhou Wang
Shuguang Hu

WILEY VCH

CIP

Authors

Prof. Fazhou Wang
Wuhan University of Technology
Wuhan, 430070
China

Prof. Shuguang Hu
Wuhan University of Technology
Wuhan, 430070
China

Cover Image: © Yuichiro Chino/Getty Images

Typesetting: Straive, Chennai, India
Printing and Binding: CPI Group (UK) Ltd, Croydon, CR0 4YY

Library of Congress Card No.: applied for

British Library Cataloguing-in-Publication Data A catalogue record for this book is available from the British Library.

Bibliographic information published by the Deutsche Nationalbibliothek
The Deutsche Nationalbibliothek lists this publication in the Deutsche Nationalbibliografie; detailed bibliographic data are available on the Internet at <http://dnb.d-nb.de>.

Print ISBN: 978-3-527-35355-2
ePDF ISBN: 978-3-527-84598-9
ePub ISBN: 978-3-527-84599-6
oBook ISBN: 978-3-527-84600-9

Published by WILEY-VCH GmbH, Boschstr. 12, 69469 Weinheim, Germany, under exclusive license granted by Chemical Industry Press Co., Ltd.

C9783527353552_020326

The manufacturer's authorized representative according to the EU General Product Safety Regulation is WILEY-VCH GmbH, Boschstr. 12, 69469 Weinheim, Germany, e-mail: Product_Safety@wiley.com.

Contents

About the Authors

Prof. Fazhou Wang currently serves as President of Wuhan University of Technology and Director of the State Key Laboratory of Silicate Materials for Architectures. Prof. Wang has long been dedicated to research on high-tech concrete materials and is the author of 200 SCI-indexed papers, more than 60 invention patents, and a monograph published in Chinese and Russian. He serves as Associate Editor of *Bulletin of the Chinese Ceramic Society* and is an Editorial Board Member of the *Journal of Building Materials* and the *Journal of the Chinese Ceramic Society*. Additionally, he serves as Deputy Chairman of the Chinese Ceramic Society and Rotating President of the China Concrete and Cement-Based Products Association. His research achievements have been applied in major national projects such as the Beijing–Shanghai High-Speed Railway, remote island infrastructure, and the Pinglu Canal Hub.

Prof. Shuguang Hu is a distinguished professor and doctoral supervisor at Wuhan University of Technology. He received his PhD from the same institution. His research focuses on advanced cement-based composite materials and high-tech concrete materials. He has been involved in over 30 national and provincial research projects, as well as more than 100 key engineering projects. He has received 6 National Technology Awards and has authored over 200 research SCI/EI-indexed papers and 7 books.

Preface

Cement and concrete, as the most widely used man-made materials worldwide, form the backbone of modern societal infrastructure. As contemporary engineering evolves toward larger scales, longer spans, deeper spaces, and harsher service environments, the requirements for concrete in terms of strength, performance, and durability have become more stringent. Concurrently, the complexities of material preparation and engineering practices are escalating, presenting formidable challenges for existing concrete technologies.

Since its origin, concrete technology has been characterized by relentless exploration and innovation. Particularly over the past three decades, the rapid progression of related sciences and technologies has empowered researchers to adopt advanced theories and methodologies, achieving transformative breakthroughs in its performance and preparation technologies. These advancements have provided essential technical support for pioneering engineering projects and high-standard infrastructure development in the modern era.

The unprecedented dynamism in technical concepts and the expanded freedom and scope of research have spurred significant advancements in concrete technology. At the same time, these factors have occasionally given rise to disputes over technical definitions and evaluation frameworks, which can hinder the overall efficiency of research efforts. It is, therefore, essential to systematically reflect on prior experiences, adhere to the inherent principles governing the evolution of concrete materials, foster consensus on scientific research trajectories, and elevate the overall research and development (R&D) capabilities in the field to drive forward concrete technology.

This book introduces the concept of high-tech concrete, a term not confined to performance metrics or compositional features but defined by the integration of advanced preparation technologies with superior material properties. The objective is to establish a concept that promotes shared understanding, encourages collaboration and knowledge exchange, and optimizes the overall efficiency of innovation in concrete technology. High-tech concrete encompasses high-strength concrete, high-performance concrete, ultra-high-strength concrete, ultra-high-performance concrete, as well as advanced composite structural concrete and innovative functional concrete materials.

The structure of this book is founded on the extensive research undertaken by the authors' team. It systematically addresses the principles of material science design, elucidates the critical characteristics and mechanisms of high-tech concrete, and introduces innovative methods for structural design and performance enhancement. Additionally, it consolidates innovative approaches and technological advancements in material R&D, establishing a comprehensive framework for high performance, multifunctional, eco-sustainable, and technologically advanced applications across the entire life cycle of concrete materials. The book further highlights technical developments and engineering applications in high-performance composite structural concrete and explores recent advancements in functional concrete technologies.

This work encapsulates over three decades of research and application outcomes achieved by the authors' team in the domain of high-tech concrete. Chapter 1 outlines evolution trajectory, characteristic innovations, and frontier challenges of high-tech concrete. Chapter 2 analyzes the composition and design methodologies of composite cementitious materials. Chapter 3 focuses on the functional materials in concrete including admixtures, polymers, fibers, nano-seeding, and internal curing media. Chapter 4 investigates the microstructure optimization strategies and performance enhancement mechanisms of ultra-high-strength concrete. Chapter 5 details the design philosophy and durability enhancement mechanisms of ultra-high-performance concrete. Chapter 6 examines engineering applications of high-performance composite structural concrete in mega-infrastructure projects. Chapter 7 explores innovative technologies including recyclable materials, resin aggregate, and CO_2-driven 3D printing concrete. This work consolidates findings from over a hundred graduate theses, national authorized invention patents, hundreds of research papers, and diverse critical engineering applications in landmark engineering projects.

The authors are profoundly grateful for the support provided by the Ministry of Science and Technology, the National Natural Science Foundation of China, provincial and ministerial science and technology departments, as well as numerous project owners, design institutions, and construction entities.

The primary research and editorial work for this book was led by Professors Fazhou Wang, Shuguang Hu, Qingjun Ding, Yongjia He, Linnu Lu, Congsheng Jiang, and Zhen He, with significant contributions from Drs. Zhichao Liu, Lu Yang, Yunpeng Liu, Yongjia He, Yunsheng Zhang, Peng Liu, Meijuan Rao, Wenqin Zhang, Chuanlin Hu, Jin Yang, Yunhua Zhang, and others. Special thanks are extended to the Chemical Industry and Engineering Society of China, Chemical Industry Press, and WILEY Press for their substantial support. The authors also acknowledge the valuable assistance and guidance of numerous experts and colleagues throughout the preparation of this book.

Despite our best efforts, this book may contain errors or omissions due to the authors' limitations. We warmly invite feedback and constructive suggestions from experts, peers, and readers to refine and enhance this work further.

Fazhou Wang
Wuhan, August 2025

1

Introduction

1.1 Overview

Concrete is the most widely used man-made material and a primary building material in the world today, playing an irreplaceable role. Compared to other common building materials such as metals and organic materials, concrete has low production energy consumption, wide availability of raw materials, simple preparation processes, and consequently low production costs. At the same time, it possesses characteristics such as durability, fire resistance, strong adaptability to engineering and the environment, and convenient application [1, 2].

Concrete technology has evolved from the early use of primitive natural cementing materials to the contemporary use of artificially produced cementing materials. The material composition has progressed from a simple triad of cementitious materials, aggregates (ranging from coarse to fine), and water to a diverse system with various additives and functional admixtures. The material performance design has advanced from basic single mechanical strength to a combination of ultra-high strength, high durability and functionality. The preparation technology has developed from the simplest hand mixing to the application of various advanced technologies and intelligent methods. In short, today's concrete has comprehensively entered a new stage known as high-tech concrete [3–6].

High-tech concrete refers to concrete prepared based on ordinary concrete technology using advanced material science theories and innovative manufacturing technologies. This ensures the prepared concrete exhibits superior material properties, excellent functionality, and unique, environmentally friendly characteristics. It effectively expands the application scope of traditional concrete, giving it a stronger vitality. The categorization of high-tech concrete includes high-strength concrete (HSC), high-performance concrete (HPC), ultra-high-strength concrete (UHSC), and ultra-high-performance concrete (UHPC), as well as special composite concrete.

The development of high-tech concrete began in the mid-20th century, accompanying the progress of modern society, large-scale infrastructure construction, and the expansion of human living spaces, leading to new and higher requirements for building materials. It also relies on the rapid development of science and

High-Tech Concrete Materials: Design, Preparation, and Applications, First Edition.
Fazhou Wang and Shuguang Hu.

technology in other fields, possessing a strong logical connection to the era. During this period, concrete experts worldwide actively participated in the technological innovation wave, seizing opportunities, actively exploring, and making continuous breakthroughs in various aspects of concrete, from strength and various performance metrics to raw materials and preparation methods, as well as the application of concrete materials, products, and engineering. This has continuously elevated the foundational and critical role of concrete in social development to a new and higher level. Today, we can see various iconic modern concrete projects everywhere: super-long, super-deep, and super-large underwater tunnels; various types of large-span bridges; massive underground structures; towering skyscrapers; transportation networks crossing mountains and rivers; ocean engineering in swift waves and harsh extreme environments; and special major projects. All of these contributions are attributed to high-tech concrete.

Undoubtedly, this is a proud era in the history of concrete development. However, it is also precisely because of the unprecedented activity in technological thinking and the freedom and breadth of research work that concrete has never experienced such a flourishing of technological achievements. This often leads to disputes over technical concepts and evaluation systems, affecting the overall efficiency of research work. The reasons for this are, on the one hand, the broader and more independent thinking of contemporary scientists, who are more innovative. On the other hand, this is influenced by the widespread introduction of advanced theories and technologies from other disciplines, resulting in more research methods and technical approaches. Additionally, due to the development of information technology, people are increasingly accustomed to rapid learning through online information, lacking the important research atmosphere of conferences, effective deliberation, and consensus formation. We should view this rationally and dialectically. It is a sign of social progress, broadening our horizons, quickly borrowing from advanced theories and methods, helping researchers effectively break free from traditional constraints, improving research efficiency and level, and promoting greater progress in the scientific and technological aspects of this discipline. Of course, at the same time, we should also pay attention to new problems, strengthen academic exchanges and collaborative research on common issues, share achievements, avoid detours, and advance the overall research level and scientific and technological achievements in this discipline.

1.2 Brief History of High-tech Concrete Development

1.2.1 Ordinary Concrete

The technical characteristics of ordinary concrete are a relatively simple material composition, a higher water–cement ratio (W/C), and the use of a relatively straightforward mixing and preparation technique. The compressive strength of concrete is generally less than 50 MPa. Ordinary concrete includes plain concrete and ordinary reinforced concrete.

1.2.1.1 Plain Concrete

Plain concrete refers to concrete that does not add any other supplementary components in addition to the three basic components of concrete: cement, aggregate and water. Modern Portland (silicate) cement, the world's first patented cement invented by British construction worker Joseph Aspdin in 1824, marked the beginning of modern cement, and the cost was low, especially with lower energy consumption. The advent of this hydraulic cementitious material endowed concrete with the needed strength and durability. Its raw materials were easily available. Consequently, it significantly expanded the application scope of concrete, establishing its indispensable position in construction materials. The technical characteristics of plain concrete involve using only the three basic ingredients and simple mixing and forming processes, resulting in a compressive strength generally ranging from 20 to 30 MPa after hardening.

For over a century, concrete technology has undergone continuous development and exploration. However, there have been no fundamental breakthroughs in the main performance strength during this time. In 1896, French scientist Feret first proposed a strength formula based on pore content. In 1919, American scientist Duff Abrams found through numerous experiments that there was a correlation between the water–cement ratio (W/C) and concrete strength, indicating an inverse relationship between concrete strength and W/C. In 1930, Swiss scientist Belomey studied and proposed the famous relationship formula between concrete strength, actual cement strength, and W/C under certain conditions, demonstrating that reducing the amount of water could enhance concrete strength. In the 1960s, American researcher T. C. Powers conducted more in-depth research, considering the main hydration product of cement (gel) and establishing the well-known T. C. Powers model. He proposed the relationship formula between strength and gel-to-void ratio, indicating the impact of capillary pore volume on strength. The basic conclusion of these research results is that a lower W/C can reduce the volume of voids (pores) in concrete materials, leading to a higher strength. However, for ordinary concrete, to meet the preparation process and ensure the workability of fresh concrete mixtures, it is necessary to add water significantly higher than the complete hydration of cement. Therefore, it is challenging to significantly improve the strength of concrete, and a higher W/C also makes the material structure relatively loose, affecting other properties of concrete [7].

1.2.1.2 Ordinary Reinforced Concrete

To improve the strength and enhance the performance of concrete, researchers began to experiment with introducing additional raw materials or improving the concrete preparation methods based on plain concrete. The goal was to achieve concrete with better strength and performance. These efforts paved the way for the development of modern concrete technology. A milestone achievement in this journey occurred in 1849 when French landscape architect Joseph Monier invented the original reinforced concrete, thereby enhancing the stability and durability of concrete structures. In 1879, Austrian Hatschek produced asbestos fiber concrete, marking the era of fiber-reinforced concrete and achieving breakthroughs in the crack resistance of concrete products. In 1910, H. F. Porter from the United States obtained a patent for steel fiber-reinforced concrete, improving the strength and overall performance of concrete. In 1928, Freyssinet from France created prestressed reinforced concrete. In 1934, the United States invented the concrete vibrator, effectively compacting and uniformizing concrete during the forming process, thereby enhancing its strength and impermeability. In 1937, E. W. Scxiptrit from the United States researched and patented the world's first water reducer, Pozzolitn, providing a technological approach to improving concrete strength by reducing the water–cement ratio (W/C). At the end of the 19th century and the beginning of the 20th century, polymers appeared in concrete, and it was discovered that the brittleness of concrete could be improved by adding organic materials. In the 1960s, the United States invented polymer-impregnated concrete. In the 1970s, China developed glass fiber-reinforced concrete. By employing these methods, the strength of concrete has been increased to varying degrees, reaching a basic level of approximately 40 MPa, with some approaching 50 MPa.

1.2.2 High-strength Concrete

Modern civil engineering construction demands concrete with higher strength, and the century-long exploration of ordinary concrete technology has laid the foundation. The development of the modern chemical industry and materials science and technology has made HSC a reality. In the late 1950s to the early 1960s, the advent of highly efficient water reducers with a water-reducing rate exceeding 20% marked the beginning of the development of HSC technology [8]. During this period, Germany invented melamine-based superplasticizers, and Japan developed naphthalene-based superplasticizers. These superplasticizers overcame significant technical challenges, such as the difficulty of working with dry and hard concrete mixtures due to reduced water content. They allowed concrete to be compacted under lower water–cement ratio conditions, resulting in denser material and, consequently, higher strength. The definition of HSC evolves with the progress of scientific and technological advancements over time. From a comprehensive perspective of technical classification, HSC, as defined here, has a strength greater than 50 MPa.

The main characteristics of HSC components include the incorporation of highly efficient water reducers and mineral admixtures. The significant breakthrough in concrete technology comes from the use of highly efficient water reducers. From the perspective of material structure, the reduction in water content significantly reduces voids and internal defects in hardened concrete. The reduction in defects not only enhances concrete strength but also increases material density. This, in turn, effectively hinders harmful external substances from entering the concrete through pores, thereby improving the durability of the concrete. HSC with added superplasticizers maintains good fresh concrete flowability and can be conveniently applied in on-site construction. This is one of the reasons why various large-scale concrete projects opt for HSC.

1.2.3 High-performance Concrete

HPC is a type of concrete characterized by high strength, durability, and workability. Unlike HSC, which emphasizes only mechanical properties, HPC must meet high-performance requirements in terms of workability, mechanical performance, and durability simultaneously [9]. HPC demands not only high workability and strength but also important properties such as resistance to chemical corrosion. These include high volume stability, high elastic modulus, low drying shrinkage, low creep, low-temperature deformation, and high impermeability. The technical characteristics of HPC involve the use of new-generation superplasticizers and ultra-fine materials based on HSC. Ultra-fine materials, with particle sizes <10 μm and specific surface areas equivalent to 600 m^2/kg or more, are commonly used in high-performance cement-based composite materials. Common varieties include fly ash, finely ground water-quenched slag powder, silica fume, finely ground zeolite powder, kaolin, diatomaceous earth, calcined shale, and expanded slag.

Ultra-fine powder is a key component in preparing HPC. Its mechanisms of action include: first, filling the voids of the cementitious material, reducing porosity, and refining pore size; second, enhancing the inducing activity of the mixed material, participating in the hydration reaction of the cementitious material, improving the density of the concrete, and reducing hydration heat [10]. The reduction in porosity is a direct factor in improving concrete strength and durability, while the increased activity of the mixed material further enhances the interface structure of the concrete. The reduction in porosity, refinement of pore size, and improvement of the interface structure can enhance the concrete's resistance to permeability, thereby improving its durability. Due to its superior comprehensive performance, HPC quickly found application and promotion in practical engineering. Currently, HPC with a strength of C60 is widely used in projects such as bridges, high-rise buildings, and airport construction. Concrete with strengths exceeding 80 and 100 MPa has also been applied in various engineering projects.

1.2.4 Ultra-high-strength Concrete

UHSC, defined as concrete with compressive strength exceeding 100 MPa, emerged as a research trend worldwide from the early 1980s to the early 21st century. The theoretical basis for the design of UHSC is the maximum packing density theory. According to this theory, the constituent materials, comprising particles of different sizes, form the tightest possible packing in optimal proportions. In other words, the interstices between millimeter-sized particles are filled with micrometer-sized particles, and the interstices between micrometer-sized particles are filled with sub-micrometer-sized particles, achieving the maximum packing density. To attain this state of maximum packing density, UHSC concrete is molded using special preparation techniques. The following types are representative and influential.

1.2.4.1 Macro-defect-free Cement

In 1981, Birchall and colleagues in the United Kingdom invented a cement–polymer composite material with unique properties [11, 12]. This composite material, named macro-defect-free (MDF) cement, is characterized by the absence of large pores and coarse crystals in the hardened body, attributed to specialized production techniques and an extremely low water–cement ratio (0.12–0.16). The primary components of MDF material include cement, water-soluble polymers, superplasticizers, and water, with the cement being either Portland cement or aluminate cement. Key points in the preparation process of MDF materials include the following: first, mixing various components to form a homogeneous slurry through specialized mixing equipment; second, shaping the mixture through processes such as hot pressing, extrusion, rolling, or injection.

The mechanical properties of MDF significantly surpass those of ordinary concrete. Its compressive strength ranges from 200 to 300 MPa, tensile strength from 50 to 120 MPa, flexural strength from 150 to 200 MPa, elastic modulus from 25 to 50 GPa, and fracture toughness up to 3 $MPa{\cdot}m^{1/2}$. The hardened body of the MDF also exhibits excellent sound absorption, antistatic properties, and crack resistance at low temperatures. Its main drawback lies in poor water stability, with a noticeable strength decrease after absorbing moisture, especially when alumina cement is used primarily due to the water absorption and swelling of the polymers in MDF.

Since the mid-1980s, the author has conducted in-depth and systematic research on the reinforcement mechanism of MDF materials, as well as the practicality of material design and preparation technology. This research has revealed the polymer–cement interface enhancement mechanism and the multiple roles of polymers in MDF materials [13–19]. The formation of a characteristic multi-phase interpenetrating network structure in the material has been identified and confirmed [20], leading to the proposal of criteria for polymer selection and optimization methods for material structure design. An accurate calculation and control method for polymer dosage has been introduced, effectively improving water resistance and enhancing volume stability by minimizing excess usage through polymer interface chemical reactions. Furthermore, breakthroughs have been achieved in improving the overall durability of MDF materials by reinforcing them with fibers [21–24].

1.2.4.2 Densified System Containing Homogenously Arranged Ultra-fine Particles

In 1982, Danish researchers, led by Bache, developed a densified system containing homogeneously arranged ultra-fine particles (DSP) [25]. The formulation involves two key aspects: first, the uniform mixing of cement with an average particle size of approximately 10 μm and highly reactive ultra-fine volcanic ash with an average particle size of approximately 0.1 μm in appropriate proportions, forming the tightest packing. This results in extremely small pore sizes, very low porosities, and disconnected pores. Second, the addition of high-efficiency water-reducing agents to achieve an extremely low water–cement ratio (0.16 or lower) further reduces porosity, prevents the agglomeration of cement particles and ultra-fine powders, and strives for their uniform distribution.

DSP primarily utilizes the principle of dense filling or filling effects to reduce porosity and enhance compactness. According to theoretical models of particle packing, coarse particles have a significant impact on packing density, while merely changing the size and quantity of fine particles has a minimal effect on the overall material packing density. Therefore, it is possible to consider substituting a significant portion of unhydrated cement in DSP with other supplementary materials to reduce cement consumption. Studies have shown that using fly ash, slag powder, and other particles to replace or partially replace silica fume in DSP can achieve cost reduction without compromising performance.

The compressive strength of typical DSP made with Portland cement and silica fume can exceed 120 MPa, and when high-strength aggregates are added, the compressive strength of DSP can reach over 250 MPa. One drawback of DSP is that the increase in the modulus of elasticity and fracture energy is limited, resulting in high brittleness and the likelihood of generating numerous microcracks during the hardening and drying shrinkage processes. Introducing a certain amount of fibers into DSP can enhance ductility and toughness while can also providing low shrinkage, low permeability, and high corrosion resistance.

1.2.4.3 Compact Reinforced Composite

In 1986, Bache and others developed a composite material containing a significant amount of fibers and closely spaced steel reinforcement, known as compact reinforced composite (CRC) [26]. The base material of CRC is composed of high-strength DSP, with its main reinforcement being similar to that of steel reinforcement in conventional reinforced concrete but with a much larger quantity (volume ratio ranging from 10% to 20%). Aggregates, such as quartz particles or particles with a high Al_2O_3 content, along with appropriate fibers (with a volume ratio of 5–10%, fine diameter, and high-strength steel fibers), are used to create a base material that is tightly bonded to the steel reinforcement, providing high tensile strength, stiffness, and ductility.

CRC typically achieves a compressive strength of up to 200 MPa, with some formulations exceeding 300 MPa. Its tensile strength is comparable to structural steel reinforcement and exhibits considerable ductility. In comparison to DSP, CRC not only boasts higher strength but also has significant deformation capacity. CRC exhibits exceptional durability, surpassing that of conventional HPC, primarily due to the extremely dense nature of its base material. It demonstrates high resistance to projectile penetration, making it suitable for military engineering facilities. CRC is well suited for the production of lightweight, slender precast components or components cast on-site. It has been applied in various scenarios, including the construction of tunnel lining blocks, the bottom plate of cantilevered balconies, supporting beams for stairs, manhole covers, and joints in on-site cast connections for precast slab-column systems.

1.2.4.4 Reactive Powder Concrete

In 1993, French scholars Richard and Cheyrezy developed reactive powder concrete (RPC) using the DSP model [27–29]. The main processes of RPC include the following: first, following the principle of maximum packing density by eliminating coarse aggregates and using fine aggregates with a maximum particle size of 400–600 μm, such as quartz sand, and adding high-activity mineral admixture silica fume to form a particle system with high uniformity and density; second, employing a large amount of high-efficiency water-reducing agent to significantly reduce the water content and decrease porosity; third, adding steel fibers to reduce material brittleness and improve toughness; fourth, applying pressure and deaeration before and during hardening to further enhance density; and finally, post-curing through hot pressing to improve the chemical reactivity and internal structure density.

RPC materials exhibit ultra-high strength, exceptional durability, and high toughness. It also possesses good volumetric stability, with compressive strength exceeding 150 MPa, flexural strength up to 50 MPa, elastic modulus exceeding 50 GPa, and fracture energy ranging from 20,000 to 40,000 J/mm^2. These performance indicators indicate that RPC materials have reached a new level in mechanical properties, surpassing previous concrete materials. RPC has seen significant progress in mix design methods, raw material composition, and preparation technology.

Despite its remarkable mechanical strength and excellent performance, RPC faces some challenges. First, due to the large amounts of cement and silica fume in the material composition, as well as the elimination of coarse aggregates, RPC experiences significant shrinkage. Second, the forming process requires compression and high-temperature curing, making it suitable only for factory prefabrication and not applicable to on-site concrete engineering. Third, the high cost, high-energy consumption, and complex preparation process seriously constrain the practical application of RPC.

1.2.5 Ultra-high-performance Concrete

To promote the practical application of UHSC, researchers have proposed the concept of UHPC based on UHSC [30, 31]. "Ultra-high performance" encompasses two aspects: exceptionally high durability and mechanical performance, with a crucial emphasis on practical applicability. When in a freshly mixed state, the material should exhibit good flowability for ease of preparation and application. Therefore, the key focus of UHPC technology development lies in simplifying the preparation process and reducing material costs. Although a unified definition has not yet been established, it is clear that the material needs to possess outstanding physical and mechanical properties, with compressive strength generally reaching or exceeding 150 MPa.

The technical characteristics of UHPC include the following: first, the utilization of a multicomponent binder system, incorporating industrial waste superfine powder to achieve energy savings and cost reduction without compromising performance. Second, the incorporation of coarse aggregates with a size of 10–20 mm enhances the material's volume stability and durability while effectively lowering costs. Third, the use of superplasticizers imparts flowability to the concrete, facilitating on-site construction for engineering projects. Fourth, the preparation of UHPC involves conventional techniques, eliminating the need for pressure molding and high-temperature curing. The research conducted by the author's team primarily focuses on optimizing material composition and understanding the formation of material structures [32–36]. UHPC can be considered the most durable engineering material, with mechanical properties approaching those of steel structures when appropriately reinforced. Additionally, it exhibits excellent wear resistance and blast resistance. Therefore, UHPC is particularly suitable for use in large-span bridges, blast-resistant structures, thin-walled structures, and environments with high abrasion and corrosion.

1.2.6 High-performance Composite Structural Concrete and Innovative Functional Concrete

High-performance composite structural concrete and innovative functional concrete are based on ordinary concrete, introducing components that significantly enhance and improve concrete properties or confer special functions. These utilize composite technology methods in material design, employing advanced manufacturing processes to optimize the material's structure, resulting in superior performance and unique functionalities. Based on their application characteristics, they can be classified into structural and functional types, with the former commonly used in various structural engineering projects. The authors and their team have dedicated years to the research of high-performance composite structural concrete and innovative functional concrete technology, achieving the following outcomes.

1) Utilizing various fibers, particles, and additives to enhance the strength, impact resistance, and crack resistance of concrete materials, such as high-strength, high-penetration resistance, blast-resistant concrete, and grout fiber concrete [37–43].
2) Using various special aggregates to give concrete specific functions, such as radiation-resistant concrete, lightweight aggregate concrete, abrasion-resistant concrete, and heat storage concrete [44–49].
3) Employing supplementary cementitious materials or composite cementitious material systems to improve the properties of the concrete matrix, such as UHSC and HPC, shrinkage-compensating concrete, high-permeability crack-resistant concrete, and large-volume low-heat and highly durable concrete, and so on [50–55].
4) Material composite and structural composite technology can be applied to enhance the structural performance of concrete, including concrete-filled steel tube composite materials, bridge deck pavement concrete, ultra-high structures with large spans, tunnel structural concrete, and highly durable concrete [35, 56–60].
5) Using various organic components to improve the toughness of concrete materials, such as high-ductility concrete, polymer-modified concrete, cement asphalt mortar, and so on [61–66].

6) Employing additional components to alter the material structure and confer new uses to concrete, such as resin aggregate concrete, CO_2 driven 3D printing concrete, and other innovative concrete technologies [67–72].

1.3 Challenges and Opportunities for Concrete Materials

Concrete, steel, and polymer materials are the most important representatives of the three main materials: inorganic non-metallic, metallic, and organic. These three major types of materials have their own properties. Table 1.1 shows their performance characteristics [6, 72].

As shown in Table 1.1, there are significant differences in the engineering performance of the three materials. Steel exhibits high strength, with elevated tensile, flexural, and shear strengths. It possesses good ductility, allowing for substantial plastic deformation and various cold-working processes such as bending, stretching, drawing, and stamping at room temperature. Additionally, steel is characterized by uniform quality and reliable performance. However, its primary drawback lies in susceptibility to corrosion, leading to poor durability. Polymer materials feature characteristics such as light weight, high elasticity, low elastic modulus, high insulation, and low thermal conductivity. Nevertheless, their disadvantages include lower strength compared to steel, poor rigidity, high thermal expansion, low heat resistance, and susceptibility to aging.

Table 1.1 Comparison of properties of concrete, steel, and polymer materials.

Performance	Steel	Polymers	Ordinary concrete	High-tech concrete
Compressive strength (MPa)	235–1600	105–400	10–50	100–300
Flexural strength (MPa)	~500	60–150	3–6	50–200
Tensile strength (MPa)	400–1900	46–200	1.27–3.11	4.5–24
Flexural/compression ratio	0.23–2.13	0.25–1.10	0.15–0.31	0.20–0.50
Elastic modulus (GPa)	190–210	0.10–2.00	22–38	37–55
Toughness	Excellent	Excellent	Poor	250 times for ordinary concrete
Poisson's ratio	0.25–0.30	0.30–0.50	0.20	0.19–0.24
Density (kg/m^3)	~7800	900–1500	1900–2500	~2500
Coefficient of thermal expansion (10^{-6}/°C)	10–15	72–150	~10	~10
Thermal conductivity [W/(m·K)]	48–52	0.08–0.50	~3	1–30
Bulk resistivity (Ω·cm)	10–15	10^{16}–10^{18}	10^{8}–10^{15}	10^{9}–10^{11}
Fracture energy (J/m^2)	10^{5}–10^{6}	600–1500	110–160	10^{4}~4×10^{4}

Note: The data presented in Table 1.1 for high-tech concrete are representative indicators and are not taken from a specific material.

Ordinary concrete materials have the lowest strength, but within a certain range, their strength can be designed and controlled from low to high to meet the varied strength performance requirements of different projects. Concrete mixtures exhibit plasticity, allowing for simple and convenient forming processes. They can be cast into various shapes and sizes of integral structures or prefabricated components as per engineering requirements, making their processing much easier than that of other materials. However, due to its low tensile strength and brittleness, concrete typically cannot withstand tensile stress and has low impact resistance and a relatively low strength-to-mass ratio. Moreover, the volume instability of concrete is a significant drawback.

From the above comparative analysis, as the most widely used artificially prepared material globally, concrete has made significant contributions to societal civilization and technological progress. However, with societal development and technological advancements, there is an increasing demand for higher performance in various concrete products. Modern engineering construction is evolving toward larger-scale, longer-span structures, deeper underground spaces, and more complex and harsh environments. These major construction projects pose challenges with high difficulty, requiring high strength and durability. The material properties of ordinary concrete and its inherent limitations are no longer able to meet the demands of this progressive era. Concurrently, rapid advancements in other material technologies pose additional competition in traditional application areas for concrete.

Nevertheless, it is heartening to observe, from the innovative development and achievements in concrete technology (as indicated in Table 1.1), that high-tech concrete materials produced using advanced techniques have made significant progress in many aspects. Some properties and functionalities have even experienced substantial breakthroughs, indicating substantial development potential in concrete materials. On the other hand, the ongoing progress in science and technology, coupled with developments in relevant disciplines and interdisciplinary fields, provides a broader perspective and valuable experiences for breakthroughs in concrete material technology. Importantly, the foreseeable future promises unprecedented opportunities for concrete materials due to China's rapid and high-quality economic development, extensive infrastructure construction, and urbanization. Therefore, building upon existing work, carefully summarizing past experiences, analyzing and clarifying development strategies, selecting and establishing research objectives, following scientifically effective research methods, and further improving research and development efficiency to achieve new breakthroughs in concrete technology are not only crucial tasks for scientists and technologists in this field but also significant opportunities for the innovative development of concrete materials.

1.4 Technical Characteristics and Research Content of High-tech Concrete

The concept of high-tech concrete (ATC) proposed by the author differs from previous definitions based on concrete performance indicators or compositional features. Instead, it is defined based on the advanced nature of concrete preparation technology and material performance, utilizing a comprehensive evaluation method. The concept of high-tech concrete is introduced from the perspective of the technical characteristics of concrete. Its purpose is to establish a more scientific and consensus-forming concrete technology system and evaluation standards. This approach aims to avoid disputes arising from different definitions of certain concrete technologies, facilitating consensus formation and enhancing the efficiency and level of innovation in concrete technology. The categorization system of high-tech concrete materials includes HSC, HPC, UHSC, UHPC, and special composite concrete. The technical characteristics and research content of high-tech concrete encompass aspects such as high performance, multifunctionality, full ecological integration, and the incorporation of new technologies.

1) High Performance: Researchers can adopt the comprehensive technological innovation to enhance various chemical, physicochemical, and mechanical properties of concrete, making it more conducive to production, structural construction, and possessing superior material and engineering performance. This further enhances the safety and durability of concrete engineering, extending its service life.
2) Multifunctionality: Researchers can adopt the material design innovation to adjust the composition of concrete materials and forming corresponding performance structures, the characteristic functions of concrete can be expanded to meet the needs of various special purposes. In addition, they can adopt the innovative composite technologies, exploring and developing composite structures of concrete with other materials, creating material composite effects, overcoming inherent shortcomings, and achieving new breakthroughs in expanding application areas.
3) Full Ecologization: Researchers can adopt the material chemistry innovation, developing and utilizing various industrial and domestic waste as substitutes for the components of ordinary concrete, achieving the goals of resource conservation, energy efficiency, and environmental friendliness in concrete production. Through innovative design concepts, developing concrete from material selection to manufacturing, use, and recycling, incorporating materials into a fully life-cycle recyclable design and preparation technology.
4) Adoption of new technologies: Researchers can adopt the interdisciplinary innovation, learning from and applying technological methods and experiences from other fields, developing new designs, new raw materials, and new manufacturing methods, seeking revolutionary leaps in the properties and functions of concrete materials. In addition, they can adopt the intelligent innovation, developing smart concrete with self-diagnostic, self-repair, self-organization, and self-adjustment functions. Ultimately, they will integrate information technology innovation, developing big data, digitalization, and networking technologies, endowing concrete with new qualities for the modern era.

References

1 Mindess, S., Young, J.F., and Darwin, D. (2005). *Cement*. Beijing: Chemical Industry Press.
2 Mehta, P. and Monteiro, P. (2005). *Concrete: Microstructure, Properties, and Materials*. New York: The McGraw Hill Companies Inc.
3 Jahren, P. and Sui, T. (2013). *Concrete and Sustainability*. Beijing: Chemical Industry Press.
4 Zhongjin, L. (2011). *Advanced Concrete Technology*. Hoboken: Wiley.
5 Hu, S. and Yang, D. (1995). Optimizing design of high-tech concrete composites. New Development in Concrete Science & Technology. Southeast University Press, Nanjing, (in Chinese).
6 Hu, S. (2009). *Advanced Cement-Based Composite Materials*. Beijing: Science Press.
7 Li, F.M. (1984). *Cement and Concrete Chemistry*. Beijing: China Architecture Publishing House (in Chinese).
8 Miao, C.W. (2008). *High Performance Admixture for Concrete*. Beijing: Chemical Industry Press (in Chinese).
9 Wu, Z. and Lian, H. (1999). *High Performance Concrete*. Beijing: China Railway Publishing House (in Chinese).
10 Steven, H.K., Beatrix, K., and Willliam, C.P. (2009). *Design and Control of Concrete Mixtures*. Beijing: Science Press.
11 Birchall, J.D., Howard, A.J., and Kendall, K. (1981). Flexural strength and porosity of cements. *Nature* 289 (5796): 388–390.
12 Birchall, J., Kendall, K., and Howard Anthony, J. (1981, January 07). Cementitious product. EP 0021682.
13 Hu, S., Guan, X., and Ding, Q. (2002). Research on optimizing components of microfine high-performance composite cementitious materials. *Cement and Concrete Research* 32 (12): 1871–1875. (in Chinese).
14 Peng, Y., Hu, S., and Ding, Q. (2009). Dense packing properties of mineral admixtures in cementitious material. *Particuology* 7 (5): 399–402. (in Chinese).
15 Ding, Q., Ye, Q., and Hu, S. (2018, April 10). An ultra-high performance concrete prepared by wet granite powder. CN 201711192782. (in Chinese).
16 Hu, S. (1994). XPS nondestructive depth analysis method and its application in cement based composite materials. *Cement and Concrete Research* 24 (8): 1509–1514. (in Chinese).
17 Hu, S. (1993). An investigation of bonding behavior on the interface of polyacrylamide-aluminous composites. *Journal of Wuhan University of Technology (Materials Science Edition)* 8 (2): 19–25. (in Chinese).
18 Hu, S. (1993). Structure analysis of polymer-cement interface bonding layer. *Journal of Wuhan University of Technology* (04): 12–17. (in Chinese).
19 Hu, S. (1994). Kinetic analysis of the hydration of $3CaO\text{-}3Al_2O_3\text{-}CaSO_4$ and the effect of adding $NaNO_3$. *Thermochemica Acta* 246: 129–140. (in Chinese).

20 Hu, S. and Yang, D. (1995). IPCN structure and its formation condition in MDF cement. New Development in Concrete Science & Technology. Southeast University Press, Nanjing, (in Chinese).

21 Ding, Q., Hu, S., and Li, Y. (1997). Research on fiber reinforced MDF cement Materials. *Concrete and Cement Products* 1997 (3): 14–17. (in Chinese).

22 Hu, S. (1996). Studies of MDF cement reinforced with glass fiber. MAETA Workshop on High Flexural Polymer-Cement Composite. Sakata, Japan, (in Chinese).

23 Ding, Q., Li, Y., and Hu, S. (1998). Research on carbon fiber reinforcement of MDF cement. *Journal of Wuhan University of Technology* 20 (3): 14–17. (in Chinese).

24 Li, B. (1998). *Research on Moisture Sensitivity and Modification of Cementitious Composite Materials Without Macroscopic Defects*. Wuhan: Wuhan University of Technology (in Chinese).

25 Bache, H. (1981). Densified cement ultra-fine particle-based materials.

26 Bache, H. (1989). Fracture mechanics in integrated design of new, ultra-strong materials and structures. Fracture Mechanics of Concrete Structures: From Theory to Applications. Chapman & Hall, London.

27 Richard, P. and Cheyrezy, M. (1995). Composition of reactive powder concretes. *Cement and Concrete Research* 25 (7): 1501–1511.

28 Richard, P. and Cheyrezy, M. (1989). Reactive powder concretes with high ductility and 200-800 MPa compressive strength. *ACI Symposium Publication* 144: 507–518.

29 Cheyrezy, M., Maret, V., and Frouin, L. (1995). Microstructural analysis of RPC (reactive powder concrete). *Cement and Concrete Research* 25 (7): 1491–1500.

30 Zhang, Y., Zhang, W., and Liu, J. (2005). *Ultra-High Performance Cementitious Composites*. Beijing: Chemical Industry Press (in Chinese).

31 Guan, X., Hu, S., and Ding, Q. (2022). Research on ultrafine high performance cement. The 5th International Symposium on Cement and Concrete. Tongji University Press, Shanghai, (in Chinese).

32 Hu, S. and Li, Y. (1999). Research on the hydration, hardening mechanism, and microstructure of high performance expansive concrete. *Cement and Concrete Research* 29 (7): 1013–1017. (in Chinese).

33 Ding, Q., Liu, X., and Liu, Y. (2016). Effect of curing regimes on microstructure of ultra high performance concrete cement pastes. First International Conference on UHPC Materials and Structures. RILEM Publications S.A.Rl.L., Paris, (in Chinese).

34 Zhang, X., Liu, Z., and Wang, F. (2019). Autogenous shrinkage behavior of ultra-high performance concrete. *Construction and Building Materials* 226: 459–468. (in Chinese).

35 Hu, S., Wu, J., Yang, W. et al. (2010). Relationship between autogenous deformation and internal relative humidity of high-strength expansive concrete. *Journal of Wuhan University of Technology (Materials Science Edition)* 25 (3): 504–508. (in Chinese).

36 Ding, Q., Yan, P., and Hu, S. (2018, April 20). A lightweight low shrinkage ultra-high performance concrete and its preparation method. CN 201711233051. (in Chinese).

37 Alford, N.M. and Birchall, J.D. (1985). Fibre toughening of MDF cement. *Journal of Materials Science* 20 (1): 37–45.

38 Wu, S., Hu, S., and Ding, Q. (1999). The effect of polymer in steel fiber reinforced cement composites. *Journal of Wuhan University of Technology* 14 (3): 23–27. (in Chinese).

39 Zhang, F., Ding, Q., and Lin, Q. (2005). Research on steel fibre reinforcement lightweight aggregate concrete for deck pavement of steebox bridge. *Concrete* (11): 47–49+60. (in Chinese).

40 Li, Y., Hu, S., and Ding, Q. (2001). Properties of ettringite type expansive agent. *Journal of Wuhan University of Technology* 16 (4): 53–56. (in Chinese).

41 Hu, S., Ding, Q., and Wang, H. (2007, May 09). A high-energy delayed expansion agent for enclosed concrete and its preparation method. CN 200510019477. (in Chinese).

42 Hu, J. and Hu, S. (2004). Effect on anti-shrinkage property of concrete with multiple anti-cracking additive. *Concrete* 171 (1): 22–23+40. (in Chinese).

43 Ma, B., Wang, X., and Hu, S. (2006, June 21). Shield tunnel lining segments and their preparation methods. CN 200510120533. (in Chinese).

44 Hu, S., Ding, Q., and Wang, F. (2004, April 28). Preparation method of high-strength lightweight aggregate concrete. CN 03125310. (in Chinese).

45 Hu, S., Yang, T., and Wang, F. (2010). Influence of mineralogical composition on the properties of lightweight aggregate. *Cement and Concrete Composites* 32 (1): 15–18. (in Chinese).

46 Ding, Q., Hu, S., and Tian, Y. (2010, February 24). Preparation method of high-strength and high toughness lightweight aggregate concrete. CN 200710053634. (in Chinese).

47 Hu, S. and Wang, F. (2006). *Lightweight Aggregate Concrete*. Beijing: Chemical Industry Press (in Chinese).

48 Nie, S., Hu, S., Wang, F. et al. (2017). Pozzolanic reaction of lightweight fine aggregate and its influence on the hydration of cement. *Construction and Building Materials* 153: 165–173. (in Chinese).

49 Shen, P., Lu, L., Wang, F. et al. (2020). Water desorption characteristics of saturated lightweight fine aggregate in ultra-high performance concrete. *Cement and Concrete Composites* 106: 103456. (in Chinese).

50 Hu, S, He, Y., and Lu, L. (2006, November 29). A composite cementitious material and its preparation method. CN 200610018820. (in Chinese).

51 Hu, S, Ding, Q., and Wang, H. (2007, February 14). High activity compensating shrinkage mineral admixture and its preparation method. CN 200510018820. (in Chinese).

52 Peng, Y., Hu, S., and Ding, Q. (2010). Preparation of reactive powder concrete using fly ash and steel slag powder. *Journal of Wuhan University of Technology (Materials Science Edition)* 25 (2): 349–354. (in Chinese).

53 Hu, S., Wang, H., and Ding, Q. (2006). The preparation of a high activated and slight expansive admixture base on steel slag and the study on its performance. *Concrete* 202 (8): 47–49. (in Chinese).

54 Huang, X., Hu, S., Wang, F. et al. (2019). The effect of supplementary cementitious materials on the permeability of chloride in steam cured high-ferrite

Portland cement concrete. *Construction and Building Materials* 197: 99–106. (in Chinese).

55 Hu, S. (2010). *Special Cement*. Wuhan: Wuhan University of Technology Press (in Chinese).

56 Hu, S,, Ding, Q., and Lu, L. (2007, May 23). A method for finely controlling the expansion of steel tube concrete. CN 200510019476. (in Chinese).

57 Hu, S., Lin, H., and Ding, Q. (2007, May 23). A paving method for steel bridge deck composite layer. CN 200410061407. (in Chinese).

58 Wang, X., Ding, Q., and Hu, S. (2009). Research on the performance of lightweight aggregate concrete for steel box girder bridge deck. *Highway* (4): 60–65. (in Chinese).

59 Li, Y. and Hu, S. (2001). The microstructure of the interfacial transition zone between steel and cement paste. *Cement and Concrete Research* 31 (3): 385–388. (in Chinese).

60 Wang, F., Liu, Y., and Hu, S. (2013). Effect of early cement hydration on the chemical stability of asphalt emulsion. *Construction and Building Materials* 42: 146–151. (in Chinese).

61 Wang, T., Hu, S., and Wang, F. (2008). Research on the main influencing factors of CA mortar strength. *Railway Engineering* (2): 109–111. (in Chinese).

62 Hu, S., Zhang, Y., and Wang, F. (2012). Effect of temperature and pressure on the degradation of cement asphalt mortar exposed to water. *Construction and Building Materials* 34: 570–574. (in Chinese).

63 Wang, T. (2008). *Research and Application of CA Mortar for High-Speed Railway Slab Ballastless Track*. Wuhan: Wuhan University of Technology (in Chinese).

64 Wang, F., Liu, Y., and Hu, S. (2013). Interaction mechanism between portland cement and cationic asphalt emulsion droplets. *Journal of Materials Science & Engineering* 31 (2): 186–190. (in Chinese).

65 Hu, S., Wang, T., and Wang, F. (2006, August 23). A high early strength self expanding CA mortar material. Wuhan CN 200610018439. (in Chinese).

66 Ding, Q., Hu, S., and Sun, Z. (2008, November 19). Preparation method of interface modified injectable organic-inorganic composite pavement materials. Wuhan CN 200910272815. (in Chinese).

67 Tian, K. (2010). Study of new cementitious microwave absorbing materials. *Journal of Building Materials* 13 (3): 295–299.

68 Ding, Q., Hu, S., and Huang, X. (2014, May 14). A type of absorbing lightweight aggregate and its preparation method. Wuhan CN 201410202947. (in Chinese).

69 Hu, S., Wu, J., Yang, W. et al. (2009). Preparation and properties of geopolymer-lightweight aggregate refractory concrete. *Journal of Central South University of Technology* 16 (6): 914–918. (in Chinese).

70 Tian, Y., Hu, S., and Wang, F. (2009). Analysis of fatigue properties of damping and high strength concrete. *Journal of Building Materials* 12 (3): 328–331. (in Chinese).

71 Hu, S., Ding, Q., and Tian, Y. (2008, May 07). Preparation method of high-strength and high damping concrete. Wuhan CN 200710053633. (in Chinese).

72 Wang, C. (2001). *Building Engineering Materials*. Beijing: Earthquake Publishing House (in Chinese).

2

Cementitious Material for High-tech Concrete

The development of concrete went through the stages of ordinary concrete, high-strength concrete, ultra-high-strength concrete, high-performance concrete, and ultra-high-performance concrete. The type and performance of cementitious materials used in concrete have also undergone a major change from Portland cement with a relatively simple composition, structure, and performance to the composite cementitious material with complex components, structure, and performance, as shown in Figure 2.1. The component of cementitious material of concrete changes from the Portland cement clinker composed of four main minerals including tricalcium high (C_3S), dicalcium silicate (C_2S), tetracalcium aluminoferrite (C_4AF), and tricalcium aluminate (C_3A) to special aluminate cement, special cement, and Portland composite cement including aluminate cement, sulfoaluminate cement, iron-rich cement, Portland slag cement, etc. The gradation and interaction of cementitious materials became more scientific to match the development of concrete and realize its high strength and performance. For instance, the cementitious material with rough grading has changed to composite cementitious materials with various wastes and functional materials. The study of the reaction in concrete has changed from the formation of hydration products caused by the reaction of the minerals in cement clinker to the synergistic effect of the reaction of minerals in cement clinker and supplementary cementitious material. Therefore, to achieve the requirements for strength and performance of high-tech concrete, the cementitious material should be a composite cementitious material with an optimized structure and composition. This composite cementitious material is characterized by an optimal grading and chemical composition. In addition, the economic cost of the raw material should be taken into account.

2.1 Characteristics of High-tech Concrete

Regarding the history of the development of concrete, cementitious materials for high-tech concrete are characterized by distinct compositions, structures, performances, and economic costs.

High-Tech Concrete Materials: Design, Preparation, and Applications, First Edition.
Fazhou Wang and Shuguang Hu.

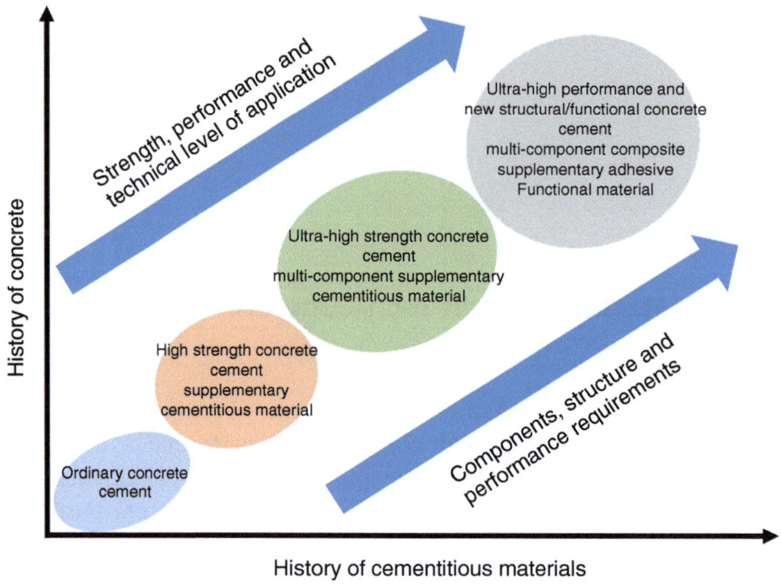

Figure 2.1 Development history of cementitious materials for high-tech concrete.

1) In terms of composition, high-tech concrete is composed of cement and supplementary cementitious materials, including fly ash, slag powder, and silica fume. Hence, the sequential hydration, pore filling and spherical effect of cementitious materials are realized, which improves the workability, mechanical properties, and durability of concrete.
2) In terms of structure, high-tech concrete is composed of multiple gradations of particles. A structure of dense packing of high-tech concrete is formed by combining ordinary cement with ultra-fine supplementary cementitious materials or using ultra-fine cement. Thus, the mechanical properties and durability of concrete are improved.
3) In terms of performance, the setting and hardening properties, mechanical properties, and durability of high-tech concrete can be adjusted according to the requirements of the application. The hydration property, workability, mechanical properties, and durability of cementitious materials can be controlled by the composition and structural design of cementitious materials.
4) In terms of the economic cost, cement in high-tech concrete is partially replaced by wastes with potential cementitious properties, including fly ash, slag powder, and steel slag powder, to improve the performance of cementitious materials. Meanwhile, the cost of the raw materials used in high-tech concrete has reduced, making it more economical.

To sum up, cementitious material for high-tech concrete is a kind of high-performance composite cementitious material with adjustable hydration performance, setting and hardening performance, mechanical properties, and durability. The performance of the cementitious materials is adjusted by changing the components and gradation of the cementitious materials by regulating the component of cement clinker and using supplementary cementitious materials. Thus, this cementitious material with low cost is applicable to concrete with high-performance requirements.

2.2 Types of Cementitious Materials for High-tech Concrete

Cementitious materials for concrete can be divided into cement and supplementary cementitious materials. Cement, as a kind of cementitious material, is widely used in various concrete and a large number of construction projects. Supplementary cementitious materials refer to materials such as fly ash, silica fume, slag powder, and so on, which improve the hydration property, workability, mechanical properties, and durability of concrete. Supplementary cementitious materials are also termed mineral admixtures. The ultra-fine modification of supplementary cementitious materials is one of the important research aspects of developing cement and concrete materials [1, 2].

2.2.1 Cement

According to the classification of construction engineering, cement can be divided into common cement and special cement. According to GB175-2023 common cements are divided into six categories. The compositions and names of common cement are shown in Table 2.1, which are various kinds of cement prepared by Portland cement clinker compounded with mineral additives (fly ash, slag powder, limestone, and other supplementary cementitious materials and gypsum) for different requirements. This type of cement is the most widely used cement, and its performance meets the requirements of high-strength, ultra-high-strength, high-performance, and ultra-high-performance concrete [1].

2.2.2 Supplementary Cementitious Materials

2.2.2.1 Fly Ash

Fly ash is a solid waste discharged from the production of coal-fired power plants. It is a fine dust obtained from the collection of boiler flue gas by electrostatic precipitators. In the process of coal combustion, the vast majority of combustibles in the furnace are burned out. Part of the refractory or non-combustible (mainly inorganic mineral components) and a small amount of unburned carbon are mixed in the high-temperature flue gas. Before the flue gas is emitted into the atmosphere, these fine particles are captured, separated and collected by the dust collector to obtain the fly ash. Generally, fly ash exhibits a spherical shape due to the small particle size of the coal dust and the tendency to form glass beads at high temperatures, as shown in Figure 2.2. The content of unburned carbon and the fineness of fly ash can be appropriately reflected according to the color of the fly ash. Fly ash with darker

Table 2.1 Classification and composition of common cement.

Types of common cement	Clinker + Gypsum (%)	Lime-stone (%)	Blast furnace slag (%)	Fly ash (%)	Pozzolanic materials (%)
Portland cement	≥95	≤5		—	—
Ordinary Portland cement	80–94	—	6–20		
Portland slag cement	30–79	—	21–70	—	—
Portland pozzolana cement	60–79	—	—	—	21–40
Portland fly ash cement	60–79	—	—	21–40	—
Portland composite cement	50–79	21–50			

Figure 2.2 Morphology of fly ash microbeads.

colors exhibits finer particle size and the content of carbon. Meanwhile, fly ash can be divided into high-lime fly ash and low-lime fly ash based on the content of Ca. Generally, the color of high-lime fly ash (Class C, CaO content >10%) is yellowish, while the color of low-lime fly ash (Class F, CaO content <10%) is grayish. Nowadays, according to GB/T 1596-2017, fly ash for supplementary cementitious materials is divided into three classes, and the requirements for physicochemical properties of fly ash are shown in Table 2.2.

2.2.2.2 Slag

Slag is a kind of industrial by-product discharged in the process of smelting iron ore, which is a kind of calcium-silicon inorganic silicate material formed via the reaction of lime, silica in iron ore, and impurities such as alumina at a molten state of 1350–1550 °C, followed by cooling. The process of cooling slag significantly affects its chemical activity. Molten slag produced through slow cooling is composed of calcium, silicon, aluminum, magnesium, and other elements, which is a stable solid

Table 2.2 Requirements for physicochemical properties of fly ash for supplementary cementitious materials.

Items		Requirements		
		Grade I	Grade II	Grade III
Fineness (45 μm sieve residue), ≤%	Class F	12.0	30.0	45.0
	Class C			
Water demand ratio, ≤%	Class F	95	105	115
	Class C			
Loss on ignition, ≤%	Class F	5.0	8.0	10.0
	Class C			
Water content, ≤%	Class F	1.0		
	Class C			
SO_3, ≤%	Class F	3.0		
	Class C			
Free-CaO, ≤%	Class F	1.0		
	Class C	4.0		
Increase in distance of Le Chatelier after boiling, ≤mm	Class C	5.0		

solution composed of calcium melilite and magnesium melilite. The only component with cementitious reactivity in this slag is β-C_2S of low content. Therefore, the cementitious reactivity of the crystalline slag is extremely low, and thus it is only utilized in autoclaved bricks and low-strength building materials. In contrast, the slag obtained by rapid cooling retains its glassy structure formed at high temperatures, and thus it is potentially hydraulic and pozzolanic activity after grinding. It is widely used in concrete as a common supplementary cementitious material.

The chemical composition of slag varies with the method of ironmaking and the type of iron ore. However, the basic compositions of slag are CaO (30–50 wt%), SiO_2 (27–42 wt%), Al_2O_3 (5–20 wt%), and MgO (1–18 wt%). Meanwhile, according to the specific surface area of slag powder, slag powder is divided into three types in China: S75, S95, and S105, and the physicochemical properties of different levels of slag powder are specified in the national standard GB/T 18046-2017, as shown in Table 2.3.

2.2.2.3 Silica Fume

Silica fume is the dust recovered by electrostatic dedusting during the production of ferrosilicon alloy or industrial silicon. Generally, quartzite gravel and pig iron are utilized as the raw material, while coke is the reducing agent for the production of ferrosilicon alloy. When the temperature rises to 1750 °C, the quartzite reacts with carbon to produce monatomic silicon. As the temperature reaches 2160 °C, about 10%–15% of the silicon is vaporized, and the vapor reacts with the oxygen in the air to produce SiO_2-rich smoke. When the smoke spreads to the low-temperature zone,

Table 2.3 Physicochemical properties of slag for cement and concrete.

Items			Grade		
			S105	S95	S75
Density (g/cm^3)	≥		2.8		
Specific surface area (m^2/kg)	≥		500	400	300
Activity index (%)	7 d	≥	95	70	55
	28 d	≥	105	95	75
Fluidity ratio (% ≥)			95		
Initial setting time (%)	≤		200		
Water content (mass fraction) (%)	≤		1.0		
SO_3 (mass fraction) (% ≤)			4.0		
Chloride content (mass fraction) (% ≤)			0.06		
Loss on ignition (mass fraction) (% ≤)			1.0		
Insoluble matter (mass fraction) (% ≤)			3.0		
Content of vitreous body (mass fraction) (% ≥)			85		
Radioactivity			$I_{Ra} \leq 1.0$ and $I_r \leq 1.0$		

the gaseous SiO_2 is cooled to form silica fume of small, amorphous spherical particles on a nanometer scale. The color of silica fume varies from white to black with the change of content of carbon, and it generally appears as gray. The refractoriness of silica fume can reach 1650 °C. The particles of silica fume are spherical, with an average particle size between 0.1 and 1 μm, as shown in Figure 2.3. The particle size of silica fume is 1/50–1/100 of the size of ordinary cement particles. The specific surface area of silica fume is 15,000–25,000 m^2/kg. Because of the large specific surface area, the pozzolanic activity of silica fume is very high. The potential pozzolanic activity of silica fume under alkaline conditions increases with the higher SiO_2 content. As the main chemical composition of the silica fume, the content of amorphous SiO_2 varies from 80% to 95%. The impurities in silica fume are a small amount of carbon and metal oxides generated by volatilization in the metallurgical process, including Fe_2O_3, Al_2O_3, MgO, CaO, Na_2O, etc. According to GB/T 27690-2011, the requirements for the performance of silica fume for supplementary cementitious materials are shown in Table 2.4.

2.2.2.4 Steel Slag and Other Supplementary Cementitious Materials

Steel slag is a by-product of smelting metal, mainly from the oxides generated by the oxidation of various elements in the metal charge, impurities brought into the furnace by eroded materials, and supplementary metal charge materials. For example, mud and slag-making materials are added into the metal charge to adjust

Figure 2.3 Morphology of silica fume particles.

Table 2.4 Requirements for the performance of silica fume for supplementary cementitious materials.

Items	Index
Solid content of liquid phase	±2% of the control value of the manufacturer
Total alkali content	≤1.5%
SiO_2 content	≥85.0%
Chloride content	≤0.1%
Water content of powder	≤3.0%
Loss on ignition	≤4.0%
Water demand ratio	≤125%
Specific surface area (BET)	≥15,000 m^2/g
Reactivity index (7 d)	≥105%
Radioactivity	$I_{Ra} \leq 1.0$ and $I_r \leq 1.0$
Performance of inhibiting alkali-aggregate reaction	Expansion ratio after 14 d reduces more than 35%
Chloride ion penetration resistance	Ratio of electric flux after 28 d ≤ 40%

the properties of the steel slag, such as limestone, dolomite, iron ore, and silica. The yield of steel slag is about 15–20% of crude steel production. Figure 2.4 shows the morphology of steel slag. The mineral composition of steel slag is mainly tricalcium silicate, followed by dicalcium silicate, RO phase, dicalcium iron, and small amounts of free calcium oxide, magnesium oxide, etc. The chemical compositions of steel slag varies due to the variation of geography and the conditions of production. Table 2.5 shows the chemical composition of steel slag discharged by several large steel plants in China.

Figure 2.4 Morphology of steel slag.

The composition of steel slag varies depending on different raw materials, steelmaking methods, production stages, steel types produced, and furnace batches. Currently, the steelmaking methods mainly used in China include converter, open hearth, and electric furnace steelmaking. Steel slag can be divided into converter steel slag, open hearth steel slag, and electric furnace steel slag according to the steelmaking method. According to different production stages, steel slag can be divided into steelmaking slag, casting slag, and splashing slag. In steelmaking slag, open hearth slag can be further divided into early slag and final slag (including refining steel slag and crude steel slag), and electric furnace steel slag can be divided into oxidizing slag and reducing slag. According to the properties of the slag, it can be divided into alkaline slag and acidic slag. According to the physical and chemical properties of steel slag, China has established technical specifications for the use of steel slag in concrete, as shown in Table 2.6.

The research on the utilization of steel slag in concrete in China has a history of more than 40 years, with rich achievements and experiences. The main applications include steel slag aggregate, steel slag cement, and steel slag mineral admixture. However, the utilization of steel slag in construction materials is limited for several reasons. First, due to the different chemical properties of the raw materials and production process, the chemical composition and mineral composition of steel slag fluctuate significantly. Second, steel slag contains a large amount of iron and iron-containing compounds, which are difficult to grind. If steel slag is ground to the same fineness as cement clinker, it has poor economic and environmental performance. Additionally, steel slag has a high density, leading to higher transportation costs. Third, compared to cement clinker, the mineral phases in steel slag tend to be overburned due to the high formation temperature, resulting in low

Table 2.5 Chemical composition of steel slag produced from different areas in China (%).

Manufacturer	SiO_2	Fe_2O_3	Al_2O_3	CaO	MgO	MnO	FeO	P_2O_5	f-CaO	$CaO/(P_2O_5 + SiO_2)$
Ma steel	12.1–16.3	2.79–7.24	2.70–6.83	44–52.7	6.9–12.4	0.62–2.5	10.5–18	0.33–4.7	0.46–4.2	1.82–3.0
An steel	16.6–22.77	1.79–7.02	1.10–1.64	16.5–38	11.2–20	1.04–3.9	7.97–37	0.13–1.0	0.7–5.49	0.78–1.8
Wuhan steel	16–18	7.76–12.8	7–8	40–50	9–12	0.5–1	8–14	0.5–1.5	1–6	2.42–2.56
Shougang	12.26	6.12	3.04	52.66	9.12	4.59	10.42	0.62	6.24	4.08
Benxi steel	16.36	1.49	2.56	50.44	13.22	2.66	11.50	0.56	1.57	2.98
Bao steel	13–17	4–10	1–3	40–49	4–7	5–6	11–22	1–1.4	2–9.6	2.66–2.86
Shanghai steel	11–20	3–7	10–18	44–55	8–13	—	0.5–1.5	1.2	2–5	2.56–3
Chengdu steel	15–17	4–9	3–4	19–33	12–14	4–5	19–22	0.47–0.9	1.23–1.84	1.2–2.7

Table 2.6 Technical specifications for steel slag used in concrete.

Items			First level	Second level
Surface area (m^2/kg)		≥	400	
Density (g/cm^3)		≥	2.8	
Water content (%)		≤	1.0	
f-CaO (mass fraction) (%)		≤	3.0	
SO_3 (mass fraction) (%)		≤	4.0	
Alkalinity coefficient		≤	1.8	
Activity index (%)	7 d	≥	65	55
	28 d	≥	80	65
Fluidity ratio (%)		≥	90	
Soundness	Boiling process		Qualified	
	Pressure evaporation process		Qualified when the content of MgO is higher than 13%	

hydration activity, which reduces its early performance when incorporated into concrete. Lastly, steel slag contains some free CaO and MgO. If not fully consumed in the early stages, these oxides will react with water in the later stages, causing volume expansion and reducing the volume stability of concrete. Therefore, the large-scale and effective utilization of steel slag has always been a major challenge that both the academic and industrial sectors need to overcome.

In addition to the supplementary cementitious materials mentioned above, there are other supplementary cementitious materials, such as limestone, quartz, and zeolite powder. Some of these materials come from natural mineral resources, while others are obtained from ore dressing by-products or through physical and chemical stimulation methods [2–6]. For example, limestone and quartz powder mainly come from limestone or quartz mineral resources. In addition, they are primarily used as raw materials for calcined cement clinker in the cement industry. By means of grinding, these materials can improve the properties of cementitious materials. Some studies indicate that when the limestone content is low, it improves the early strength of the cementitious materials, but increasing the content will significantly reduce the cement strength [7, 8]. Quartz sand can be ground into particles of different sizes and used as inert aggregate to replace cement. It is commonly used in the preparation of high-tech concrete or reactive powder concrete. By adding a small amount of fine quartz powder into conventional concrete, the content of cement and heat release in the stage of early hydration can be reduced, and the concrete's abrasive resistance can be improved.

2.3 Mechanism and Properties of Cementitious Materials for High-tech Concrete

The system of cementitious materials for high-tech concrete is a type of composite cementitious material that improves the performance of cement and concrete materials through the interaction of cement clinker with different supplementary cementitious materials. This section introduces the development of single-component (single admixture), two-component (compound admixture), and multi-component supplementary cementitious materials for high-tech concrete based on the composite types of cement and supplementary cementitious materials, and their interaction mechanisms and performance.

2.3.1 Mechanism and Properties of Single-component Supplementary Cementitious Material

2.3.1.1 Fly Ash–cement System

1) Effect of Fly Ash

   Fly ash reacts very slowly with water and generates a small amount of heat of hydration, making it difficult to achieve the necessary cementitious properties in construction projects. However, when fly ash partially replaces cement to form a fly ash–cement composite cementitious system, the addition of fly ash into cement results in three major effects [9].

   1) **Morphological Effect:** Fly ash particles are spherical and can act as rolling ball lubricants in cement paste, reducing the sliding resistance between cement particles and thereby improving the flow properties of the paste.
   2) **Microaggregate Effect:** The strength of fly ash particles collected through high-temperature cooling is relatively high. When incorporated into cement concrete, some of the unhydrated fly ash particles are not hydrated. These unhydrated particles bear external stresses with the hardened cement paste, which is known as the microaggregate effect.
   3) **Pozzolanic Effect:** The active chemical components, including SiO_2 and Al_2O_3 in fly ash, react chemically with $Ca(OH)_2$ produced by cement hydration, forming secondary hydration products like calcium silicate hydrates and calcium aluminate hydrates. Some of these hydration products deposit on the surface of fly ash particles, while others fill the voids in cement hydration products or concrete, reducing the pore diameter and densifying the structure. Moreover, the consumption of $Ca(OH)_2$ near the interfacial transition zone by fly ash significantly improves the stability of the interfacial transition zone of concrete.

2) Influence of Fly Ash on Hydration

   In the initial minutes of cement hydration, there is a rapid dissolution of various mineral phases, leading to the appearance of a short and sharp heat release peak, reaching maximum hydration heat rate rapidly. When fly ash is added as a mineral admixture to cement, the rate of this heat release peak significantly increases,

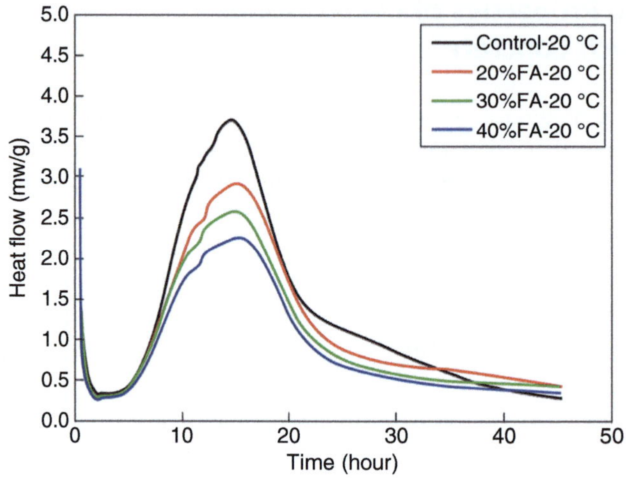

Figure 2.5 Effect of different amounts of fly ash on cement hydration.

and the degree of increase is proportional to the amount of fly ash added. This increase is mainly due to the dilution effect of fly ash, which increases the effective water-to-cement ratio of cement, enhancing the dissolution rate of various mineral phases in the early stages of cement hydration. Additionally, as mentioned earlier, fly ash can provide nucleation sites, promoting the formation of early hydration products, as shown in Figure 2.5.

During the second stage of hydration, known as the induction period, the addition of fly ash prolongs the induction period, and this prolongation effect is more pronounced with increasing fly ash content. This is because fly ash particles readily adsorb Ca^{2+} from the cement paste, reducing the concentration of Ca^{2+} in the liquid phase and lowering the saturation of $Ca(OH)_2$ in the solution, thereby delaying its crystal nucleation and growth. In addition, due to the reduced availability of Ca^{2+}, the Ca/Si ratio of the C-S-H gel formed during hydration is lower, resulting in poorer structural stability of the C-S-H gel, requiring a longer time for its transformation into a stable C-S-H gel [10, 11].

In summary, the effect of fly ash on the hydration process of the cement–fly ash binary system can be summed up in the following ways: (1) By replacing an equivalent amount of cement with fly ash, the cement content is reduced, increasing the effective water-to-cement ratio and leading to an increase in the rate of early heat release during hydration. However, due to the overall reduction in cement quantity, the total heat release is decreased. (2) The adsorption of Ca^{2+} by fly ash particles reduces the concentration of Ca^{2+} in the liquid phase, delaying the formation of stable C-S-H and the crystallization of $Ca(OH)_2$, thereby postponing the hydration of clinker minerals. (3) At a certain mixing ratio, fly ash can exhibit its pozzolanic effect in the later stages of hydration, where the active components of fly ash react with $Ca(OH)_2$ in the paste, promoting the hydration of the cementitious material [12].

3) The presence of free calcium oxide (CaO) in fly ash significantly affects the rate and amount of heat released during cement hydration. High-calcium fly ash exhibits higher reactivity, particularly in alkaline cement environments, leading to a more pronounced accelerating effect on hydration reactions and faster heat release rates. The higher the amount of fly ash participating in reactions, the greater the heat release. Additionally, high-calcium fly ash has a lower aggregation of silicate anions, making it more prone to chemical reactions. Therefore, high-calcium fly ash exhibits higher pozzolanic activity and serves as a better mineral admixture for reducing water content.
   The size and lattice integrity of free calcium oxide crystals are crucial factors influencing the hydration performance of fly ash. In high-calcium fly ash, smaller CaO crystal grains with significant lattice distortion exhibit higher hydration activity. Therefore, in cement-based materials with high-calcium fly ash at the same water-to-cement ratio, the dispersibility of the cementitious material in water is better, leading to a larger contact area with water and higher initial hydration heat release rates.
4) Influence of Fly Ash on the Hydration Product C-S-H
   An increase in the amount of fly ash leads to a diversification of the nano-scale morphology of C-S-H gel particles. Observations of hardened cementitious samples containing fly ash using an atomic force microscope (AFM) (Figure 2.6) revealed that in the composite cementitious system with 30% fly ash, the size of C-S-H gel particles ranged between 20 and 50 nm, with an average particle size of 47 nm, predominantly spherical or elliptical in shape. In high-density regions enriched with C-S-H gel particles, elongated structures composed of 5–10 particles were observed. Furthermore, the particle size in the high-density region was comparable to that in the low-density region. Figure 2.7 shows the particle size of C-S-H gel in a composite cementitious system with 50% fly ash, distributed between 50 and 100 nm, with an average particle size of 78 nm, predominantly flat in shape, with unclear boundaries between particles in high-density regions. The alteration in the morphology of C-S-H is primarily attributed to the reduction in the alkalinity of the pore solution in cementitious materials due to the increased fly ash content, which affects the nucleation and growth of calcium silicate hydrates. Additionally, the hydrated calcium silicate formed by the hydration

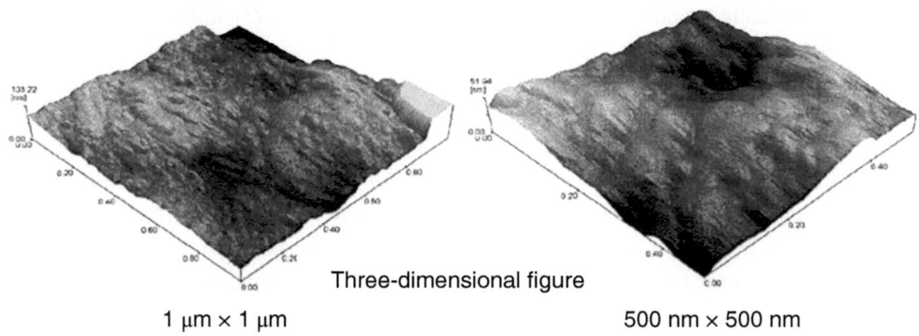

Figure 2.6 Morphology of C-S-H in composite gelling system doped with 30% fly ash.

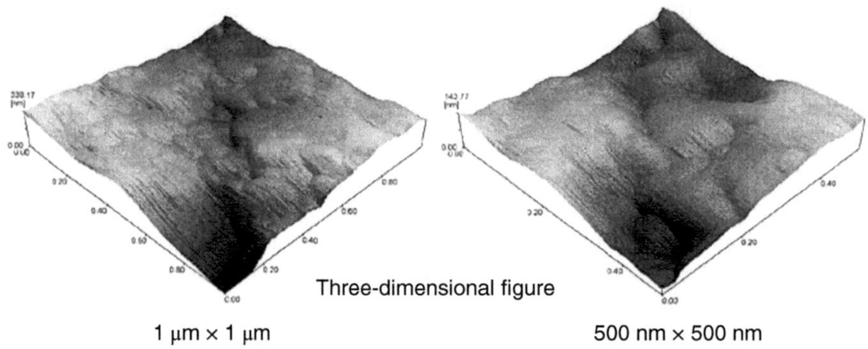

Figure 2.7 Morphology of C-S-H in compound gelling system doped with 50% fly ash.

reaction of fly ash in alkaline conditions may exhibit a nano-scale morphology different from that generated by cement silicate minerals, leading to overall morphological changes.

Based on the hydration reaction performance and product structure of the fly ash–cement composite system, its influence on the mechanical properties of cementitious materials is mainly reflected in the following aspects: the compressive strength increases initially and then decreases with an increase in fly ash content. When the fly ash content is excessive, the actual amount of cement significantly decreases, resulting in a reduction in total heat release during hydration and a decrease in the generation of $Ca(OH)_2$, a necessary product for the pozzolanic effect, thereby inhibiting the development of strength. The fineness of fly ash also has a significant impact on strength. Typically, fly ash with a larger specific surface area is more susceptible to stimulation in an alkaline cement environment, thus favoring the development of strength in the cementitious system.

2.3.1.2 Slag Powder–cement System

Slag powder affects the chemical and physical performance of cementitious materials. The effects on chemical properties are mainly characterized by potential hydration reactivity and pozzolanic effect. The effects on physical properties can be divided into microaggregate effect, dilution effect, and microcrystalline nucleation effect [13, 14].

1) Mechanism of effect on the chemical properties of slag powder
   The potential hydraulic stiffness of slag powder is determined by its chemical composition, and its calcium-rich component can dissolve Ca^{2+} under the condition of non-cement-provided $Ca(OH)_2$ by itself and form gel-like products with its own dissolved silicon and aluminum ions to form the cement strength. Meanwhile, the cleavage of slag powder vitreous also needs to consume $Ca(OH)_2$ in the cement system, which will reduce the directional arrangement of $Ca(OH)_2$ in the interfacial transition zone of the aggregate, reduce the grain size, and improve the microstructure of hardened cement paste.

However, there are many factors affecting the activity of slag powder, such as chemical composition, vitreous content, and structure, among which, fineness is the key factor. It is generally believed that the pozzolanic activity of slag powder increases with its fineness. The prolongation of the grinding time not only increases the specific surface area of the material but also improves the number of active crystalline surfaces located at the edges and corners, which improves the hydration reactivity of slag powder. Generally speaking, the activity of slag powder particles with a size above 20 μm is very poor, and the reaction speed is very slow. However, when the particle size is reduced to less than 2 μm, it can be fully reacted within 24 h in the Portland cement system with suitable alkalinity. Therefore, the regulation of slag powder particle size is a key parameter for the preparation of slag cement. Figures 2.8 and 2.9 show the scanning electron microscope (SEM) morphology and particle size distribution of slag powders with two different particle sizes, respectively.

Figure 2.8 Morphology of slag powder with different finenesses under SEM.

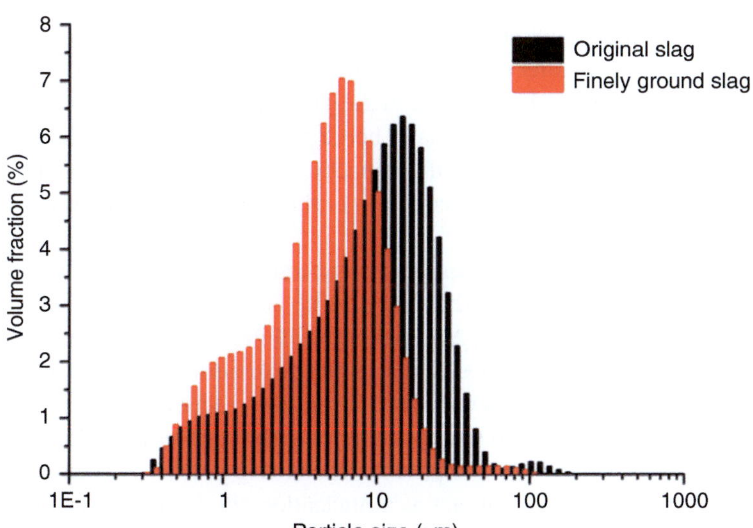

Figure 2.9 Particle size distribution of two kinds of slag powder with different fineness.

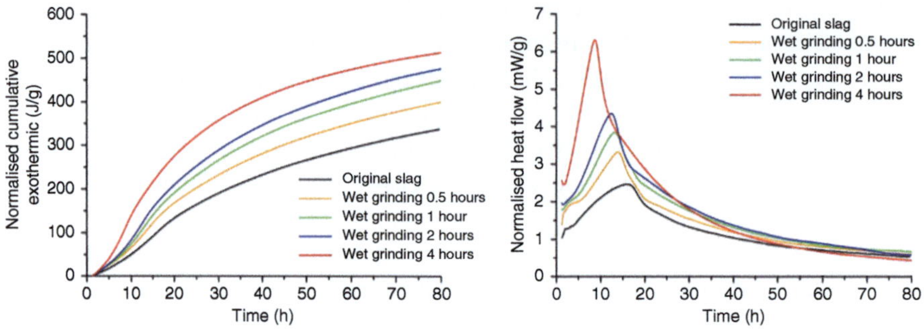

Figure 2.10 Cumulative heat of hydration and exothermic rate of hydration of two particle size slag powder under supersaturated $Ca(OH)_2$.

Figure 2.10 show the hydration exotherm of slag powder with different particle sizes under 20 °C supersaturated $Ca(OH)_2$ solution. The graph on the left reflects the cumulative exotherm, which shows that the total hydration exotherm increases with the increase of grinding time in three days, indicating that the increase of fineness improves the reaction activity of the activated slag powder. According to the figure on the right, it can be seen that the peak exothermic rate of slag powder not only increases with the extension of grinding time but also the reaction time is constantly advanced. Therefore, the improvement of the fineness of slag powder increases its activity and reaction degree, but in the actual production, the balance between the material properties and economical consideration is also an important aspect to be considered in the preparation of slag cement.

2) Slag powder physical action mechanism

During the process of slag powder reacting with cement, it will also produce physical effects on the slurry. If the cementitious material mixed with slag powder is regarded as a multi-functional system, according to the principle of the closest packing, its particle size is smaller than the cement particles. After replacing part of the cement, these small particles are filled in the gaps between the cement particles. Hence, the cementitious material has a better gradation, forming a dense filling structure and a fine level of the self-compacting accumulation system. In addition, the filling effect increases the cohesion of the mix and prevents water segregation of the concrete [10].

The incorporation of slag powder reduces the cement content, resulting in a thinner distribution of cement particles in the slurry, which is conducive to adequate cement hydration. As shown in Figure 2.11, under the dilution effect of slag powder, each gram of cement in the slurry exhibits a higher exothermic capacity, and the hydration efficiency of cement is enhanced before the reaction of slag powder starts at 15 h. The cement hydration efficiency is enhanced by the dilution effect of slag powder. Since the average particle size of slag powder for cement is smaller than that of cement, it can act as a nucleation site for the heterogeneous nucleation of cement hydration products in the early slurry, which exhibits a microcrystalline nucleation effect.

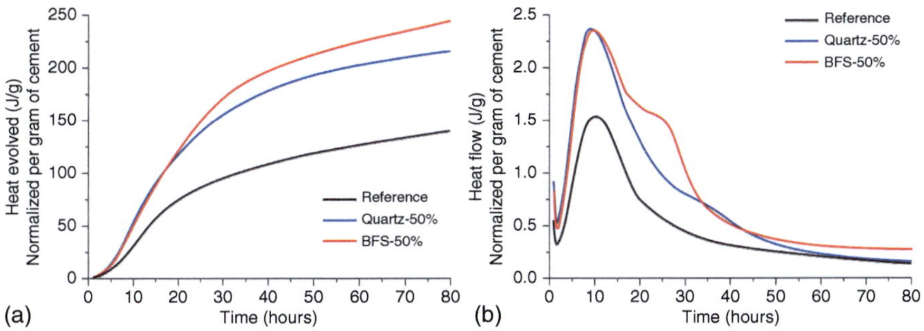

Figure 2.11 The exothermic behavior of cement paste with 50% mineral admixtures. (a) Cumulative exothermic and (b) exothermic rate curves for reference, quartz powder, and slag powder systems.

2.3.1.3 Silica Fume–cement System

Silica fume as an admixture in concrete has been widely promoted in the industry. The inclusion of silica fume can improve not only the recycling rate of industrial waste, resulting in good economic and environmental benefits, but also the durability of concrete, optimizing the mechanical properties of concrete. However, it should be noted that the density of silica fume is low, and it is easy to cause dust pollution in the process of transportation or use. To reduce this aspect of harm, in recent years, most of the silica fume products on the market have been changed to densified silica fume. Densified silica fume refers to the original silica fume rolled into tiny particles by compressed air, which can increase the volumetric density of silica fume to 900 kg/m^3. At present, in China, the utilization of silica fume is still concentrated in the mineral admixture for concrete, refractory materials, grouting works, and other construction works, as well as the application in the industry of electrically conductive composite and exterior wall thermal insulation materials.

1) Silica fume effect

   As a supplementary cementitious material for concrete, silica fume has four basic effects [13]: pozzolanic activity effect, filling effect, solution chemistry, and rheological effect. The pozzolanic activity effect is consistent with the previously mentioned fly ash pozzolanic effect, which refers to the reaction between the more active SiO_2 in the material and $Ca(OH)_2$ generated by the hydration of Portland cement to generate a C-S-H gel. In practical application, the silica fume with SiO_2 content greater than 90% should be selected, and the content of active SiO_2 in its components can be more than 40%, thus having higher pozzolanic activity.

   The filling effect of silica fume refers to the small particle size of silica fume, which can be filled in the smaller pores in the cement paste, thus greatly improving the compactness of the paste. When a certain amount of silica fume is added to concrete, silica fume can greatly reduce the proportion of harmful pores in the hardened concrete, and the effect is equivalent to filling the fine aggregate in the coarse aggregate between the voids. In terms of the structure, the addition of silica fume optimizes the particle gradation in concrete and improves the degree of compactness of the paste, so the concrete mixed with silica fume has better mechanical properties.

The chemical effect of silica fume refers to the fact that the C-S-H gel generated by the pozzolanic effect absorbs more metal ions in solution, such as aluminum ions (Al^{3+}), alkali metal ions (K^+, Na^+), and so on, which leads to a decrease in the pH of the slurry pore solution due to the trapping of these alkali metal ions. For concrete containing a large amount of alkali-silica reactive aggregates, this effect of silica fume can reduce the alkali-silica reaction to a certain extent and, at the same time, inhibit the penetration rate of chloride ions, which improves the durability of the concrete as a whole.

The rheological effect of silica fume comes from its very small particle size, which can play a better lubricating effect in the cement particles, improve the cohesion and water retention of the concrete mixture, and reduce the segregation of the concrete and the phenomenon of water seepage. However, the extremely fine particle size of silica fume also results in a significantly larger specific surface area, which leads to substantial adsorption of free water during concrete mixing. Therefore, in practical applications, it is necessary to employ high-performance water-reducing agents to optimize the rheological properties of the concrete. Additionally, the dosage of silica fume should be carefully controlled within an appropriate range to ensure both optimal performance and cost-effectiveness.

2) Influence of silica fume on the hydration of cement

Due to the large specific surface area of the particles, the incorporation of silica fume adsorbs a large amount of free water in the cement paste, resulting in a decrease in the effective water–cement ratio of the cement, thus slowing down the rate of dissolution of the mineral phases of the cement, and prolonging the hydration induction period. Therefore, when applying a silica fume–cement composite cementitious system, a high-performance water reducer is usually added to offset the high water absorption of silica fume to ensure the normal hydration development of the cementitious system.

Unlike fly ash, silica fume has high hydration activity and can be hydrated in large quantities at an early stage. Moreover, the particle size of silica fume particles is small, which makes it easier to act as a nucleation site for hydration products. Therefore, the exothermic rate of the main peak of the hydration exotherm of the cementitious system will be accelerated with the incorporation of silica fume and the peak value will increase. Under the appropriate amount of silica fume doping, the silica fume–cement system exhibits a higher heat release rate at the main hydration peak compared to plain cement, though its total cumulative heat release is generally lower over the full hydration period.

It has been shown [14] that silica fume, when participating in the hydration reaction of cementitious materials, can rapidly diffuse into the cement paste to form a particle layer of silica fume material phase, which is characterized by silica-rich and calcium-poor, and can further react with $Ca(OH)_2$ produced by cement hydration to generate C-S-H gels. At the same time, the decrease in the concentration of Ca^{2+} leads to a decrease in the degree of saturation in the solution, which encourages the dissolution of the cement minerals that have not yet been hydrated, and therefore promotes the exothermic hydration of the cement. At higher water–cement ratios, the silica fume particles are fully diffused and the more Ca^{2+} adsorbed, the more obvious this diffusion effect is, while at low water–cement ratios, the silica fume particles adsorb a large number of water molecules, resulting in cement particles not being adequately dissolved in the slurry and therefore hydration cannot proceed smoothly. Sánchez De Rojas et al. have shown [15] that, in the cement–cement silica fume binary cementitious system, the surface of cement particles encapsulates about 100,000 silica fume particles, which explains that at low water–cement ratios, silica fume prevents the contact of cement particles with water and hinders the hydration reaction. In the case of a high water–cement ratio, the silica fume particles are sufficiently dispersed, and it is difficult for this phenomenon of hindering hydration to occur, so the hindering effect of silica fume on the hydration reaction becomes smaller with the increase in the water-to-cement ratio.

In summary, in the case of a high water-cementitious ratio or mixed with the appropriate dose of a water-reducing agent, the silica fume accelerates the hydration reaction of cementitious materials, and in the case of a low water-to-cement ratio, it reduces the hydration reaction rate of cementitious materials. In addition, under the condition of a low water-to-cement ratio, silica fume shows different promoting or hindering effects in different stages of the hydration process, for example, the incorporation of silica fume prolongs the induction period of the hydration reaction and reduces the exothermic rate of hydration in the acceleration period, but promotes the growth of the exothermic rate of hydration in the deceleration period. The incorporation of silica fume has an obvious improvement effect on the mechanical properties of silica fume–cement composite cementitious materials, but in the early stage, this improvement effect is not particularly obvious. As the hydration reaction continues, the silica fume begins to react with the cement hydration product $Ca(OH)_2$ to promote the further hydration of the cement, so in the later stage, especially after seven days, the mechanical properties of the composite cementitious material have a greater improvement, and within a certain range with the increase in the amount of silica fume doping, the greater the magnitude of the improvement.

2.3.1.4 Steel Slag–cement System

The mineral composition of steel slag depends mainly on the chemical composition. According to the alkalinity $CaO/(SiO_2 + P_2O_5)$, it can be divided into olivine slag, roserite pyroxene slag, dicalcium silicate slag, and tricalcium silicate slag. The slow-cooled and sharply cooled steel slag has a crystal structure and contains no vitreous body, which is essentially different from granulated blast furnace slag. However, the rapid cooling treatment can retain more C_3S phase in the steel slag, reduce the free calcium oxide content, and improve its gelling activity but also, to a certain extent, its grindability [16–19]. Figure 2.12 lists the XRD pattern of a kind of steel slag, in which the diffraction peaks of tricalcium silicate and dicalcium silicate can be found, but the difference lies in the fact that the generation temperature of the steel slag is more than 1560 °C, whereas the forging temperature of the Portland cement

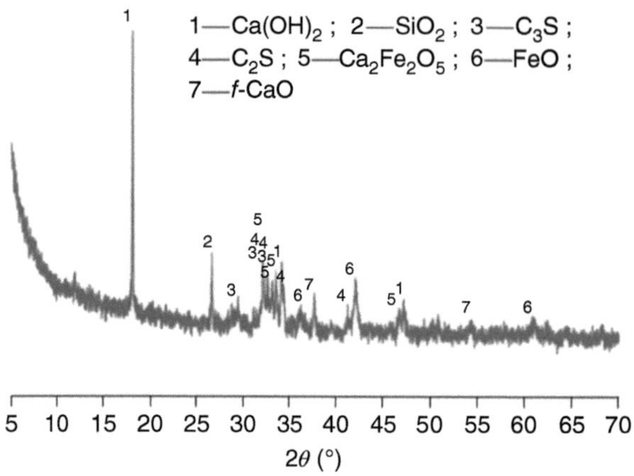

Figure 2.12 XRD pattern of steel slag.

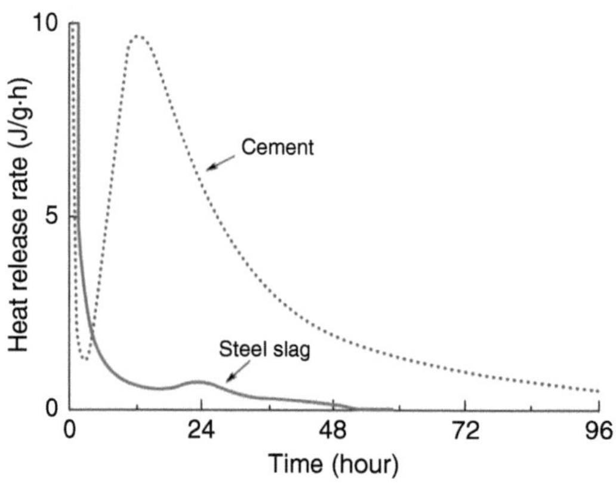

Figure 2.13 Hydration heat release curve of steel slag and cement. *Source*: Grutzeck et al. [14] / American Concrete Institute.

clinker is around 1450 °C. Steel slag has a high generation temperature, dense mineral crystals and large crystal grains, and a slow hydration rate [20]. The heat of hydration of the reaction is shown in Figure 2.13, and the hydration process of steel slag and cement is similar, which can be roughly divided into induction, acceleration, and stabilization periods, but the difference is that steel slag is less reactive, and its peak hydration exothermic rate is only 1/12 of that of cement [21].

Based on the special physicochemical properties of steel slag, it is necessary to activate the finished steel slag to meet the higher requirements of cement concrete manufacturing, mainly using mechanical activation and chemical activation.

The principle of physical activation is to use mechanical methods to improve the fineness of steel slag so that the average particle size is maintained at 10–30 μm, and the grinding process is not only the process of particle size reduction, accompanied by changes in the crystal structure of the material and the physical and chemical properties of the surface. The bonding energy of the crystal will also change; the lattice energy decreases rapidly, and lattice dislocations, defects, and recrystallization occur in the position of the loss of lattice energy. The decrease in lattice size results in an increase in the contact area between minerals and water in the steel slag. The increase in lattice strain improves the force between minerals and water. The crystalline structure of minerals is distorted, and the decrease in crystallinity makes the bonding bonds of mineral crystals decrease, and water molecules can easily enter into the minerals, accelerating the hydration reaction [22–24].

The principle of chemical activation is to promote the alkaline excitation of steel slag by using active exciters to accelerate the dissolution and migration of Ca^{2+}, AlO_4^{5-}, Al^{3+}, and SiO_4^{4-} ions in its glassy phase, which reacts with the hydration products of the cement to generate secondary hydration products, thus decreasing the $Ca(OH)_2$ concentration of the reaction system and inducing the clinker minerals to be further hydrated to maintain the alkalinity. In addition, increasing the curing temperature will further enhance the dissolution and migration rate of raw material ions and the formation rate of hydration products so that the hydration reaction of the system is accelerated and the strength of steel slag cement is effectively improved.

2.3.1.5 Other Pozzolanic Supplement Cementitious Material–cement System

Other supplementary cementitious materials with pozzolanic activity and cement physicochemical reactions are similar to the previously mentioned materials. The difference mainly lies in the structure of the hydration products, the reaction mechanism, and the physical effect of the existence of some differences, which will not be described specifically. The role of relatively inert supplementary cementitious materials, such as limestone powder, quartz powder, and so on, in the cementitious materials can be attributed to the microaggregate filling effect, microcrystalline nucleation effect, and active effect. For example, according to the concrete aggregate grading design of the appropriate fineness of limestone powder, with the help of the closest packing principle, to fully play its microaggregate filling effect, so that the compactness of concrete is improved. The microcrystalline nucleation effect is consistent with other admixtures, which can be used as the nucleation site of cement hydration products to promote cement hydration. The activity effect lies in the fact that although the activity of limestone powder is extremely low and negligible in the early stage, it has been shown [25] that limestone can react with the intermediate phase of the cement in the later stage to produce calcium carboaluminate hydrate, and it can delay the transformation of ettringite to monosulfate-type calcium sulfoaluminate hydrate. It should be pointed out that other inert admixtures such as quartz powder and natural pozzolanic materials do not have obvious potential hydraulicity or cementing properties, but their incorporation into cement concrete is significant, not only to reduce the amount of cement and carbon emissions but also to optimize the aggregate gradation of ultra-high-performance concrete or activated powder concrete, which is of great help to the overall performance improvement.

2.3.2 Mechanism and Properties of Two-component Supplementary Cementitious Material

Compared to single-component supplementary cementitious materials, the use of dual-component materials offers more pronounced advantages, as they can leverage synergistic effects between different supplementary cementitious materials. This approach is more widely applied in the preparation of high-strength and high-performance concrete. This section focuses on the interaction mechanisms and performance impacts of dual-component systems, such as silica fume, fly ash, slag powder, and steel slag powder, with Portland cement.

2.3.2.1 Fly Ash–silica Fume–cement System

For concrete materials with high cementitious material usage and low water–cement ratios, fly ash microspheres have a distinct advantage in preparing high-tech concrete materials in combination with silica fume. In the trinary system of cement–silica fume–fly ash microspheres, the types of hydration products are essentially the same as in the cement–silica fume system, both containing tobermorite-like C-S-H gel and $Ca(OH)_2$, but with a lower content of $Ca(OH)_2$. Typically, for a silica fume content of 15% and a higher content of fly ash microspheres, $Ca(OH)_2$ can be essentially consumed. Concurrently, as the content of fly ash microspheres in the system gradually increases, the $Ca(OH)_2$ generated in the system will be completely consumed [20]. In the fly ash–silica fume–cementitious material system, the XRD diffraction peak intensity of quartz will gradually increase with the increase of fly ash microsphere content, which can be attributed to the higher reactivity of silica fume compared to fly ash microspheres. Notably, with 15% silica fume content, most of the $Ca(OH)_2$ produced by cement hydration will react with silica fume, leading to some fly ash microspheres not undergoing pozzolanic reactions. Therefore, as the content of fly ash microspheres increases, the relative content of quartz will increase, and the intensity of the quartz diffraction peak will also increase. It should be noted that as the hydration age extends, cement will continue to hydrate and produce $Ca(OH)_2$, consuming fly ash microspheres. Based on this difference in pozzolanic reaction performance with silica fume, the sequential hydration reactions of cementitious materials can be designed to improve their early and later performance. SEM-EDS observations (Figure 2.14) [26] show that fly ash microspheres significantly affect the morphology of C-S-H phase. With the increase of fly ash microsphere content, the morphology of tobermorite crystals changes from flaky to rod-like, with gradually decreasing size, and the Ca/Si ratio gradually decreases with the increase of fly ash microsphere content, to 1.03, 0.94, and 0.85, respectively [20].

The interaction mechanism of the fly ash–silica fume–cement system can be summarized as follows: In the early stage of the reaction, in addition to cement hydration, silica fume also hydrates extensively, while promoting the hydration of cement clinker; when a certain amount of cement is consumed, the amount of $Ca(OH)_2$ generated begins to increase and reacts with the active SiO_2 in silica fume, reducing the amount of $Ca(OH)_2$ in the hydrated cement paste, thereby forming hydration products that reduce the volume of large pores in the hardened cement paste, increasing the number of small pores, reducing connected pores, and significantly enhancing the microhardness at the aggregate-cement paste interface. After the initial consumption of silica fume, the $Ca(OH)_2$ content gradually increases, and in the later stage, fly ash is activated by $Ca(OH)_2$, undergoing a secondary pozzolanic reaction,

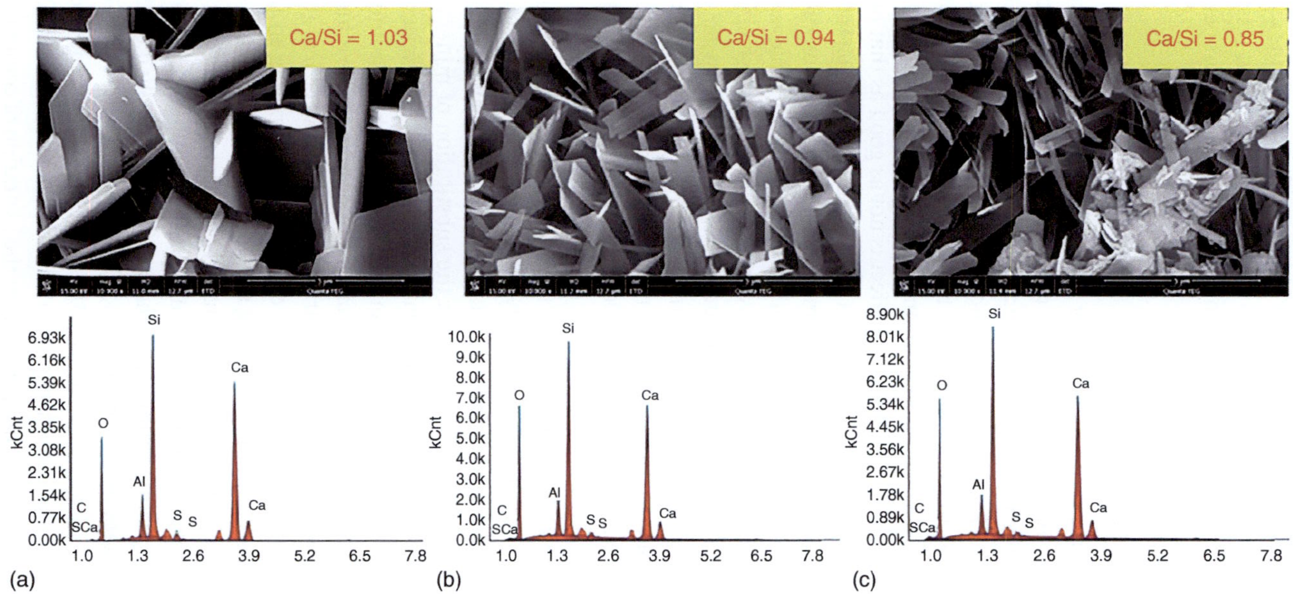

Figure 2.14 Different fly ash–silica fume–cement systems. (a) 10% Fly ash–silica fume cement, (b) 20% Fly ash–silica fume cement, and (c) 30% Fly ash–silica fume cement.

generating more C-S-H gel, further refining the pore structure of the matrix, and thereby improving the performance of the matrix. In addition, fly ash in the system can effectively dilute the water–cement ratio of cement and the water consumption of silica fume, ultimately effectively improving the workability of the composite cementitious material. The gradation effect of silica fume, fly ash, and cement particles in particle size can also effectively improve their mechanical properties and durability.

2.3.2.2 Fly Ash–slag Powder–cement System

Fly ash and slag powder are the most widely used active mineral admixtures in composite cementitious materials. They have different characteristics in composite cementitious materials. The addition of fly ash affects the early strength of the cementitious matrix, which decreases with increasing content, and has a greater potential for strength growth in the later stage; the addition of slag powder does not affect the early strength of the matrix, a small amount of slag powder does not reduce the heat of hydration, and its water-reducing effect is not as good as that of fly ash. When the content of slag powder is large, it can cause greater shrinkage in the later stage, leading to cracking and affecting performance. The simultaneous addition of fly ash and slag powder in composite cementitious materials can achieve complementary advantages and better overall effects. When dual-mixed, fly ash can exhibit better "micro filler effect" and "microaggregate effect" than slag powder and can achieve complementary effects on work performance. On the one hand, it improves the fluidity of the paste, and on the other hand, the dual-mixed fly ash and slag powder can improve the time-dependent loss of fluidity when single-mixed with fly ash or slag powder [27–29].

The hydration reaction characteristics of the fly ash–slag powder–cement system can mainly be described as follows. In the early stage of hydration, it is mainly the hydration of cement clinker, and fly ash and slag powder only play a diluting role. Although it promotes the hydration of clinker, the substitution of mineral admixtures leads to a reduction in the amount of clinker used, which will cause the generated $Ca(OH)_2$ to be less than that of pure cement, reducing the alkalinity of the system and affecting the early strength of the composite system. In the middle stage of hydration of the composite system, the hydration of slag powder mainly affects the alkalinity of the system, and the amount of fly ash participating in the reaction at this time is very small, mainly playing a diluting role. In the later stage of hydration, the pozzolanic reaction of fly ash mainly controls the alkalinity of the system. Therefore, the reaction mechanisms of each stage of the dual-component composite supplementary cementitious material system are controlled by the reactions that occur in each stage of the components, which are closely related to the type of reacting components, the degree of activity, the order of reaction, and the degree of hydration. Accordingly, the early strength contribution of the cementitious material hydration hardening body mainly comes from the hydration of cement, and slag powder and fly ash play a physical filling and compaction role to provide strength, forming a point contact skeleton with the hydration products of cement clinker to produce strength, rather than forming a network structure by the interweaving and

Table 2.7 Influence of compound mineral admixture content on the alkalinity of the pore solution of cementing material [23].

Code	Cement (%)	Fly ash (%)	Slag powder (%)	Alkalinity of pore solution (pH)			
				3 d	7 d	28 d	90 d
1	100	0	0	12.688	12.678	12.965	12.681
2	80	10	10	12.573	12.461	12.673	12.461
3	60	20	20	12.025	12.206	12.311	12.144
4	40	30	30	11.823	12.451	12.185	12.516

overlapping of the products generated by the hydration reaction. As the hydration reaction proceeds, slag powder and fly ash undergo secondary hydration reactions, and the hydration products interweave to form a compact hardened matrix.

Table 2.7 shows the pH values of pore solutions in the fly ash–slag powder–cement system at the same age, which can be found to gradually decrease with the increase of mineral admixture content. However, the change of pH values with age for different composite cementitious materials is more complex. Samples 3 and 4 have gradually increasing pH values in the early stages of hydration age and slightly decreasing values in the later stages. However, sample 2 has a higher pH value in the early stage, a significant decrease in the middle stage, and then gradually increases, and then decreases again in the later stage. This experimental result well verifies the hydration reaction law of the fly ash–slag powder–cement system. Therefore, the optimization of the type and content of admixtures is not only a superimposed effect of the complementary advantages of particle filling and compaction on macroscopic performance but also reflects the formation mechanism of microstructure. Of course, the improvement effect of mineral admixtures in concrete is not only the central quality effect on physical properties but also the interconnection and interweaving of various hydration products and unhydrated particles in the fine structure to form a spatial network structure of different levels of central quality and medium. There is also a significant feature in the hydration product structure of the slag powder–fly ash–cement system. With the addition of slag powder and fly ash, the diluting effect of fine admixtures and secondary hydration reactions both accelerate the hydration of cement clinker, resulting in a larger degree of hydration within a 60 d hydration age, and the system will have more type III C-S-H gel similar to later products, which can be attributed to the secondary hydration products formed by the secondary hydration reaction of mineral admixtures (generally occurring in the middle and later stages). In general, in the early stage of the reaction, the $Ca(OH)_2$ generated by the hydration of cement clinker and the gypsum in cement act as alkaline activators and sulfate activators for slag powder, interact with the active components in slag powder to generate hydration products, and fill in the cement matrix; in the later stage, the $Ca(OH)_2$ released by the hydration of cement clinker diffuses to the surface of the fly ash spherical glass particle, promoting the secondary hydration reaction of fly ash, making the later strength of the composite cementitious material rise, showing a superimposed effect [30].

The mechanisms by which fly ash and slag powder improve the shrinkage performance of concrete generally lie in their "dilution effect" on cement, which reduces the total amount of cementitious materials involved in the hydration reaction in the early stages and increases the number of coarse capillary pores and free water within the pores in the early concrete, thereby reducing the early shrinkage of concrete. In the later stages of hydration, although fly ash and slag powder will also undergo "secondary hydration," the pores in the concrete will be refined, but due to the sufficient strength of the concrete in the later stages, the chemical shrinkage caused by the hydration of fly ash and slag powder and the shrinkage caused by the loss of water from fine capillary pores are relatively small. The effectiveness of fly ash in improving the early shrinkage of concrete is better than that of slag powder, which is related to its low early activity, as the low-activity fly ash can better reflect the "dilution effect" of the admixture.

2.3.2.3 Steel Slag–silica Fume–cement System

Steel slag, as one of the active mineral admixtures used in composite cementitious materials, has an improving effect on the workability and strength of composite cementitious materials when added in small amounts. However, when the amount added is large, the strength of the composite cementitious materials at various ages will be significantly reduced. To increase the amount of steel slag in composite cementitious materials, the activity of the composite cementitious materials themselves must be increased. The combination of silica fume and steel slag can effectively improve this drawback. The addition of silica fume significantly improves the strength of the composite cementitious materials at various ages, and the strength continues to increase with the increase in the amount added. The main mechanisms are as follows: (1) The particle size of silica fume is two orders of magnitude smaller than that of steel slag and cement, and the relative size of the particles significantly affects the packing density of the system, enhancing the physical filling effect of the system and compensating for the early strength reduction caused by steel slag. (2) In the steel slag–silica fume–cement system, the activity of steel slag is relatively weak, and the early hydration rate is slow. Due to the presence of a certain amount of active components in its composition, the hydration process is similar to that of Portland cement, and the hydration products contain a considerable amount of $Ca(OH)_2$. During the hydration process of the composite cementitious material, when silica fume comes into contact with water in the system, a gel rich in SiO_2 and poor in calcium forms an adhering layer on the surface of silica fume particles. After a certain period, the rich SiO_2 and poor calcium gel adhering layer begin to dissolve in the $Ca(OH)_2$ produced by cement and steel slag, reacting to form C-S-H gel, which significantly increases the later strength of the composite cementitious material [31].

2.3.2.4 Steel Slag–fly Ash (Slag Powder)–cement System

The pozzolanic effect of fly ash can consume the free calcium oxide in steel slag, overcoming the poor stability problems of steel slag powder, while the free calcium oxide and its hydration product $Ca(OH)_2$ in steel slag powder can serve as activators for the secondary hydration reaction of fly ash. On the other hand, the early hydration activity of steel slag powder is higher than that of Class I fly ash, and the combination of the two can improve the early strength of fly ash concrete. Moreover, since Class I fly ash can significantly improve the workability of concrete more than steel slag powder, the combination of fly ash and steel slag powder can also improve the workability of concrete with single steel slag powder.

1) Hydration reaction performance of steel slag–fly ash (slag powder)–cement system
   The heat of hydration of cement is produced by the hydration of various clinker minerals. The heat of hydration of cement generally depends on the mineral composition of clinker, fineness, water–cement ratio, and the type and amount of cementitious materials mixed with clinker. It is generally believed that admixtures such as fly ash can significantly reduce the heat of hydration of cement. The hydration heat release period of cement is very long, but most of the heat is released within three days after adding water, especially in the early stages of setting and hardening of cement paste, which is consistent with the acceleration period of cement hydration. Figure 2.15 shows the influence of steel slag powder, slag powder, and fly ash on the hydration heat performance of cement when each is used to replace 30% of cement in single and combined additions [32]. From the figure, it can be seen that the hydration heat release of each sample is mainly concentrated before three days, and after three days, the hydration heat release of each sample tends to slow down with the extension of time. After adding 30% of various admixtures, the hydration heat release of cement at various

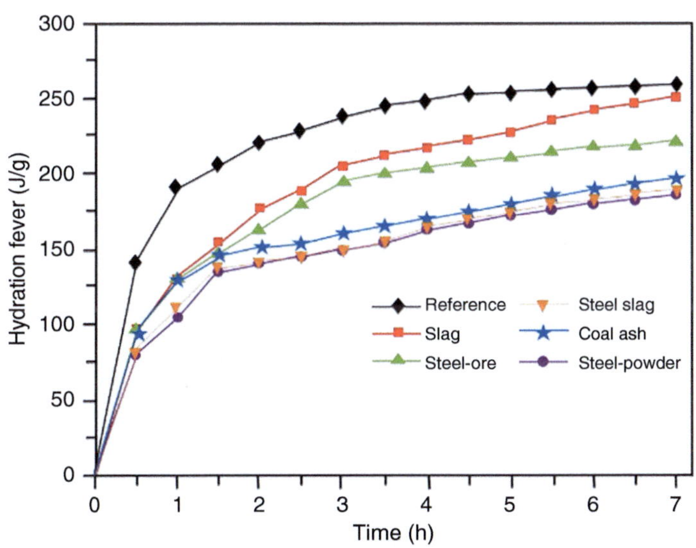

Figure 2.15 Hydration heat performance of cement with 30% steel slag powder, slag powder, and fly ash singly or in combination.

ages has a significantly different degree of decrease, which is of great significance for large-volume concrete. Comparatively, the highest hydration heat is for pure Portland cement without admixtures. The samples with slag powder and the composite of slag powder and steel slag powder have relatively lower hydration heat, and the hydration heat of the two is close; the lowest hydration heat is for the samples with single steel slag, fly ash, and the composite of steel slag and fly ash. The three admixtures, steel slag powder, slag powder, and fly ash, all reduce the hydration heat of cement to varying degrees when used alone or in combination. However, the hydration heat of 30% steel slag powder is reduced the most after seven days, by 24%, and the hydration heat of steel slag powder–fly ash is reduced by 25% after seven days. Considering the early activity of the three admixtures, it can be considered that using steel slag powder to prepare large-volume concrete with relatively high early strength requirements has more advantages.

The quantitative analysis results of $Ca(OH)_2$ in the hydration products of the above different samples at different hydration ages, calculated by comprehensive thermal analysis, are shown in Figure 2.16. For the three admixtures, in the case of single addition, the amount of $Ca(OH)_2$ in the hydration products of the slag powder paste at various ages is the least, and the amount in the fly ash paste is relatively more, while the amount of $Ca(OH)_2$ in the hydration products of the steel slag powder paste at various hydration ages is relatively more. In the case of combined addition, the amount of $Ca(OH)_2$ in the hydration products at each hydration age is between the amounts of $Ca(OH)_2$ in the hydration products of each admixture when used alone, that is, the composite of slag powder or fly ash with steel slag powder can increase the amount of $Ca(OH)_2$ in the hydration products of single slag powder or fly ash, and the same is true for steel slag powder

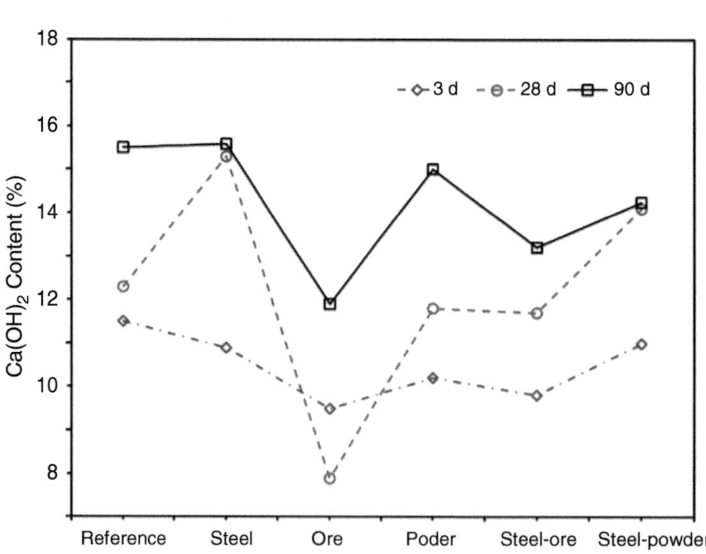

Figure 2.16 Quantitative analysis of $Ca(OH)_2$ in cement with singly or combined steel slag powder, slag powder, and fly ash.

with composite slag powder or fly ash, which can reduce the amount of $Ca(OH)_2$ in the hydration products of single steel slag powder. For slag powder and fly ash, their secondary hydration reactions require the consumption of a certain amount of $Ca(OH)_2$, so the amount of $Ca(OH)_2$ in the hydration products of the paste containing slag powder or fly ash is relatively less; as the hydration time extends, on the one hand, the hydration of Portland cement will increase the amount of $Ca(OH)_2$ in the composite system, and on the other hand, the hydration of slag powder and fly ash will reduce the amount of $Ca(OH)_2$ in the composite system. In fact, the content of $Ca(OH)_2$ in the hydration products changes with the hydration time, depending on the relative amounts and reaction rates of cement and slag powder or fly ash. Slag powder has higher early activity, and the early absorption of $Ca(OH)_2$ is dominant, so the amount of $Ca(OH)_2$ in the hydration products decreases with the extension of time before 28 d, but at the age of 90 d, the hydration of Portland cement to generate $Ca(OH)_2$ is dominant, so the amount of $Ca(OH)_2$ in the hydration products increases significantly; compared with slag powder, the secondary hydration activity of fly ash is lower, so the hydration of Portland cement to generate $Ca(OH)_2$ is dominant throughout the hydration process, but because of its dilution effect, the amount of $Ca(OH)_2$ in the hydration products is also reduced. Unlike slag powder and fly ash, the hydration reaction of steel slag powder can generate $Ca(OH)_2$, so the content of $Ca(OH)_2$ in the hydration products is the highest. The combination of steel slag powder with fly ash and slag powder can combine the characteristics of steel slag powder generating $Ca(OH)_2$ in the hydration reaction with the characteristics of slag powder and fly ash absorbing $Ca(OH)_2$ in the secondary hydration, and the amount of $Ca(OH)_2$ in the hydration products is between the three admixtures.

2) Performance of the Steel Slag–Fly Ash (Slag Powder)–Cement System

Research and application of steel slag–slag cement have shown that steel slag powder and slag powder have a good composite effect when used as mineral admixtures. It is generally believed that slag powder can consume excess free calcium oxide in steel slag powder, overcoming the stability issues associated with steel slag powder, while the hydration product $Ca(OH)_2$ of steel slag powder serves as an effective activator for slag powder. The combination of the two can improve the early strength development of concrete with single slag powder and increase the later strength growth rate of concrete with single steel slag powder, thus achieving coordinated strength development of concrete.

In Figure 2.17, the solid line represents the performance of concrete prepared with a composite of steel slag powder and slag powder, while the dashed line compares the performance of concrete prepared with steel slag alone. The experimental results of single and combined admixture concrete performance show that the workability and mechanical properties of the composite concrete are similar in trend to those of concrete prepared with single steel slag powder, depending on the amount of admixture and hydration time. The concrete workability test results indicate that after adding 50% slag powder to steel slag powder, the initial slump of the concrete is slightly improved, but the slump after 60 minutes is almost the same as that of concrete prepared with steel slag

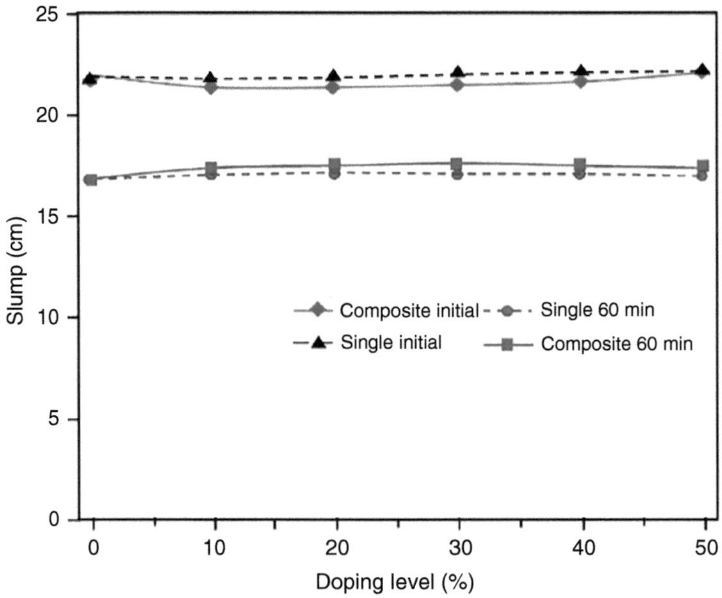

Figure 2.17 Performance of workability and mechanical properties of cement systems with single or combined steel slag powder and slag powder.

powder alone. This means that the combination of steel slag powder and slag powder can slightly improve the workability of freshly mixed concrete, but it does not reduce the slump loss of concrete. The mechanical performance results of concrete show that the optimal mix ratio is 30% steel slag powder combined with 50% slag powder, with a slight reduction in the 7 d strength of concrete, but no significant impact on the 28 d and 90 d strengths.

The combination of steel slag powder and slag powder increases the shrinkage rate of concrete, while the combination of steel slag powder and fly ash can reduce the shrinkage rate of concrete at various ages [33]. There has been extensive research on the impact of slag powder and fly ash on the drying shrinkage performance of concrete but less research on the combination of steel slag with other supplementary cementitious materials. Insights can be gained from the analysis of the effects of slag powder and fly ash on concrete shrinkage [34, 35]. Although steel slag powder has higher early activity, its active component content is still significantly lower than that of Portland cement. Therefore, due to the "dilution effect" of steel slag powder on cement, the early shrinkage of concrete is reduced. In the later stages of hydration, the porosity and pore size of steel slag concrete are lower than those of control concrete, resulting in relatively higher shrinkage stress caused by the loss of water from capillary pores. Additionally, the carbonation resistance of steel slag concrete is relatively poor, and the larger carbonation shrinkage may also increase the total shrinkage in the later stages, leading to relatively greater shrinkage in steel slag concrete in the later stages.

2.3.2.5 Mechanism of Multi-component Supplementary Cementitious Material Systems

The type of admixture changes the hydration reaction process, making the formation of hydration products in the system more complex. The generation of hydration products and the reactions that occur are closely related to the hydration environment. In multi-component composite systems with admixtures, due to the denser structure and lower water–cement ratio, there is less free reaction space between hydration products and reactants. Particles mainly generate new hydration products through mutual contact. Therefore, in the early stages of hydration, the transformation of hydration products with changes in alkalinity and other factors is mainly achieved through dissolution-precipitation processes, while in later stages, it is primarily through "solid-state reactions." The mechanism by which multi-component mineral admixtures improve the durability of concrete can be summarized as follows: as the pH gradually decreases, the repeated cycling of product dissolution, supersaturation, and recrystallization processes will change the porosity, elastic modulus, strength, and bonding properties, ultimately having a negative effect on the stability of the volume. The process described here is likely the true cause of sulfate attack and degradation of concrete materials. After the addition of mineral admixtures, the physical and chemical actions of the admixtures cause the hydration products to transition from a cycle of dissolution, supersaturation, and recrystallization to mutual contact reactions, preventing sulfate ions and other ions from the outside environment from causing erosion to the concrete material through reactions between solutions. Therefore, the compacting action of mineral admixtures not only blocks the erosion channels of sulfate ions but also changes the reaction mechanisms between internal components, preventing sulfate erosion reactions from occurring. Due to the "solid-state reactions" between mineral admixtures and hydration products, a gradual transition with a very small gradient is formed between product A and another particle B, weakening the effect of the interfacial transition zone and significantly improving the compositional structure and properties of the material. The amount and type of mineral admixtures have a significant impact on the morphology and size distribution of hydration products in composite cementitious materials, which in turn affects the final performance of the composite cementitious materials. This can be roughly summarized as follows.

1) In the composite cementitious system with mineral admixtures, the crystal sizes of $Ca(OH)_2$ and AFt in both the transition zone and the matrix are smaller than those in the corresponding areas of ordinary Portland cement. When the amount and type of mineral admixtures are the same, the crystal sizes of $Ca(OH)_2$ and AFt in both the transition zone and the matrix differ significantly.
2) The reason for the refinement of $Ca(OH)_2$ crystals by compound mineral admixture may be that dual and triple mineral admixtures improve the particle size distribution of the binding material, making the system stacking more compact and reasonable (including the interfacial transition zone), resulting in a denser structure of the hardened cement paste (including the matrix and interface transition zone), thus leaving $Ca(OH)_2$ and AFt with insufficient space to grow. The high pozzolanic activity of mineral admixtures, especially ultra-fine slag powder and

silica fume, consumes most of the edges, corners, and faces of the $Ca(OH)_2$ crystal in the hydration products, causing it to lose its complete shape and refine the crystal size. At the same time, they also play a role similar to the "nucleation effect," providing a location for crystal nucleation, allowing $Ca(OH)_2$ crystal nuclei to form rapidly, but the nuclei do not have enough time to grow, affecting the morphology and size of the hydration product $Ca(OH)_2$ and AFt crystals.

3) As the composite of mineral admixtures and the extension of the hydration age, fibrous and acicular gels transition to reticulated gels and flocculent gels earlier due to the secondary pozzolanic effect of mineral admixtures consuming $Ca(OH)_2$, causing the needle-like products to decompose and promoting more flocculent products from cement hydration.

2.3.3 Mechanism and Performance of Multi-component Supplementary Cementitious Materials

2.3.3.1 Selection Principles for Multi-component Supplementary Cementitious Materials

Domestically and internationally, a large amount of industrial waste residues are mixed into concrete as mineral modifiers. The initial intention is to utilize industrial waste residues to save cement and reduce costs. With the in-depth study of high-tech concrete theory, industrial waste admixtures are added to cementitious materials. In addition to effectively reducing the excessive consumption of resources and energy, as well as environmental and ecological destruction, the addition of mineral admixtures can significantly improve the structure and performance of cement-based materials, enhance later strength, improve durability, such as corrosion resistance and impermeability, and significantly inhibit alkali-aggregate reactions. These issues have attracted great interest from professionals at home and abroad. At the same time, the use of mineral admixtures can lead to low early strength of materials and deterioration of concrete structures and some properties due to excessive dosage. To address these issues, some scholars have used the method of compounding two or more types of industrial waste residues and combined them with high-efficiency water reducers and other admixtures to prepare high-tech concrete. This method, based on the "superposition effect" of composite materials, mixes different types of admixtures into concrete at appropriate ratios and total amounts, allowing them to complement each other's strengths and weaknesses. It can not only adjust the water demand but also improve the compressive strength of concrete, as well as increase the flexural strength, reduce shrinkage, and enhance durability [36].

In the selection of mineral admixtures in composite cementitious materials, the following principles are generally followed:

1) Economic Principle
   Commonly used admixtures in multi-component composite cementitious materials include silica fume, fly ash, slag powder, steel slag powder, etc. (Figure 2.18). Silica fume and slag powder are currently recognized as the most effective active materials, but they are limited in quantity. For example, China

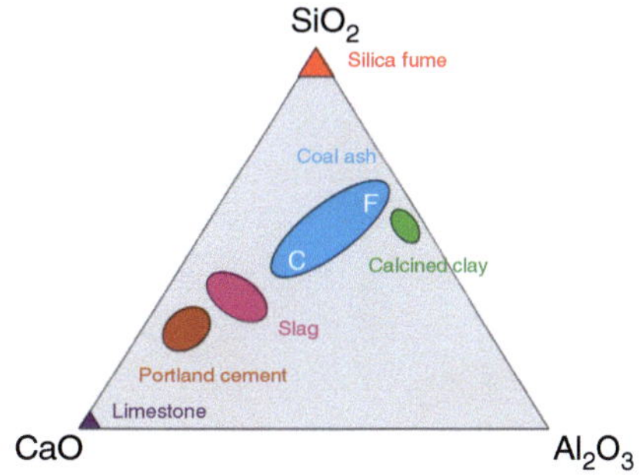

Figure 2.18 CaO-SiO_2-Al_2O_3 phase diagram of various types of supplementary cementitious materials.

lacks silica fume resources, and the price is expensive (2500–3000 yuan/ton), which can no longer meet the needs of concrete engineering. Other active materials, such as pozzolan, diatomite, zeolite powder, pumice powder, and rice husk ash, also have problems with insufficient resource distribution and total amount. Industrial waste residues that are widely distributed and abundant, such as steel slag, have large fluctuations in their composition and lack in-depth theoretical research and immature utilization technologies.

2) Close-Packing Principle

The particle packing density of the cement-based material system has an important impact on the performance of the hardened paste. The higher the particle packing density of the mixture system, the faster the hydration reaction process, the stronger the microstructure of the system, and the higher its mechanical performance. Therefore, optimizing the particle grading of the constituent materials and improving their packing density is the key to preparing ultra-high-strength and ultra-high-performance concrete.

Dense packing density is a basic principle in the design of high-tech concrete and is also applicable to the design of ultra-high-strength and ultra-high-performance concrete. Much of the research on packing density was completed by French researchers from the 1970s to the 1980s. In the design of ultra-high-strength and ultra-high-performance concrete mix proportions, packing density can be applied to the design of cementitious materials [37, 38]. Increasing density is similar to reducing the porosity of hardened cement paste. The capillary porosity rate controls the strength and durability of the hardened paste. Measures to reduce porosity mainly include reducing the water–cement ratio and particle packing. For example, ultra-fine powders smaller than cement particles, such as silica fume, can be used as filling materials to fill the pores in hardened cement paste. The dense packing effect of two-dimensional ordinary concrete and ultra-high-strength, ultra-high-performance concrete is shown in Figure 2.19. From Figure 2.19(a), it can be seen that the gaps between the particles in ordinary concrete are larger and the packing is loose; from Figure 2.19(b), it can be seen that the gaps between the particles in ultra-high-strength, ultra-high-performance concrete are smaller, and the packing is denser [39, 40].

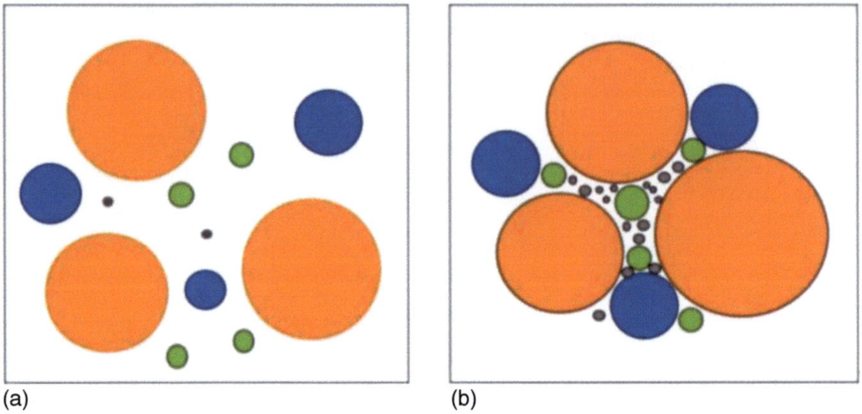

Figure 2.19 The compactness of concrete packing. (a) Ordinary concrete and (b) ultra-high strength, ultra-high performance concrete.

3) Hydration Matching Principle of Mineral Components

The hydration process of cement-based materials can be considered as a hardening process controlled by nucleation, crystallization, and crystal growth. These actions control the entire hydration and hardening process of cement. In multi-component cementitious materials, the mineral admixtures used each have different hydration reactivity characteristics. When the promotion of hydration by mineral admixtures dominates, the degree of hydration of each component is deeper; when the inhibitory effect of mineral admixtures dominates, the degree of hydration of each component decreases. As the dosage increases, the hydration reaction rate, reaction mechanism, and hydration process of each component will change significantly, thereby affecting the formation of the microstructure of hardened cement paste and its properties. Improper selection of mineral admixtures can severely affect the final performance; therefore, when designing multi-component cementitious materials, the interaction mechanisms between components should be fully considered.

The composite design of various mineral admixtures introduces more secondary center materials into the cement paste, which can reduce the spacing between secondary center materials, improve the particle gradation of secondary center materials and secondary media, and enhance the degree of secondary center material effect, significantly improving the structure of harden cement paste. On the other hand, $Ca(OH)_2$ crystals have a high degree of crystalline orientation at the interface, which is considered a "negative center material" compared to the dense hydration product structure [41–44]. The secondary hydration reaction of active fine admixtures not only consumes a large amount of $Ca(OH)_2$ but also generates more and more closely bound C-S-H gel, calcium aluminate, calcium sulfoaluminate, and other hydration products. These hydration products, as secondary nucleation centers, fill the pores of the cement paste together with unreacted fine particles, refining the pore size. As the hydration age develops, the number of harmful large pores in the cement paste decreases, and harmless or less harmful small pores or micropores increase, that is, the pore structure is improved, and the cement paste becomes denser. The composite of mineral admixtures of different particle size scales can effectively increase the number of center materials, reduce the spacing between center materials, increase particle dense stacking, and enhance the degree of center material effect. At the same time, it can also reduce negative center materials such as micropores and large $Ca(OH)_2$ crystals, reduce porosity, improve pore structure, and thus significantly enhance the strength of hardened cement paste.

2.3.3.2 Mechanism of Multi-component Supplementary Cementitious Materials

For most mineral admixtures, their effects when added to concrete generally include microaggregate effect, morphological effect, pozzolanic effect, and interfacial effect, but different mineral admixtures may exhibit positive or negative effects depending on the form of the effect, which mainly depends on the physical state and chemical composition of the mineral admixtures. If the physical properties and dosage ratios of the admixtures are properly controlled, the comprehensive positive effects exhibited by multi-component composite mineral admixtures added to the concrete are greater than those of single mineral admixtures. This can be attributed to the synergistic effect and interactions of multi-component composite mineral admixtures.

1) Composite of Microaggregate Effects
   In concrete powder, the particle size of cement is the largest, followed by ground slag powder and fly ash, with silica fume being the smallest (Figure 2.20). If the powders in the cementitious materials are mixed in appropriate proportions, it is possible to form a good continuous micro scale of powder materials in concrete. During the hydration process, the cementitious material particles of different particle sizes fill each other, reducing the gaps between particles, thereby further

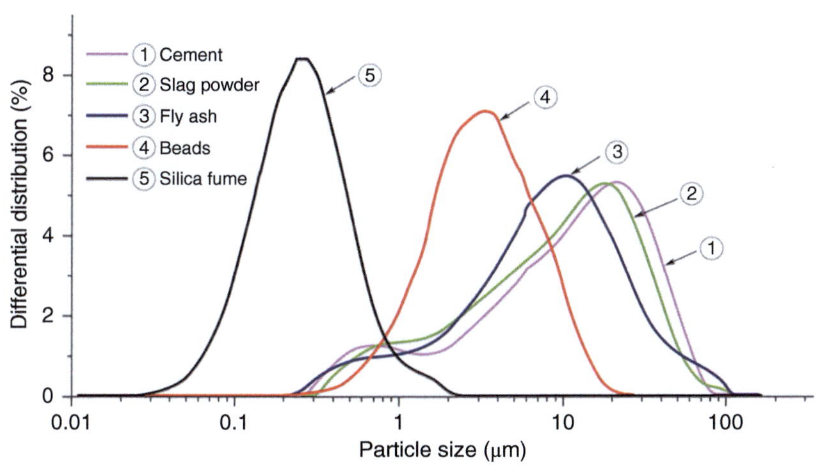

Figure 2.20 Particle size distribution of different supplementary cementing materials and cements [28].

reducing the total porosity of the composite cementitious material system after setting and hardening, which may reduce the permeability of concrete.

2) Composite of Morphological Effects
   The particle shape, fineness, and distribution of admixtures have varying degrees of impact on the hydration extent, depth, and properties after the hardening of concrete. Due to the irregular and rough surface of slag powder particles, their addition to concrete may reduce the workability of fresh concrete. Silica fume has a very small particle size and a large specific surface area, which requires a significant amount of water to maintain workability. However, the combination of slag powder with fly ash and slag powder with silica fume with fly ash can compensate for this loss, playing a certain plasticizing and water-reducing effect, which is beneficial for the formation of a dense concrete structure. Therefore, the composite of multi-component cementitious materials can have a complementary effect on morphology, improving the macroscopic performance of concrete from the microscopic structure.
3) Composite of Interface Effects
   During the mixing and casting of concrete, a layer of water film is formed around the aggregates, resulting in a higher water–cement ratio near the aggregates than in areas away from the aggregates. This leads to large capillary pore volumes in the interface transition zone, preferential orientation of $Ca(OH)_2$ crystals, and a large number of microcracks. Therefore, the interfacial transition zone of concrete is usually the weakest link in the chain of concrete properties. The addition of one or several types of slag powder, fly ash, and silica fume can reduce the formation of $Ca(OH)_2$ in concrete and inhibit the growth of $Ca(OH)_2$ crystals in the interface zone. At the same time, slag powder, fly ash, and silica fume have small particle sizes and good water retention, which can inhibit the formation of water films around aggregates, thereby improving the structure of the interface transition zone and enhancing the bonding strength at the colloid-aggregate interface (Figure 2.21). Therefore, regardless of whether slag

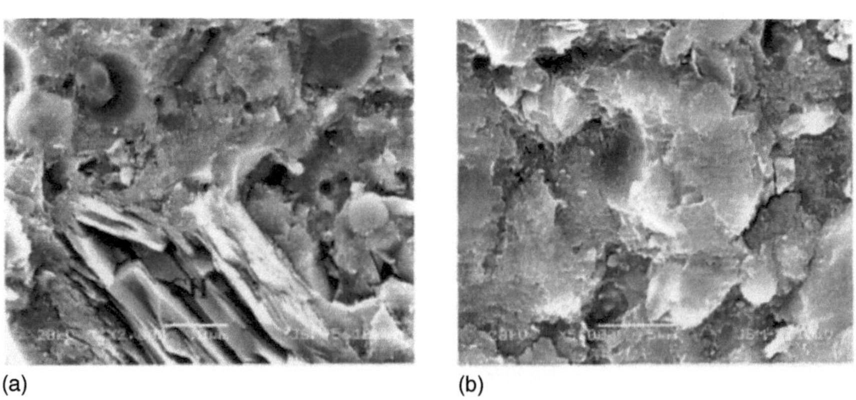

(a) (b)

Figure 2.21 Schematic diagram of the improvement of interface performance by the consumption of $Ca(OH)_2$ crystals by supplementary cementitious materials. (a) Effect of single mixing of fly ash and (b) multi-compound blending effect.

powder, fly ash, and silica fume are added in binary or multi-component forms, their interfacial effects are positive.

4) Composite of Pozzolanic Effects
Materials such as slag powder, fly ash, and silica fume all exhibit pozzolanic reactivity in composite cementitious materials. These mineral admixtures, when combined to form cementitious materials, can mutually enhance each other's reactivity during the hydration process, producing a composite cementitious effect. In the composite cementitious system, cement clinker always hydrates first, forming C-S-H and $Ca(OH)_2$. The $Ca(OH)_2$ and gypsum in the cement can activate the hydration of slag powder, fly ash, and silica fume. For silica fume, due to its higher hydration activity and surface energy compared to slag powder and fly ash, it reacts quickly in the cement paste, helping to increase the C-S-H gel. For slag powder, the released CaO can promote the formation of C-S-H gel and AFt (when gypsum is present) around fly ash particles, thus promoting the dissolution of aluminum and silicon phases in fly ash particles, increasing the concentration of aluminum and silicon in the hydration liquid phase, which in turn accelerates the hydration process of slag powder and silica fume (Figure 2.22) [32].

In addition, high-quality industrial wastes such as slag powder and pozzolanic material powder, after mixing with cement and water, first undergo hydration of the cement clinker minerals, followed by the reaction of the released $Ca(OH)_2$ during the hydration process of the clinker minerals with the active components in the mixed materials, generating hydration products. Since the cement-based system is alkaline, a certain degree of alkalinity is required to maintain its structural stability, such as the protective performance of concrete, resistance to carbonation, and the stability of various hydration products, with $Ca(OH)_2$ being the main source of alkalinity in the system. With the addition of admixtures, secondary reactions occur, consuming $Ca(OH)_2$ and reducing alkalinity, which improves the engineering performance of the material but has a significant impact on the stability of the cementing and crystalline phases, especially when secondary reactions occur after the hardening of the cement paste, the transformation of hydration products will inevitably lead to poor volume stability of concrete. Moreover, the alkalinity in the pore solution changes the composition and properties of the cement pore solution, thereby also affecting the growth environment of cement hydration and hydration products. The change in alkalinity in the pore solution system also has a certain impact on the composition, structure, morphology, and stability of hydration products [45].

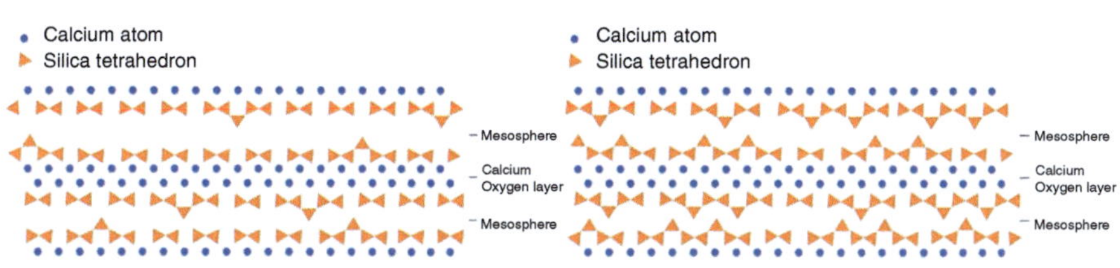

Figure 2.22 Schematic diagram of the formation of different types of C-S-H gel structures in the supplementary cementitious material–cement system.

In summary, multi-component composites can make greater use of supplementary cementitious materials compared to single and double composites. In addition, the synergistic effect of ternary mineral admixtures can achieve better overall performance than single and double composites.

2.4 Prospects for Design and Development of Composite Cementitious Materials

Multi-component composite cementitious systems are primarily designed and prepared for use in ultra-high-strength and ultra-high-performance concrete. Considering the material development strategy, technical requirements, and economic viability, the design classification of these systems can be evaluated according to the three-tiered criteria shown in Figure 2.23 [46].

The first tier mainly focuses on performance and economic viability, aiming to design materials with high strength, excellent durability, and good volume stability. It emphasizes the technical and economic rationality of the materials, with a high cost-effectiveness ratio being the primary goal of this level of design. The second tier builds upon the first, taking into account resource and energy consumption, as well as environmental protection issues. The design goal at this level is for concrete materials to have high technical indicators, high economic viability, and good environmental compatibility. To achieve this goal, it is necessary to extensively use industrial waste as active fine admixtures in concrete materials, realizing the recycling of industrial waste such as steel slag and fly ash in concrete, while also minimizing the use of cement and silica fume to reduce greenhouse gas CO_2 emissions during cement production. The second tier falls within the category of environmentally-friendly concrete.

The third tier is the highest level, with the design philosophy aimed at enabling concrete materials to fully leverage their performance and functional technical

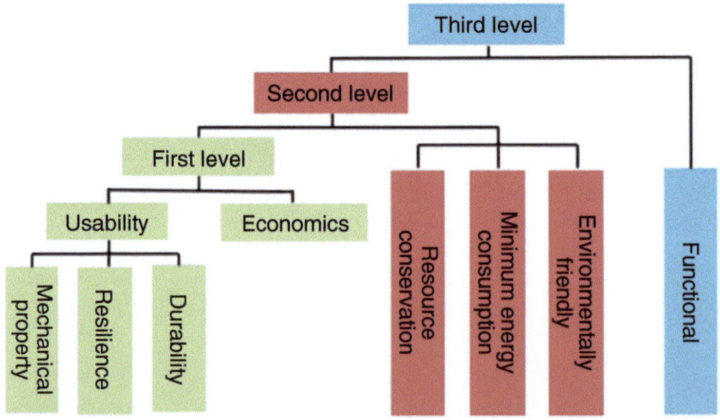

Figure 2.23 Reference standards for the design level of multi-component composite cementitious materials.

advantages. The main purpose of this level of design is to ensure that the material meets the performance requirements of engineering technology while also possessing high functionality (such as seismic resistance, radiation protection, and so on), achieving a harmonious unity of high performance, high functionality, and environmental compatibility.

In recent years, both domestic and international researchers have adopted the multi-component composite supplementary cementitious material model for the second and third-tier designs, preparing high-tech concrete materials such as ultra-high-strength concrete materials, ultra-high-performance concrete materials, and cement-based composite materials without macroscopic defects. The designed and developed multi-component composite cementitious material components are divided into the following types: (1) cement–steel slag–fly ash–silica fume composite cementitious system; (2) cement–slag powder–fly ash–silica fume composite cementitious system; (3) cement–steel slag–fly ash–slag powder composite cementitious system; (4) cement–fly ash–slag powder–limestone powder composite cementitious system; (5) cement–silica fume–slag powder–quartz fine powder composite cementitious system; and (6) cement–calcined clay–limestone powder–slag powder system. Of course, in addition to the above multi-component cementitious systems, with the continuous in-depth research on high-tech concrete materials, various types and functions of cementitious systems are still being developed and studied. However, the mechanism and performance of supplementary cementitious materials play to their respective advantages, achieving a "superimposed" effect. Due to space limitations, this section will not elaborate on each one.

It should be pointed out that in the face of the huge consumption of resources and energy in the traditional cement industry and the society nowadays that advocates energy saving, emission reduction, and a low-carbon economy in all industries, the production technology of the traditional cementitious materials industry has reached a bottleneck and its sustainable development capability is questioned. It is imperative for the traditional cementitious materials industry to move toward sustainable development. The use of mineral admixtures not only effectively solves the production pressure and environmental pressure of traditional cementitious materials but also effectively reuses waste from other industries. Although single and dual incorporation of mineral admixtures can effectively improve the workability, mechanical properties, and durability of new cementitious materials, with the development of concrete technology, dual incorporation of cementitious materials still cannot meet the requirements in some aspects. Effective compounding based on the characteristics of various mineral admixtures, such as mineral phase composition, hydration characteristics, and particle size matching, to achieve a "superimposed" effect is expected to further develop concrete technology. Recent research has also found that multi-component supplementary cementitious materials with triple or even multi-component incorporation have superior comprehensive performance compared to single and dual incorporation, achieving significant improvement in concrete performance, which has a significant impact on the development of the entire concrete industry.

References

1 Hu, S. (2010). *Special Cement*. Wuhan: Wuhan University of Technology Press (in Chinese).

2 Hu, S. and Li, Y. (1997). Research progress of limestone silicate cement. *Cement Guide for New Epoch* 3 (5): 11–13. (in Chinese).

3 Ding, Q., Li, Y., and Hu, S. (1998). Research on magnesium slag as cement admixture. *Cement Engineering* (3): 37. (in Chinese).

4 Wei, J., Li, Z., Chen, P. et al. (2013). Research on calcined carbonaceous shale used as cement mixtures. *Cement Engineering* (6): 22–23+22. (in Chinese).

5 Ming, Y., Chen, P., Li, L. et al. (2012). Experimental study on preparation of composite cement by double mixing natural zeolite and ethylene furnace slag. *Cement Engineering* (1): 26–29. (in Chinese).

6 Chen, X. (2014). Research on the preparation of composite cement with high content of coal slag and manganese slag. *Cement Engineering* (2): 10–12+41. (in Chinese).

7 Li, Y. and Hu, S. (1997). Development and application of high content limestone silicate cement. *Cement Engineering* (05): 29–31. (in Chinese).

8 Li, Y., Ding, Q., and Hu, S. (2007). Utilization of limestoneas mineral admixturein cementand concrete. *Journal of Wuhan University of Technology* (3): 35–41. (in Chinese).

9 Chen, Y. and Xu, Z. (2008). *The Scientific Basis for the Preparation and Application of High-performance Cement*. Beijing: Chemical Industry Press (in Chinese).

10 Liang, W., He, Y., Ji, H. et al. (2009). Research on the durability of high content fly ash faceplate concrete. *Concrete and Cement Products* (02): 14–16. (in Chinese).

11 Hu, S., Gen, J., Lu, L. et al. (2015). The effect of ultra-fine fly ash on the hydration process of different fineness portland cement under low water cement ratio. *Cement* (01): 6–8. (in Chinese).

12 Han, J. (2011). *Study on the Degree of Polymerization of C-S-H Gel, the Hydration Product of Cement with High Content of Fly Ash*. Wuhan: Wuhan University of Technology (in Chinese).

13 Hu, S. (2009). *Advanced Cement-Based Composite Materials*. Beijing: Science Press (in Chinese).

14 Grutzeck, M., Atkinson, I., and Roy, D. (1983). Mechanism of hydration of CSF in calcium hydroxide solutions. The First International Conference on the Use of Fly Ash Silica Fume, Slag and Natural Pozzolans in Concrete, 643–664.

15 Sánchez De Rojas, M.I., Luxán, M.P., Frías, M. et al. (1993). The influence of different additions on portland cement hydration heat. *Cement and Concrete Research* 23 (1): 46–54.

16 Hu, S., Peng, Y., Chen, K. et al. (2009). Preparation and deformation performance of reactive powder concrete (RPC) with steel slag powder. *Journal of Wuhan University of Technology* 31 (01): 26–29+33. (in Chinese).

17 Hu, S., Wang, H., and Ding, Q. (2006). The preparation of a high activated and slight expansive admixture base on steel slag and the study on its performance. *Concrete* (08): 47–49. (in Chinese).

18 Hu, S., He, Y., Ding, Q. et al. (2008, June 04). A production process of steel slag powder. CN 200510019018. 7. (in Chinese).

19 Peng, Y. (2009). *A Study on the Composition, Structure, and Properties of Steel Slag Activated Powder Concrete*. Wuhan: Wuhan University of Technology (in Chinese).

20 Xiong, F., Liu, X., He, Y. et al. (2009). Investigation on ant-cracking ability of mortar with high volume steel slag powder. *Journal of Wuhan University of Technology* 31 (05): 1–4. (in Chinese).

21 Hu, S., Ding, Q., Wang, H. et al. (2006, November 29). Steel slag metakaolin composite cementitious material and its preparation method. CN 200510018695. 7. (in Chinese).

22 Peng, Y., Chen, K., and Hu, S. (2011). Influence of the characteristics of steel slag powder particle on compressive strength of reactive powder concrete. *Journal of Building Materials* 140 (4): 541–545. (in Chinese).

23 Qiang, W. and Peiyu, Y. (2010). Characteristics of hydration products of steel slag. *Journal of the Chinese Ceramic Society* (09): 133–136. (in Chinese).

24 Guan, S. and Hu, S. (2007). Mechanism of curing schemes influence on volume change of steel slag compound cement mortar. *Cement Engineering* (3): 25–28. (in Chinese).

25 Kakali, G., Tsivilis, S., Aggeli, E. et al. (2000). Hydration products of C3A, C3S and portland cement in the presence of $CaCO_3$. *Cement and Concrete Research* 30 (7): 1073–1077.

26 Liu, Y. (2018). *The Influence of Steaming Time and Mineral Admixtures on the Microstructure and Mechanical Properties of UHPC*. Wuhan: Wuhan University of Technology (in Chinese).

27 Liu, M., Wang, F., and Ding, Q. (2010). Study on the hydro- thermal dissipation law of massive concrete mixing with slag powder and fly ash. *Concrete* (01): 21–24. (in Chinese).

28 Hu, S., He, Y., Lu, L. et al. (2008, June 04). A composite cementitious material and its preparation method. CN 200610019244.X. (in Chinese).

29 Ding, Q., Geng, J., Hu, S. et al. (2009). Different effects of fly ash and slag on anti-rebar corrosion ability of concrete with chloride ion. *Wuhan University Journal of Natural Sciences* (4): 355–361.

30 Zhang, L. (2002). *Study on the Hydration Characteristics and Product Properties of Multi-Component Cement-Based Materials*. Wuhan: Wuhan University of Technology (in Chinese).

31 He, L. (2016). *Research on the Preparation of Medium and Low Strength Steel Tube Concrete Using Steel Slag as Cementitious Material and Fine Aggregate*. Wuhan: Wuhan University of Technology (in Chinese).

32 Guan, S. (2008). *Research on the Activity and Cementitious Properties of Steel Slag Powder and Its Concrete Properties*. Wuhan: Wuhan University of Technology (in Chinese).

33 Hu, S., Ding, Q., Wang, H. et al. (2007, February 14). High activity compensating shrinkage mineral admixture and its preparation method. CN 200510018820. 4. (in Chinese).

34 Zhang, Y., Sun, W., and Hu, S. (2001). Influence of mineral admixtures on the shrinkage of binder paste in HPC. *Building Technique Development* (11): 24-26+60. (in Chinese).

35 Peng, Y., Hu, S., and Ding, Q. (2010). Preparation of reactive powder concrete using fly ash and steel slag powder. *Journal of Wuhan University of Technology-Materials Science Edition.* (2): 349–354. (in Chinese).

36 Hu, S., Guan, X., and Ding, Q. (2002). Research on optimizing components of microfine high-performance composite cementitious materials. *Cement and Concrete Research* 32: 1871–1875.

37 Peng, Y., Hu, S., and Ding, Q. (2009). Dense packing properties of mineral admixtures in cementitious material. *Particuology* 7 (5): 399–402.

38 Zhu, M., Zhang, H., and Hu, S. (2011). Research on the characters of ball grinded tricalcium silicate grains and fractal theory. *Journal of Wuhan University of Technology* 33 (2): 56–58. (in Chinese).

39 Wang, F. (2014). *Preparation and Performance Study of Low Viscosity Ultra High Strength (c100) Concrete.* Wuhan: Wuhan University of Technology (in Chinese).

40 Hu, S., Zhang, Y., and Ding, Q. (2000). Research on the hydration heat of binder pastein high performance concrete. *Journal of Building Materials* (3): 202. (in Chinese).

41 Wang, F., Qi, G., Liu, X. et al. (2016, January 13). An inorganic admixture for adjusting the strength and viscosity of ultra-high strength concrete. CN 201410071585. 6. (in Chinese).

42 Zhang, L., Hu, S., and Ding, Q. (2002). The research on the microstructure and its development of cement-based systems. *Journal of Wuhan University of Technology* (06): 11–14. (in Chinese).

43 Lu, L., He, Y., Ding, Q. et al. (2004). Hydration heat release behavior of multi-component cement-based materials. *Cement* (9): 1–3. (in Chinese).

44 Mao, R. (2017). *Study on Microstructure and Mechanical Properties of Reactive Powder Concrete.* Wuhan: Wuhan University of Technology (in Chinese).

45 Zhang, G. (2016). *The Microstructure Formation and Evolution Mechanism of C-S-H Gel in Portland Cement Pastes with Granulated Blast Furnace Slag Under the Attack of Corrosive Ions.* Wuhan: Wuhan University of Technology (in Chinese).

46 Hu, S. and Wang, F. (2006). *Lightweight Aggregate Concrete.* Beijing: Chemical Industry Press (in Chinese).

3

Functional Materials for High-tech Concrete

In essence, the difference between modern concrete and traditional concrete is the extensive use of other raw materials in the material design and preparation process of concrete, except for cement, aggregates, and water. It can be seen that the development history and technical path of high-tech concrete, the application of material science design theory and technical methods, and the exploration of various functional materials are its prominent features. Through the introduction of functional materials, various properties of cementitious composites can be significantly enhanced, the efficiency of material preparation can be improved, and new functions can be imparted. In particular, it can eliminate the inherent defects of materials and expand the space for new applications. Therefore, understanding and utilizing functional materials is essential to researching and developing high-tech concrete.

In the previous chapter, the properties and effects of various supplementary cementitious materials (SCM) with hydration activity were described. Studies have shown that the role of SCM is realized by participating in the hydration reaction of cement minerals. This chapter will introduce the commonly used functional materials for high-tech concrete, especially some advanced functional materials that have been adopted and researched by the author's team in recent years, including high-performance composite expansion materials, polymer-modified materials, fiber-reinforced materials, ultra-fine particle materials, as well as nano-seeds, internal curing materials, and functional aggregates.

3.1 Concrete Admixtures

There are numerous studies on concrete admixtures, and their research and application results are mature and stable. Here, we will give a brief description of the concepts of admixtures from the view of material systematics, focusing on the high-performance composite expansive materials, which have an important role and technical characteristics in the research and development of high-tech concrete by the author's team.

High-Tech Concrete Materials: Design, Preparation, and Applications, First Edition.
Fazhou Wang and Shuguang Hu.

3.1.1 Types of Concrete Admixtures

Concrete admixture is a kind of material that is added before or during the mixing process of concrete to improve the properties of freshly-mixed and hardened concrete. Concrete admixture has become one of the indispensable and important components of modern concrete and is an important technology for concrete modification. The most important function of concrete admixtures is to improve the workability of freshly mixed concrete, the physical and mechanical properties, and the durability of hardened concrete. At the same time, the research and application of admixtures also promote the development of concrete production and construction techniques and the emergence of new concrete types.

Concrete admixtures can be roughly divided into four categories according to their main function [1]:

(1) Admixtures that improve the rheological properties of freshly mixed concrete, including a variety of water-reducing agents, water-retaining agents, plasticity-retaining agents, flocculants, pumping agents, etc.
(2) Admixtures that regulate the setting time and hardening properties of concrete, including retarders, accelerators, and quick-setting agents.
(3) Admixtures that improve the physical and mechanical properties and durability of concrete, including early-strength agents, toughening agents, air-entraining agents, defoamers, waterproofing agents, rust inhibitors, etc.
(4) Admixtures that improve other properties of concrete, including expansion agents, antifreeze agents, mold inhibitors, moisture-proofing agents, sterilizing agents, coloring agents, and so on.

Admixtures are highly variable in their chemical composition; some admixtures have more than one function. Therefore, attention needs to be paid to its characteristics when using.

The main role of concrete admixtures [2] are as follows: First, improving the performance of concrete, mortar, and cement paste, which mainly includes

(1) Enhancing workability without increasing water usage, or reducing water usage while maintaining the same workability.
(2) Shortening or lengthening the initial setting time.
(3) Reducing or avoiding shrinkage or producing minor expansions.
(4) Changing the rate of water secretion or the amount of water secretion or both.
(5) Reducing segregation.
(6) Improving permeability and pumpability.
(7) Reducing the rate of slump loss.

Second, improving the properties of hardened concrete, mortar, and cement paste, which mainly includes

(1) Retarding or reducing the heat of hydration.
(2) Accelerating the rate of early strength growth.
(3) Increasing strength (compression, tension, or flexure).

(4) Improving durability or resistance to severe exposure conditions, including the application of antifreeze salts.
(5) Reducing capillary water movement.
(6) Reducing liquid phase permeability.
(7) Control of alkali reaction with certain aggregate components to produce expansion.
(8) Formulating porous concrete.
(9) Improving the bond between concrete and reinforcement.
(10) Increasing the bond between old and new concrete.
(11) Improving resistance to impact and abrasion.
(12) Stopping the corrosion of metals buried in concrete.
(13) Formulating colored concrete materials.

3.1.2 The Functionalization Principle of Concrete Admixtures

Admixtures can be divided into two categories in terms of their principle of action: one type of admixture acts on the cement–water system simultaneously by affecting the surface tension of the water and through adsorption on the surface of cement particles, and the other type of admixture dissociates into the ionic species and affects the chemical reaction between cement and water within minutes to hours of addition.

After adding admixtures to concrete, physical effects will occur, such as forming adsorption films on the surface of cement particles, changing the potential, and generating different attractive or repulsive forces. Some will disrupt the flocculation structure, enhance the stability of the cement dispersion system, and improve the conditions for cement hydration; others can form large molecular structures, altering the adsorption state on the surface of cement particles. Some will reduce the surface tension and surface energy of water. A few also participate directly in chemical reactions, generating new compounds with cementitious components.

The use of high-performance water-reducing agents allows for the thorough dispersion of cement particles, significantly reducing water usage and fully utilizing the reactivity of cement. This results in denser cement paste, with improved microstructures of the pore structure and interfacial transition zone, thereby greatly enhancing the physical and mechanical properties of concrete. In terms of impermeability, chloride ion diffusion, carbonation, sulfate attack resistance, and impact and wear resistance, these concretes outperform those without admixtures. They not only increase strength and workability but also enhance the durability of concrete. It can be said that only by mixing high-performance water-reducing admixture, the preparation of high-performance concrete with high construction ability, high strength, and high durability can be realized [3, 4].

3.1.3 The Main Types of Expansive Agents and the Mechanism of Action in Concrete

A great number of research and application practices show that the application and role of concrete expansion agents in high-tech concrete materials are very important and unique, which is different from the general concept and role of concrete admixtures. Functionally, concrete expansion agents can effectively regulate the stability of concrete volume, improve durability, and generate designable composite effects. Qualitatively, a concrete expansion agent is a functional material that lies between admixtures and supplementary cementitious materials. It directly participates in hydration reactions, and its reaction products greatly affect the structure and properties of concrete. Based on its characteristics and importance, this section provides a focused introduction to it.

The principle of concrete expansion is to produce volume expansion through chemical reactions in concrete. Hence, it can effectively compensate for volume shrinkage, densifying the material structure, preventing material cracking, and so on. Each expansion type is different in terms of raw material composition, expansion principle, performance characteristics, compensation technology, and method of use. Types of concrete expansion materials are as follows [5]:

3.1.3.1 Ettringite-type Expansion

The chemical reaction that causes the volume expansion of hardened cement paste is due to the formation of a certain amount of ettringite phase with the molecular formula $3CaO{\cdot}Al_2O_3{\cdot}3CaSO_4{\cdot}32H_2O$ (i.e., trisulfide-type calcium sulfoaluminate [AFt]).

At present, the main types of expansion sources for ettringite include calcium aluminate (barium, strontium), calcium sulfoaluminate, and alunite. The corresponding expansion raw materials are aluminate cement clinker, calcium sulfoaluminate (barium and strontium) clinker, calcined alumina, and natural alumina. Its main expansion reaction formulas are as follows:

$$3CA + 3CaSO_4 + 38H_2O \rightarrow C_3A \cdot 3CaSO_4 \cdot 32H_2O + 2(Al_2O_3 \cdot 3H_2O) \quad (3.1)$$

$$3CA_2 + 3CaSO_4 + 47H_2O \rightarrow C_3A \cdot 3CaSO_4 \cdot 32H_2O + 5(Al_2O_3 \cdot 3H_2O) \quad (3.2)$$

$$C_4A_3\overline{S} + 2(CaSO_4 \cdot 2H_2O) + 31H_2O \rightarrow C_3A \cdot 3CaSO_4 \cdot 32H_2O + 2Al(OH)_3 \quad (3.3)$$

$$K_2SO_4 \cdot Al_2(SO_4)_3 \cdot 4Al(OH)_3 + 13Ca(OH)_2 + 5CaSO_4 + 78H_2O \rightarrow 3AFt + 2KOH \quad (3.4)$$

Depending on the type of various aluminum phase reactants, the solid-phase volume increases by about 107–133% after reacting with gypsum to form ettringite, depending on the type of reaction:

(1) Calcium Sulfoaluminate Expansion
The main expansion source in this type of reaction is anhydrous calcium sulfoaluminate ($3CaO{\cdot}3Al_2O_3{\cdot}CaSO_4$, $C_4A_3\overline{S}$), and the expansion behavior of this type of expansive agents is closely related to the aluminum–silicon ratio (N) of the aluminate clinker, gypsum dosing, and particle fineness after grinding. The aluminum–silicon ratio represents the relative content of $C_4A_3\overline{S}$ and $2CaO{\cdot}SiO_2$

in sulfoaluminate clinker; the larger the N is, the higher the mineral content in clinker $C_4A_3\overline{S}$. The higher the mineral content, the greater the volume expansion. When the gypsum doping increases, the expansion rate increases, and the expansion stability period is extended. Fineness also has a greater impact on the expansion behavior of calcium sulfoaluminate expansion agent; when the particle is finer, the free expansion rate decreases. However, when the particle is too coarse, the expansion stabilization period of the expansion agent is longer, and it is easy to form ettringite in the later stage and produce the destructive effect, so the fineness of calcium sulfoaluminate expansion agent should be controlled in a suitable range.

(2) Calcium Aluminate Expansion
The main source of expansion in this type of reaction is the reaction of aluminum-containing calcium aluminate minerals, which produces volume expansion, along with the generation of AFt. At the same time, the aluminum gel, produced during the hydration of calcium aluminate expansion agent helps densify the hardened matrix. The reaction rate of monocalcium aluminate (CA) is higher than that of dicalcium aluminate (CA_2). Hence, the expansion mode of aluminate expansion agents with different clinker mineral compositions is different. The early-age expansion of expansion agent is larger when the content of CA is higher, and the stabilization period is early. While the mid- and late-phase expansion rate is larger when the content of CA_2 is higher, the expansion is unstable, and it is prone to cause harmful expansion. At the same time, the crystallization of light-burnt aluminate cement clinker mineral is incomplete, the grain size is fine, and thus the reaction speed with gypsum is faster, and the stabilization period is shorter. While the grain of over-burnt aluminate cement clinker mineral is coarse, the crystallization is complete, the reaction with gypsum to generate calcium alumina is slower, and the stabilization period is longer. The main factors affecting the expansion behavior of aluminate expansion agents are the mineral composition of aluminate cement clinker, calcination system, the gypsum content, and dosage, as well as the fineness of the expansion agent.

(3) Alunite Expansion
The mineral phase of alunite contains a large amount of $K_2SO_4{\cdot}Al_2(SO_4)_3{\cdot}4Al(OH)_3$ minerals, which has high solubility in the solution with the presence of $Ca(OH)_2$ and gypsum, and can dissolve a large amount of Al^{3+} and $SO_4{}^{2-}$, which react with $Ca(OH)_2$ and gypsum to form AFt. The above reaction produces calcium aluminate at a slower rate, thus the expansion stabilization period is a little longer, and the expansion rate of some alumina expansion agents can even increase during the middle and late stages of hydration hardening. Generally, the mixing of some fly ash and other SCMS in this type of expanding agent can help to improve the stability of expansion. The expansion behavior of the alumina expansion agent is also related to the variety and content of raw materials, alumina, and gypsum, as well as the fineness of the product. Due to the removal of a large amount of water of crystallization, the degree of crystallization of calcined alumina is greatly reduced. Additionally, as an

expansion agent component, it is easier to dissociate in water to provide active Al_2O_3, so that the generation of calcium aluminate speeds up. Due to the high alkali content in the alumina expansion agent, when applied to cementitious materials containing alkali-reactive aggregates, it will increase the possibility of alkali-aggregate reaction, and thus its scope of use is somewhat limited.

3.1.3.2 Calcium Hydroxide Expansion

Calcium hydroxide type expansion uses calcium oxide as the source of expansion, and this type of expansion agent is made by calcining limestone and other raw materials. Free lime (f-CaO) is the active ingredient in calcium oxide type expansion, which generates $Ca(OH)_2$ upon hydration, leading to volume expansion. This type of expansion is characterized by a fast expansion rate and large expansion, but the limit of the self-stress value generated by the cementitious materials is small. At present, it is mainly used for foundation grouting of large equipment, and grouting of foundation bolts, so that the shrinkage of cementitious material is reduced and the volume stability is increased.

The use of CaO as an expansion agent is limited by the fact that the stability of $Ca(OH)_2$ is strongly influenced by pressure, which is different from ettringite. Moreover, $Ca(OH)_2$ differs from $Mg(OH)_2$ and hydrated calcium sulfoaluminate in that it is poorly gelled. Research shows that owing to the high solubility of $Ca(OH)_2$, it will dissolve under the action of pressurized water, and in the seepage project, a white substance will appear around the cracks, which is detected as $CaCO_3$, the carbonation product of $Ca(OH)_2$. It can be seen that CaO, as an expansion agent, is not suitable for waterproofing applications. In addition, when $Ca(OH)_2$ content in cementitious materials is high, it is easy to cause poor durability of cementitious materials under the action of aggressive media rich in ions such as Cl^-, SO_4^{2-}, Mg^{2+}, Na^+, and so on. Therefore, CaO-containing expansion agents should not be used in structures exposed to seawater or other aggressive aqueous environments.

3.1.3.3 Magnesium Hydroxide Expansion

The main component of magnesium hydroxide type expansion agent is MgO, which generates $Mg(OH)_2$ during hydration and leads to volume expansion. This type of expansion agent can be produced by calcining magnesite ($MgCO_3$) into MgO at about 1000 °C. Dolomite or dolomitic limestone, the main component of which is $CaMg(CO_3)_2$, can be used to produce MgO expansion agent by controlling the MgO content.

Magnesium oxide expansion agent has delayed expansion performance, which has a low hydration rate at room temperature. In the ambient temperature of 40–60 °C, the expansion rate of MgO hydrated to $Mg(OH)_2$ is obviously accelerated, and its expansion is stable after 1–2 months of hydration reaction. Therefore, it is generally applicable to the dam rock foundation backfill of large volume cementitious materials. The higher internal temperature rise of large volume cementitious materials in 3–7 d can accelerate the hydration reaction of MgO. At this time, the magnesium oxide expansion agent starts to expand and stabilizes in one year. Its expansion happens to occur in the stage of contraction of the temperature drop of cementitious materials. It may cause structural damage due to delayed expansion when used in the design of cementitious materials for civil and industrial construction used at room temperature.

3.1.3.4 Ferric Hydroxide and Ferrous Hydroxide Expansion

Iron powder-based expansion agents use iron hydroxide as the expansion source, and its main components are iron filings and oxidizing agents. Iron filings come from metal cutting waste. Oxidizing agents include perchlorates, permanganates, and so on. Ionic catalysts are usually chloride salts such as ferric chloride or calcium chloride, and there are other additives, such as water reducers, used as dispersants. First of all, the iron filings waste was treated to remove oil, carbon, and organic matter. The method is to put the iron filings in a kiln and burn them, followed by rapid cooling and magnetic separation to remove potentially reactive metal ions, such as zinc, aluminum, and so on. The iron powder expansion agent is made by adding the above raw materials to the treated iron filings, and the iron powder expansion agent can be divided into two categories: iron monomer and pre-mixed with a specified amount of cement and fine aggregates.

Iron powder expansion agents mainly cause expansion using the expansion effect of iron oxidation and rust. The surface of iron raw material in the expansion agent is oxidized under the action of the catalyst, and these iron oxides are gradually dissolved in the cementitious materials, and Fe^{2+} and Fe^{3+} are generated with OH^- in the cement hydration system to form colloidal $Fe(OH)_2$ and $Fe(OH)_3$, which results in volume expansion. The main characteristics of this type of expansion agent are good heat resistance, early expansion stability, and suitability for dry and high-temperature environments, but the expansion is not too high. At present, this expansion agent is used in small quantities, mainly as a shrinkage compensator for machine base plates, ground screw filling in hot workshops, floor base repair and caulking, and so on. Usually, it is used by mixing with mortar or cementitious materials at the construction site, and it should be used immediately.

3.1.3.5 Gaseous Expansion

The aluminum powder expansion agent, which is made by adding metallic aluminum powder to dispersants and other components, is a typical and commonly used type of gaseous expansion agent. During the hydration process of cement, it reacts to produce hydrogen, which causes expansion. In addition, hydrogen peroxide (H_2O_2) and other substances can also be as an expansion agents. This type of expansion agent is usually used to prepare lightweight wall materials such as aerated cementitious materials.

3.1.3.6 Composite Expansion

Since it is often difficult to realize the coordinated development of strength and shrinkage deformation of cementitious materials when a single type of expansion agent is used for the preparation of cementitious materials or shrinkage-compensating cementitious materials, it is necessary to carry out the composite design of multiple expansion component phases and make a composite expansion agent to obtain a good expansion performance. The author's team has carried out some work in the research and technological development of calcium sulfoaluminate-calcium oxide composite expansion agents [6]. The main raw materials for the production of this type of composite expansion agent are bauxite, limestone, and gypsum, which are calcined at about 1300 °C and then milled. This type of expansion agent, which is a kind of composite expansion agent with a double expansion source, contains both aluminate or sulfoaluminate minerals that can generate AFt, and f-CaO that can generate $Ca(OH)_2$. A serious of composite expanders can be developed according to design requirements. For example, EA (Expansion Agent) expanders use lime as the primary expansion source and ettringite as the secondary expansion source, while HF (high-performance) expanders use ettringite as the primary expansion source and lime as the secondary expansion source. According to the expansion characteristics, lime produces early expansion, and ettringite produces medium-term expansion. Other composite expansion agents include calcium sulfoaluminate-magnesium oxide composite expansion agents, calcium sulfoaluminate-alunite composite expansion agents, and calcium oxide-magnesium oxide composite expansion agents. Some of these kinds of expansion agents are composited by pre-prepared components, and some are made by calcining and finely grinding the proportionally prepared raw materials.

In addition, there are also composite expansion agents that are composed of a particular expansion agent component and other admixtures. They can adapt to the need to prepare cementitious materials for special properties, such as water reduction, early strength, antifreeze, pumping, slow setting, air-entraining properties, and so on. An example of this kind of composite expansion agent is the JQ-type cracking and compacting agent, which consists expansion agent, water-reducing admixture, early strength agent, and other components. It not only plays the role of expansion but also has the role of water reduction, early strength, enhancement, and seepage control.

3.1.4 Design and Preparation of High-energy Blended Expansive Agent

3.1.4.1 Component Design of Composite Expansion Agent

The author's team has researched and developed a high-energy blended expansive agent (HBEA) for the need of high-energy expansion and fine expansion control in some concrete composite structure projects. The principle of design HBEA is to adopt the high-energy expansion design scheme with large expansion energy, long expansion stability period, and sustainable expansion, adopt the technical route of performance matching with different expansion sources, and the composite composition is composed of the sulfoaluminate mineral $C_4A_3\overline{S}$, calcined RO (the solid solution consists of divalent alkaline-earth metal oxides such as mainly FeO-MgO-MnO) minerals (CaO and/or MgO), and RO solid fusion phases (predominantly overcooked solid fusion phases of CaO, MgO, and FeO) as the expansion composition with the expansion sources being AFt, $Ca(OH)_2$, and $Mg(OH)_2$, respectively. Among them, the calcined RO expansion reaction is mainly in the early stage, the C_4A_3 expansive reaction is mainly in the middle and late stages, and the RO solid-melt phase expansion reaction is mainly in the late stage. Using the matching of their different hydration and expansion characteristics, the requirement of adjusting the expansion properties for balanced and sustainable development can be achieved. The HBEA concrete expansion reaction is a coordinated matching effect, the reaction equation is shown in (3.3) as follows:

$$2Al(OH)_3\ (gel) + 3Ca(OH)_2 + 3(CaSO_4 \cdot 2H_2O) + 20\,H_2O \rightarrow AFt \tag{3.5}$$

$$CaO + H_2O \rightarrow Ca(OH)_2 \tag{3.6}$$

$$MgO + H_2O \rightarrow Mg(OH)_2 \tag{3.7}$$

3.1.4.2 Preparation and Effects of Composite Expansion Agent

The range of raw material compositions for HBEA is m(calcined RO minerals): m(sulfoaluminate cement clinker):m(RO solid fusion phase):m(gypsum) = (15–32%):(28–45%):(8–24%):(25–36%), in which the mass composition of sulfoaluminate cement clinker minerals ranges from $C_4A_3\overline{S}$: 50–82%, C_4AF: 3–13%, C_2S: 5–37%. $K_m[MgO/(FeO + MnO)] > 1$ in the over-burned RO phase. Gypsum is natural hard gypsum or calcined gypsum. RO solid fused phase is ground to a specific surface area of 400–500 m^2/kg, and the rest of the powder is ground to a specific surface area of ≥280 m^2/kg. HBEA has the advantages of high expansion efficiency and long expansion time, low alkali content, and low cost. In the configuration of high-energy expansion concrete, HBEA dosage generally accounts for 10–18% of the total cementitious material dosage, the self-stress of the concrete is 1.0–5.0 MPa, its 14 d restrained expansion rate is 0.03–0.05%, 90 d restrained expansion rate is 0.02–0.04%, and the restrained expansion rate begins to stabilize in 90 d. Table 3.1 shows the experimental results of high-energy expansion concrete configured with HBEA. The HBEA composition of this experiment is sulfoaluminate clinker 37%, over-heated RO phase 20%, hard gypsum 28%, and RO solid fusion phase 15%.

3.2 Polymer Materials

An important component of ultra-high-strength concrete (UHSC) is polymers, and the main mechanism of polymer action is analyzed in this chapter. From the existing research results, the addition of polymer in cementitious materials caused a series of changes from material processing characteristics, and hydration process to material properties, this change includes both the physicochemical action of not only the polymer itself in the cementitious materials but also the essential changes in the interface and structure of the material caused by the chemical action between the two phases [7–9].

3.2.1 Types of Polymers

The authors studied four types of polymers commonly used for cement and concrete (mortar) modification [10], i.e., water-soluble polymers, polymer emulsions

Table 3.1 Experimental results on the performance of concrete with high-energy composite expansion materials.

Cement (kg/m³)	HBEA (kg/m³)	*W/C*	Sand rate (%)	FDN (%)	28 d strength (MPa)	Restrained expansion rate/10^{-4}	
						14 d	90 d
300	0	0.46	39	0.8	36.7	−3.3	−5.0
270	30	0.46	39	0.8	36.4	3.0	1.9
246	54	0.46	39	0.8	36.3	3.9	2.4
460	0	0.39	40	1.1	52.1	−3.4	−5.2
414	46	0.39	40	1.1	52.7	3.6	2.8
377	83	0.39	40	1.1	52.9	4.5	3.5
520	0	0.33	39	1.3	59.2	−3.8	−5.0
468	52	0.33	39	1.3	61.1	4.0	3.0
426	94	0.33	39	1.3	58.4	4.3	3.4

Note: P.O. 42.5 cement was used, a standard experimental method.

(or dispersions), redispersible polymer powders, and liquid polymers. The general requirements for polymers used in cement and concrete are as follows:

(1) No negative effect on cement hydration.
(2) High stability against highly reactive ions released during cement hydration, such as Ca^{2+} and Al^{3+}.
(3) High mechanical stability, e.g., no emulsion breakage under high shear during metering, transportation, and mixing.
(4) Very good storage stability.
(5) Low air-entraining properties.
(6) The ability to form polymers that are compatible with the cement hydration products and aggregates in a hardened body.

In the hardened products, it can form a film layer with good adhesion to the hydration products of cement and aggregates, and the minimum film temperature should be low.

The polymer film formed should have excellent water resistance, alkali resistance, and weather resistance.

The polymers used for UHSC are mostly water-soluble polymers, such as polyvinyl alcohol (PVA), polyacrylamide (PAM), cellulose ether, and modified starch, and also a small amount of epoxy resin and unsaturated polyester.

3.2.2 Role of Polymers

The content presented in this section does not deal with the essential changes in the interface and structure of the material due to the chemical interaction between the two phases, but only provides some analyses concerning the plasticizing, pore-reducing, and toughening effects of the polymer itself in cement pastes and the process of structural development of the material, focusing on the physical and physicochemical mechanisms of polymer action in this type of material and discussing the relationship between these effects and the chemical action.

3.2.2.1 Plasticizing Effect of Water-soluble Polymer

The water-soluble polymer with polar hydrophilic groups of surface organic active substances, will make the electrical properties of the surface of the cement particles to change, as well as the production of the hydration layer of the three-dimensional protective effect, and so on, thus making the cement slurry in the cement particles dispersion, the surface to be wetting, slurry viscosity decreased, the relative sliding of particles easier to carry out, the resistance of the internal friction between the particles to reduce the flow of the slurry properties can be improved, easy to mold the material processing. The plasticizing effect of water-soluble polymer are as follows:

(1) Adsorption-Dispersion Effect
In the hydration process of cement, when the cement is mixed with water, there may still be some flocculent structure produced. This is a kind of structural product made of cement particles bonded together with each other to enclose part of the free water. These flocculated structures are wrapped with a lot of mixing water, thus reducing the effective water in the newly mixed slurry and affecting the fluidity of the cementitious materials mix. The plasticization theory about the water-reducing agent suggests that when the water-reducing agent is added to the cement slurry, the hydrophobic groups in the water-reducing agent are directionally adsorbed on the surface of the cement particles, while the hydrophilic groups are directed to the aqueous solution, which constitutes a monomolecular adsorption membrane. Due to the directional adsorption of surfactants, cement particles acquire surface charges of the same polarity. The resulting electrostatic repulsion not only stabilizes the cement-water suspension but also disrupts the flocculated structure formed during initial hydration, releasing trapped free water and thereby reducing water demand. Water-soluble polymer contains hydroxyl (–OH), carboxyl (–COOH), amino ($-NH_2$), acyl amino ($-CONH_2$), and other hydrophilic groups with polarity. Structurally, they are effective surface-active substances. Therefore, they can play the role of adsorption-dispersion in the cement paste.

(2) Wetting Effect
After mixing cement with water, the surface of its particles is wetted by water, and the wetting condition will have a great influence on the performance of fresh cement paste. Therefore, if the surfactant that can make the whole system's interfacial tension decrease is mixed, the cement particles could be dispersed effectively, and the wetting effect will also make the hydration area of the cement particles increase. On the other hand, related to wetting is the penetration of water molecules into the capillary pores of the cement particles. The stronger

the penetration, the faster the cement hydration. The addition of water-reducing agent can increase the penetration of water molecules into the capillary pores within a certain period. Water-soluble polymers with polar hydrophilic groups can play a wetting role in the cement paste.

(3) Lubrication

Water-soluble polymer has two types of different properties of the groups (branched chain), with a clear hydrophobic or hydrophilic effect. The highly polar hydrophilic groups are negatively charged and oriented on the surface of the cement particles, which can easily bonded with water molecules in the form of hydrogen bonds. This hydrogen bonding force is much greater than the molecular attraction between the molecule and the cement particles. The force of this hydrogen bonding is much greater than the molecular attractive force between the molecule and the cement particles. When the surface of the cement particles adsorbed a sufficient number of polymers, with the help of polar groups and water molecules in the role of hydrogen bonding, coupled with the water molecules between the hydrogen bonding of the bonding, the surface of the cement particles forms a layer of stable solvent solubilized water film, which prevents the cement particles in direct contact with each other. Simultaneously, the addition of polymers generally leads to the introduction of a certain number of fine bubbles. These bubbles are enveloped by the molecular films that are adsorbed in a directional manner by the polymers. Since the charge of the adsorbed film on the bubbles is the same as that of the cement particle adsorption film, there is an electrostatic repulsion between the bubbles themselves and between the bubbles and the cement particles. This repulsion contributes to the dispersion of cement particles and enhances their sliding ability, an effect referred to as the ball-bearing effect.

3.2.2.2 Pore-reducing Effect of Polymers

The pore-reducing effect of polymers is of twofold significance in low water–cement ratio polymer cementitious composites. Since the addition of polymer as a plasticizer can greatly reduce the amount of water used in the preparation of cement paste, the porosity formed by this excess water can be effectively reduced. At the same time, due to the increase in plasticity, some of the pores that may have been formed in the system are also more easily eliminated. On the other hand, the polymer itself has a very good plasticity, it can also be easier to fill in some of the voids of the inorganic matrix and play the role of filling and closing the pores, and this role for the overall performance of the material is very influential.

For brittle cementitious materials, it can be assumed that the pores are the cracks in Griffith's theory, and if the pore size can be effectively reduced, the fracture resistance of the material can be improved. Figure 3.1 shows the structural condition of the material in situ under both conditions, with and without the addition of polymer. The cement used for the experimental samples was Portland cement, the polymer was PAM, and the water–solid ratio was 0.4, all of which were mixed manually and slightly flattened with a spatula when made into a curing block, with the temperature of the curing room at 60 °C and the curing time at 28 d. The cement specimens

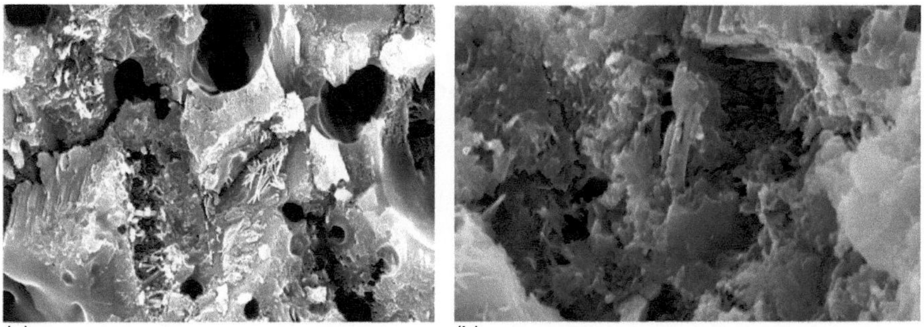

(a) (b)

Figure 3.1 Effect of polymer emulsions on the pore structure of cement pastes (a) No polymer added and (b) Polymer added.

were removed after hydration to age and polished to take a smoother section for scanning electron microscope (SEM) observation.

Figure 3.1(a) shows the hardened cement paste without PAM addition, from which a large number of pores can be seen, with the size of the pores on the order of several tens of micrometers (the magnification in the figure is 153 times). From the distribution, it is also extremely heterogeneous. Figure 3.1(b) shows the hardened cement paste with the addition of polymer, from which more obvious pores are not easily observed, and some small pores are also relatively uniform in distribution and irregular in shape, with sizes 1 to 2 orders of magnitude smaller than those of Figure 3.1(a) (magnification of 1000 times in the figure). In addition, from the comparison of the two photos, there are many round pits in Figure 3.1(a) that are remnants of holes in the section. However, in Figure 3.1(b), the surface is rougher and the residue left after the destruction of the pores is not visible, which indicates that the porosity and macropores are reduced due to the incorporation of polymer. When the destruction is carried out with external force, no aggregation of holes can be seen on the section, thus making the slurry more resistant to fracture.

3.2.2.3 The Toughening Effect of Polymers on Hardened Cement paste

One of the outstanding properties of low water–cement ratio polymer cementitious composites is the substantial increase in the toughness of the material, thus improving the inherent disadvantage of the brittleness of conventional cementitious materials. From the perspective of composites, it is reasonable to assume that this toughening comes from polymers with high toughness [5, 10]. Toughness is a measure of the energy absorbed per unit volume of material during fracture, and it is an important property that characterizes a material's ability to resist fracture. The more energy absorbed during fracture, the higher the toughness of the material and the stronger the fracture resistance. Conversely, the less energy absorbed in the fracture process, the lower the toughness of the material, and the poorer the ability to resist fracture. Figure 3.2 and Table 3.2 show the effect of different polymer dosages on the toughness of cementitious materials.

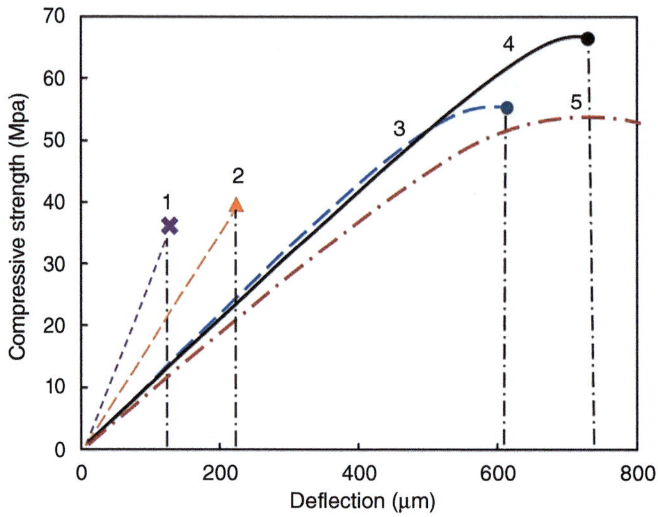

Figure 3.2 Effect of different polymer dosage on material toughness.

Table 3.2 Stress-deflection tests of materials.

Number	Polymer content (%)	*P* (N)	*b* (mm)	*h* (mm)	*σ* (MPa)	*f* (μm)
1	0	30.55	4.09	3.65	33.42	147
2	4	38.00	3.25	4.32	37.59	236
3	8	37.80	4.13	3.70	53.63	614
4	12	68.61	3.69	4.07	67.35	750
5	15	60.29	4.05	4.11	52.88	—

P is the external force applied in the test, *b* is the width of the specimen, *σ* is the stress, *h* is the height of the specimen, *L* is the span of the specimen in the test, and *f* is the perturbation value.

The results show that with the gradual increase in the amount of polymer, the toughness of the material continues to improve, and this toughening mechanism originates from the combined effect of three aspects:

(1) The Role of Polymer Energy Absorption
In general, when the material is strained by force, it will produce numerous very fine cracks in its interior. When the material is doped with polymer, there will inevitably be a significant portion of polymer cross-fill across the cracks, preventing the rapid development of cracks. The polymer consumes energy during its deformation, thus increasing the toughness of the material. The polymer is on the face of the growing crack such that the energy absorbed in the impact is equal to the sum of the fracture energy of the inorganic matrix and the work done to break the polymer. From the point of view of tensile yielding, it must be assumed that a large number of microcracks are formed,

each of which is held back by the polymer and separated from neighboring microcracks. Thus, large tensile strains can be realized by the opening of the microcracks, the elongation of the polymer body, and the bending of the material. In practical low-water–cement ratio polymer cementitious composites, this blocking, bridging action of the polymer body is connected as a whole because of the interpenetration of the two-phase network that is formed.

(2) Cracking Effect of Polymer
Polymer in the composite material is like a stress concentrator, it can concentrate a large amount of stress and produce a large number of cracks, consuming a large amount of energy. In composites, polymers play the role of both initiating cracks and controlling the development of cracks. Under tension, cracks are initiated at the points of maximum principal stress, and the cracks propagate along the plane of maximum principal stress. When the stress concentration at the crack tip decreases below the critical level required for crack growth, or when a large polymer particle or other obstacle is encountered, crack growth is terminated. The result is the appearance of a large number of small cracks, which is exactly the opposite of the formation of a small number of large cracks when no polymer is present in the same cement paste. Due to the uniform distribution and high density of polymer-induced microcracks within the material, the material is able to achieve higher strain energy densities through decentralized energy dissipation prior to fracture, a mechanism that directly explains its excellent tensile strength and impact resistance.

Figure 3.3 shows SEM photographs of crack extension in the body of the material with and without polymer doping. Figure 3.3(a) shows a large crack across the body of the material produced by the external stress of a plain hardened cement paste. Figure 3.3(b) is an SEM photograph of the hardened cement paste under strain after 10% PAM is added. The magnification of this figure is 10 times larger than that of Figure 3.3(a), but the large cracks across the material body cannot be seen in the figure; on the contrary, there are a large number of small cracks in the photograph. The doping of PAM has an obvious toughening effect on the hardened cement.

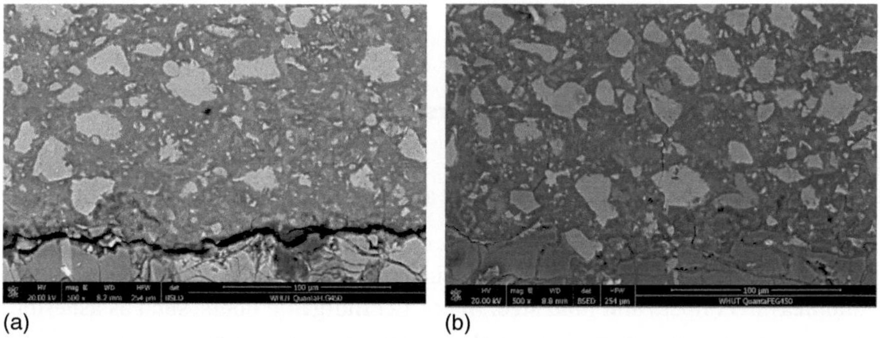

(a) (b)

Figure 3.3 Influence of polymers on crack expansion in hardened cement pastes (a) No polymer added and (b) Polymer added.

(3) Shear Yielding of the Polymer to the Matrix.
The toughening effect of polymers can also be produced by shear yielding in the matrix. In composites, due to certain exogenous factors, polymers can build up a hydrostatic tension around the matrix they are in contact with. The reason for this hydrostatic tension is, first, the thermal shrinkage difference; second, the dry shrinkage difference; and third, the mechanical effect. In the early stage of molding, the polymer absorbs water and swells, increasing its volume. During the molding, hydration, and maintenance of the material, the polymer will gradually produce dehydration. In the use of the material, in a dry environment, the polymer will lose water and shrink, and this shrinkage is numerically much larger than the dry shrinkage of the cement phase and therefore will produce hydrostatic tension on the surrounding cement phase. The specific manifestation of the hydrostatic tension effect is the generation of a certain degree of prestressing in the hardened cement paste as if it were an artificially made prestressing material, and when an external force is applied to the material, it is necessary to first overcome this pre-generated stress effect before the rest of the stresses can be applied to the material, which increases the value of the stresses that produce the damage, and an increase in the toughness of the material [11, 12].

The toughening in the actual cementitious material comes from the combined effects of all three. It should be noted that these effects must be better realized under suitable compatibility conditions between the organic and inorganic phases. The compatibility factor here mainly refers to the fact that the two phases should be well bonded to each other. When the two phases are completely excluded from each other, there is a clear interface between the phases, and the phases can be completely separated from each other. At this time, the material is left inside the weak parts, and the mechanical properties of the material (including toughness) are bound to deteriorate. The other extreme is when the two phases are not excluded, and like a "family," it is bound to make the formation of molecular dispersion between the two phases, the polymer body no longer exists, and the polymer of these physical and physicochemical actions will not exist.

3.3 Fiber Material

3.3.1 Types and Characteristics of Fibers

3.3.1.1 Types of Fibers

There are a wide variety of fibers that can be applied to cementitious materials, often with the following classifications:

(1) According to their properties, they can be divided into: (1) Metal fibers, such as stainless steel fibers and mild steel fibers. (2) Inorganic fibers, such as asbestos fibers, glass fibers, boron fibers, carbon fibers, etc. (3) Synthetic fibers, such as nylon fibers, polyester fibers, polypropylene fibers, etc. (4) Plant fibers, such as bamboo fibers, hemp fibers, etc.

(2) According to their elastic modulus, they can be divided into: (1) High-modulus fibers, such as steel fibers, carbon fibers, glass fibers, etc. (2) Low-modulus fibers, such as polypropylene fibers, certain plant fibers, etc. High-modulus fibers mainly improve the impact resistance, thermal explosion resistance, tensile strength, rigidity, and crack resistance of composite materials, while low-modulus fibers mainly improve the toughness, strain capacity, and impact resistance of cement composite materials related to toughness.
(3) According to their length, they can be divided into non-continuous short fibers and continuous long fibers, such as glass fiber untwisted roving, polypropylene fiber film, etc. At present, the fibers used for preparing fiber-reinforced concrete are mainly short fibers. Steel fibers, glass fibers, polypropylene fibers, and carbon fibers are commonly used to improve the mechanical and other application properties of cement mortar and concrete, including tensile strength, compressive strength, elastic modulus, crack resistance, durability, fatigue load life, impact and wear resistance, dry shrinkage and expansion resistance, fire resistance, and other thermal properties. Sometimes, long fibers or fiber products are also used, such as fiberglass mesh cloth and fiberglass felt.

The research and application results indicate that the factors affecting the composite effect mainly include the type, size, aspect ratio, volume fraction, orientation, shape, and surface condition of the fiber material.

3.3.1.2 Surface Treatment of Fibers

For fiber-reinforced materials, the mechanical behavior depends not only on the nature of the fibers and concrete but also on the bonding properties between them, which are represented by the interfacial layer between the fibers and the cement matrix. In the Portland cement system, the composition of the interfacial layer includes a double-layer membrane, $Ca(OH)_2$ crystal-enriched zone, and porous zone, and the total thickness of the interfacial layer varies from about 10 μm to more than 50 μm. To improve the bond strength between fibers and cement matrix, the thickness of the interfacial layer must be reduced as much as possible to improve the interfacial bond between fibers and cement matrix [13, 14].

Interfacial bonding can be improved by means of fiber treatment, and the improvement measures mainly include macro-treatment, micro-treatment, mechanical densification, and other methods. Macro-treatment is to improve the fiber surface shape, the formation of bundled multifilament, wavy, twisted, and buckled at both ends of the morphology, thus increasing the contact area of the fiber and the matrix, increasing the friction between the two, thereby enhancing the interfacial bonding properties of the fiber and the matrix. The micro-treatment is to increase the specific surface area and roughness of the fiber by ion treatment of the fiber surface, thus improving the wettability and chemical properties of the fiber to increase the friction and promote the chemical reaction between the fiber and the matrix, so that the bond strength of the interface between the fiber and the matrix is greatly improved. Mechanical densification is through the use of cylinders in the fiber mechanical densification processing, increasing the elastic modulus of the fiber at the same time to increase the roughness of the fiber surface, thus improving the fiber and cementitious interface bonding performance.

3.3.2 The Role of Fibers

3.3.2.1 Crack-blocking Effect of Fibers

There are inherent volumetric shrinkage in cement and concrete during hydration and hardening. These shrinkage deformations include plastic shrinkage, chemical shrinkage, drying shrinkage, autogenous shrinkage, and temperature shrinkage. Cracking of concrete materials is often caused by the fact that the stress generated by the volumetric deformation of the concrete itself under the action of restraint is greater than the tensile strength of the material itself. For ordinary concrete, cracking occurs when the tensile stress generated by shrinkage deformation is greater than the tensile strength of the concrete. Fibers are uniformly distributed inside the concrete and form a strong bond with the cement matrix, thus forming a whole with the matrix. In this system, the fibers play a role similar to that of reinforcing bars, which can significantly reduce the number of primary microcracks within the concrete and effectively inhibit the emergence and expansion of cracks. This reinforcing mechanism changes the damage mode of concrete from typical brittle fracture to approximate ductile damage. In addition, the fiber can form a three-dimensional system and effectively block the channel of water loss, reduce or delay the loss of water, and reduce the capillary shrinkage stress, while the fiber can also prevent the settling of aggregates and improve the uniformity of concrete. Therefore, the addition of fibers can greatly improve the crack resistance of concrete.

3.3.2.2 Reinforcing Effect of Fibers

Fiber-reinforced cementitious composites can be viewed as consisting of two phases: one is the matrix phase, i.e., the cementitious material, and the other is the dispersed phase, i.e., the fibers uniformly dispersed in the matrix phase. In the early stage of loading (tension and bending), the cement matrix and the fibers bear the external force together, and the former is the main force. After the cracking of the matrix, the fibers across the cracks become the main bearer of the external forces, i.e., the main resistance to external forces by the bridging force of the fibers. In this way, the high-strength, high-modulus dispersed-phase fibers constrain the matrix deformation and impede the dislocation motion of the matrix. When the crack continues to expand around the fibers, across the crack, the fibers will transfer stress to the uncracked concrete, and the crack tip stress concentration can not only be relieved but also likely disappear. The greater the ability of the dispersed phase to impede the dislocation motion of the matrix, the more pronounced the enhancement effect. The addition of fibers significantly improves the mechanical properties of concrete such as tensile, flexural, and shear resistance, as well as the long-term mechanical properties such as crack resistance and abrasion resistance, especially the high modulus of elasticity of the fibers can also greatly enhance the impact resistance of concrete [15–17]. The reinforcement of fiber-cementitious composites depends on the structure and properties of the fibers, matrix, and fiber–matrix interface, and their volume content.

3.3.2.3 Toughening Effect of Fibers

When the fiber is subjected to bending and tension, the matrix under tension cracks, the fiber will play a role in bearing tension and keep the matrix cracks slowly; the matrix cracks also maintain a certain residual stress. With the expansion of the crack, the matrix cracks between the residual stress gradually decrease, and the fiber has a large deformation capacity that can continue to bear the cross-section of the tensile force until the fiber is pulled off or pulled out from the matrix, the process is gradual, and the fiber in the process has played a significant toughening effect. As the load on the composite increases, the fibers will transfer additional stresses to the matrix through bonding stresses. If these bond stresses do not exceed the strength of the cement matrix, more cracks will appear in the matrix. This process of increased cracking will continue until either the fibers break or the bond strength fails and the fibers are pulled out. This process occurs gradually, and the fibers then have a significant toughening effect during this process. In the cracked section, the matrix is not able to withstand any stretching, and the fibers carry the full load of the composite.

3.4 Ultra-fine Powder

Cementitious composites are typical porous materials, and the performance is closely related to their pore structure. It has been shown that in concrete, with the increase of the packing compactness of the cementitious material particle system, the hydration reaction can be accelerated and the microstructure of the system can be enhanced. For ultra-high-strength cementitious materials with extremely low water–cement ratios, this will inevitably lead to difficulties in casting. Therefore, it is of great significance to study the properties of ultra-fine powder and their functioning principles in the cementitious material system, so as to make the system have a high degree of stacking densities, to improve and optimize the structure of the materials, and to enhance the performance of ultra-high-strength cementitious composites [18–22].

3.4.1 Commonly Used Ultra-fine Powder

The so-called ultra-fine powder is a powder with a particle size of less than 10 μm with a specific surface area equivalent to 600 m^2/kg and above. Commonly used ultra-fine powders include fly ash, ground fine water quenching slag powder, silica fume, ground fine zeolite powder, and mineral microfine powders of metakaolin, diatomaceous earth, burnt shale, boiled slag, and so on.

(1) High-Quality Fly Ash
Fly ash is the ash collected from the flue gas of thermal power plant boilers fueled by pulverized coal, consisting of a large number of spherical glass beads and a

small number of crystalline substances such as mullite quartz. Its active components are mainly reactive SiO_2 and Al_2O_3. High-quality fly ash generally refers to the graded ash with a particle size of less than 10 μm, its specific surface area is about 750 m^2/kg, and the loss on burning is 1–2%.

(2) Ultra-fine Slag Powder
Ground slag powder is an ultra-fine powder of granulated blast furnace slag ground to a specific surface area of about 750 m^2/kg and a particle size of less than 10 μm. Ground slag powder is the vast majority of unstable vitreous body, storage of high chemical energy, and high activity. The active component is generally considered to be reactive SiO_2 and Al_2O_3; they can react with $Ca(OH)_2$ even at room temperature to produce strength.

(3) Silica Fume
Silica fume is a kind of ash collected from the chimney when smelting ferrosilicon alloy or silicon in a ferroalloy factory. It is the SiO gas in the furnace oxidized into SiO_2 rapidly by air, condensed into particles, and then collected by the filter. Silica fume consists of very fine, amorphous, spherical particles, and the content of its main active ingredient SiO_2 accounts for about 90%. The average particle size is about 0.1 μm, and the specific surface area is about 15,000–20,000 m^2/kg, which is 50–100 times smaller than that of cement particles.

(4) Other Ultra-Fine Powder
From the literature, the ultra-fine powders that have been used after special treatment and finishing include ultra-fine zeolite powder, metakaolin, diatomaceous earth, ultra-fine powder, burnt shale, and boiling slag.

3.4.2 The Role of Ultra-fine Powder

The role of ultra-fine powder presented in this section includes all the generic properties of ultra-fine powders often applied in concrete. The hydration properties of SCM are systematically described in Chapter 2, and some of these SCMs, because they are often prepared as ultra-fine powders, have not only the hydration properties but also the same properties of all ultra-fine powders.

3.4.2.1 Filling Effect

As the particle size of ultra-fine powder is smaller than the cement particles, it fills the gaps between the cement particles and improves the compactness of the cementitious material. The compactness and strength of cementitious material after hardening with water are also increased. Cementitious material particle combination and void ratio changes, as shown in Figure 3.4. With cement particles sized at 10.4 μm and fly ash particles at 10.09 μm, the void ratio of the cementitious material shows almost no change, regardless of the mixing ratio of cement and fly ash. But when the particle size of fly ash is reduced to 0.95 μm, 70% of the cement and 30% of the fly ash composite is used, the void rate of cementitious material decreases from the original 40% to about 20%. The cement slurry structure after adding water is shown in Figure 3.5(a). There are a large number of free water,

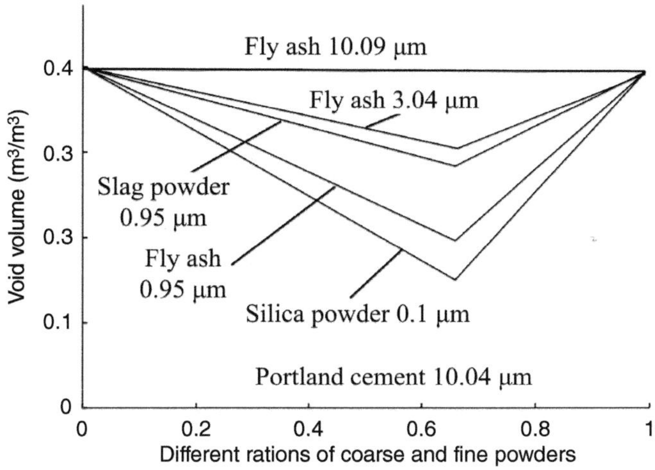

Figure 3.4 Variation of particle assemblage and void fraction.

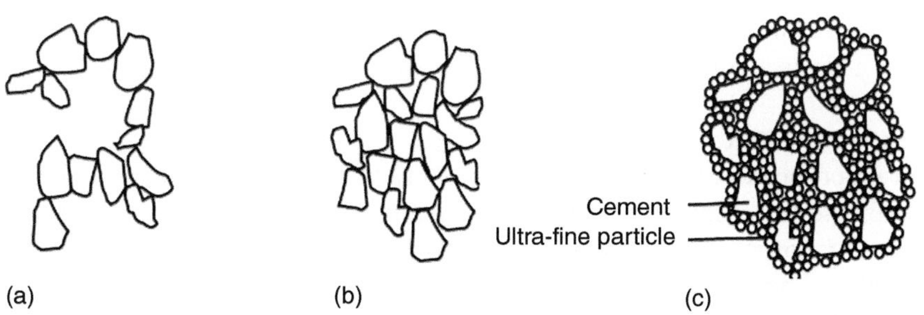

Figure 3.5 Filling role of ultra-fine powder in the cement paste structure: (a) Portland cement paste, (b) cement paste with high efficient water-reducing agent, and (c) cement paste with silica fume.

bound in the flocculent structure formed by the cement particles; this structure of the slurry fluidity is poor with a high porosity, and performs badly. The hardened structure of cement containing high efficiency water-reducing agent is shown in Figure 3.5(b). Due to the discharge of free water, cement particles have very tight connection, and the fluidity and performance after hardening are better than the portland cement paste (Figure 3.5(b)). Cement slurry containing highly efficient water-reducing agent and ultra-fine powder is shown in Figure 3.5(c); the ultra-fine powder fills the voids between cement particles and has higher strength and durability after hardening.

3.4.2.2 Fluidifying Effect

Part of the ultra-fine powder replaces cement in equal quantities can produce the fluidifying effect, but not all of the ultra-fine powder has the fluidifying effect. Although the partial replacement of cement by ultra-fine powders can fill the cement voids and discharge part of the water in the cement slurry, some ultra-fine powders, due to the high specific surface area or their porous structure that absorb water or require more free water wetting the surface, do not necessarily improve the fluidity of the slurry. This kind of ultra-fine powder includes natural zeolite powder and silica fume. It should be noted that the dosage of the high-efficiency water-reducing agent influences the fluidization of ultra-fine powder particles. This effect may arise from the adsorption of the agent molecules onto the powder surfaces, forming an electrostatic double layer. The resulting repulsive forces exceed interparticle gravitational attraction, enhancing particle dispersion and, consequently, improving slurry mobility.

3.4.2.3 Structural Optimization Effect

The structural optimization effect includes improving the interface structure between cement and coarse aggregate in concrete and improving the pore structure of cement in concrete. The interface between coarse aggregate and cement in ordinary concrete has a large amount of $Ca(OH)_2$, and these crystals are arranged directionally at the interface, which is one of the main reasons for the low strength and durability of concrete. Therefore, improving the interfacial structure between aggregate and cement by mixing part of the minerals plus ultra-fine powder in concrete is a key technology that must be solved for high-performance concrete.

3.4.2.4 Strength Effect

Different proportions as well as different types of ultra-fine powders have a great influence on the rate of strength growth. The strengthening mechanism of ultra-fine powder is as follows: in the cementitious material to which ultra-fine powder is added, as it can replace part of the cement, it reduces the content of tricalcium aluminate brought by the cement and correspondingly reduces part of $Ca(OH)_2$ generation. At the same time, the active ingredients SiO_2 and part of Al_2O_3 in the powder have a secondary hydration reaction with the cement hydration product $Ca(OH)_2$, generating a large number of amorphous C-S-H gel, filling the pores in the cementitious material, making the cementitious material more dense, and thus improving the mechanical properties:

$$xCa(OH)_2 + Al_2O_3 + mH_2O \rightarrow xCaO{\cdot}Al_2O_3{\cdot}nH_2O \tag{3.8}$$

$$xCa(OH)_2 + SiO_2 + mH_2O \rightarrow xCaO{\cdot}SiO_2{\cdot}nH_2O \tag{3.9}$$

3.4.2.5 Durability Effect

The durability effect of ultra-fine powder on concrete can be summarized into two aspects. First, it prevents the occurrence of durability issues in concrete. Generally speaking, mixing 30% fly ash, 40% slag powder, 20–25% natural zeolite powder, 15–20% metakaolin superfine powder, and 10% silica fume in concrete can effectively inhibit the harmful expansion of alkali–silica reaction. Second, it enhances the resistance of concrete to durability issues. All concrete structures are subject to the deteriorating effects of the external environment, such as temperature, humidity, salt damage, frost damage, saline corrosion, and so on. The addition of ultra-fine powder in concrete can make the electrical flux of concrete decrease significantly.

Tests have shown that after replacing part of the cement in concrete with different types of ultra-fine powder, the anti-chloride ion penetration performance of concrete is significantly improved.

3.5 Nano-seeds

3.5.1 Nano-seeds and Their Mechanism of Action

The use of mineral admixtures (fly ash, slag powder, calcined clay, etc.) for partial replacement of cement clinker is one of the most effective means for the green and sustainable development of the cement industry. Compared with cement clinker, the reactivity of mineral admixtures is greatly reduced, resulting in the slow development of the mechanical properties of cementitious materials with large quantities of mineral admixtures, which greatly limits the amount of mineral admixtures replacing cement clinker. Therefore, it is necessary to develop early strengthening techniques to enhance the incorporation of mineral admixtures in concrete. The traditional means of early strengthening mainly include thermal and chemical activation, but the steam (pressure) maintenance of high-energy consumption, early strengthening agent calcium chloride will be corrosive to the steel reinforcement, triethanolamine dosage is difficult to control, and there is a huge negative impact on the late mechanical and durability properties of cementitious materials. In recent years, the introduction of nano-materials in cementitious materials to accelerate the hydration of cementitious materials is a new technology called "Nano-Seeds" technology. Nano-materials in the appropriate dosage not only improve the early mechanical behavior of cementitious materials but also have no obvious retrogression phenomenon on the development of later mechanical properties, showing excellent advantages and application potential [23].

So far, the nano-materials used as nano-seeds in cementitious materials mainly include the following two categories: traditional nano-seeds and novel nano-seeds. Traditional nano-seeds can be further divided into inert and active nano-seeds, while new nano-seeds refer to calcium silicate hydrate (C-S-H gel) nano-seeds.

(1) Inert Nano-Seeds
Inert nano-seeds are mainly oxides such as nano-titanium dioxide, nano-zirconium dioxide, nano-iron trioxide, and carbon nano-tubes. The inert nano-seeds accelerate the hydration of cementitious materials mainly through physical effects (high specific surface area of nano-materials) but not through chemical reactions.

(2) Active Nano-Seeds
Active nano-seeds include a wide range of species, including not only some oxides such as nano-silica, nano-alumina, nano-metakaolin, and other amorphous silica-aluminum nano-materials but also calcium-rich nano-materials, such as nano-calcium carbonate, nano-$Ca(OH)_2$, and other calcium-rich nano-materials. The active nano-seeds can not only exert their physical effects

but also participate in a series of chemical reactions in the process of promoting the hydration of cementitious materials.

(3) C-S-H Seeding

As a new type of nano-materials, C-S-H seeding has been widely used in pure cement systems, cementitious systems with large dosages of admixtures, steam-free cementitious systems, alkali-inspired cementitious systems, and cementitious systems incorporated with functional additives. Since C-S-H seeding is similar to C-S-H gel in chemical composition and structure, it is considered to be the best nucleation site and catalyst for the rapid formation of C-S-H gel. Therefore, the promotion effect of C-S-H seeding on cement hydration is significantly better than that of conventional nano-crystalline seeding.

The formation of C-S-H gel is an autocatalytic process with the presence of spherical amorphous precursors in the precipitation process, following the two-step non-classical nucleation theory. The surface of C-S-H seeding contains numerous broken bonds and structural defects. Its high surface free energy enables the adsorption of ions and molecules, thereby modifying the cement hydration process. This mechanism mitigates the high-concentration shielding effect at the original mineral interfaces, as well as the crystallization pressure induced by localized precipitation and crystallization. It makes the hydration products grow rapidly and synchronously, uniformly dispersed in the whole system, thus obtaining a dense and uniform cementite structure.

3.5.2 Preparation of C-S-H Nano-seeding

C-S-H seeding preparation methods mainly include the pozzolanic effect, sol–gel method, chemical co-precipitation method, and so on.

The pozzolanic effect refers to the method of using $Ca(OH)_2$ (or calcium oxide obtained by calcining calcium carbonate), silicon source (amorphous silica, quartz glass, etc.), and water according to a certain calcium–silicon ratio, water–solid ratio, and reacted for a certain period of time at a certain temperature to obtain C-S-H nanocrystalline seeds. Advantages are wide source of raw materials and low price. Disadvantages are as follows: the reaction time can be as long as several weeks or months, and it is difficult to control the grain size of the crystal nucleus.

In the sol–gel method, ethyl orthosilicate is generally used as the precursor of the silicon source and calcium ethoxide is used as the precursor of the calcium source obtained by the sol–gel method. Advantages are as follows: the reaction time only takes a few minutes or hours, and the result is an amorphous nucleus. Disadvantages are complicated preparation process and high cost of raw materials.

In the chemical co-precipitation method, it is generally obtained by calcium nitrate as a calcium source and sodium silicate as a silicon source and obtained by stirring the reaction for a period of time under a certain atmosphere and temperature, and it is also the most widely used method for the preparation of C-S-H seeding. Advantages are simplicity, rapid reaction, low cost, and short reaction time of a few minutes to a few days. Disadvantage is the difficulty in controlling the grain size of the crystal seed.

The smaller the particle size of nano-seeds, the greater the tendency of agglomeration between particles, and once agglomerated, the effect of nano-seeds will be greatly reduced. Currently, the most commonly adopted method is to adjust the particle size of C-S-H seeding by using highly efficient polycarboxylate superplasticizers (PCE) as dispersants. However, the molecular structure of different PCEs varies greatly, and the dispersing effect of C-S-H seeding varies significantly under different synthesis conditions. With the same acid–ether ratio of PCE, the grafted carboxyl group in PCE has a smaller particle size and a stronger seeding effect than the sulfonic acid group, and with the same acid–ether ratio of PCE, the more structural units in the side chain of PCE, the more spatial site-barrier it can provide. In the case of the same PCE acid–ether ratio, the more PCE side-chain structural units, the stronger the spatial site-blocking effect, the smaller the size of crystal seed particles, the stronger the effect of crystal seeding, the larger the PCE acid–ether ratio, the stronger the ability of PCE adsorption on the C-S-H crystal seed, the smaller the size of crystal seed particles, and the stronger the effect of crystal seeding. There is a threshold value for the anionic charge density of the PCE main chain under different pH conditions. The greater the anion charge density, the greater the ability to adsorb PCE, the smaller the crystal seed particle size, and the stronger the crystal seed effect.

3.5.3 Influence of C-S-H Seeding on the Hydration Process of Cementitious Materials

The C-S-H seeding acts as the optimal heterogeneous nucleation site during the dissolution-migration-deposition process of tricalcium silicate ions in contact with water, which leads to the rapid formation of C-S-H gels at an early stage, and thus significantly shortens the duration of the induction period in the hydration stage of tricalcium silicate. The earlier the hydration temperature peak appears, the greater the slope of the accelerated period. The incorporation of C-S-H seeding nano-seeds in the cement–fly ash system could significantly accelerate the consumption of $Ca(OH)_2$ in the slurry, which led to a significant advancement of the reaction time of pozzolanic initiation in fly ash. The earlier appearance of calcium carbonate diffraction peaks in the slag powder/calcined clay composite cement system and the higher peak intensity indicate that it should have a promoting effect on the aluminum phase of the mineral admixture.

Our team proposed for the first time the use of sodium sulfate as a green alkali activator and the composite use of C-S-H seeding to improve the compressive strength and hydration rate of fly ash cement pastes with 50% of low-calcium fly ash addition at the early stage (1d). As shown in Figure 3.6 and Table 3.3, the combined use of C-S-H seeding and sodium sulfate enhanced the 1 d compressive strength, cumulative exothermic heat of hydration, and the degree of fly ash reaction of fly ash cement pastes with high fly ash dosage to a greater extent than the sum of sodium sulfate and C-S-H seeding alone, which showed synergistic effects [24]. In addition, as shown in Figure 3.7, the combined use of C-S-H seeding and sodium sulfate synergistically promoted the early compressive strength and the cumulative exothermic heat of

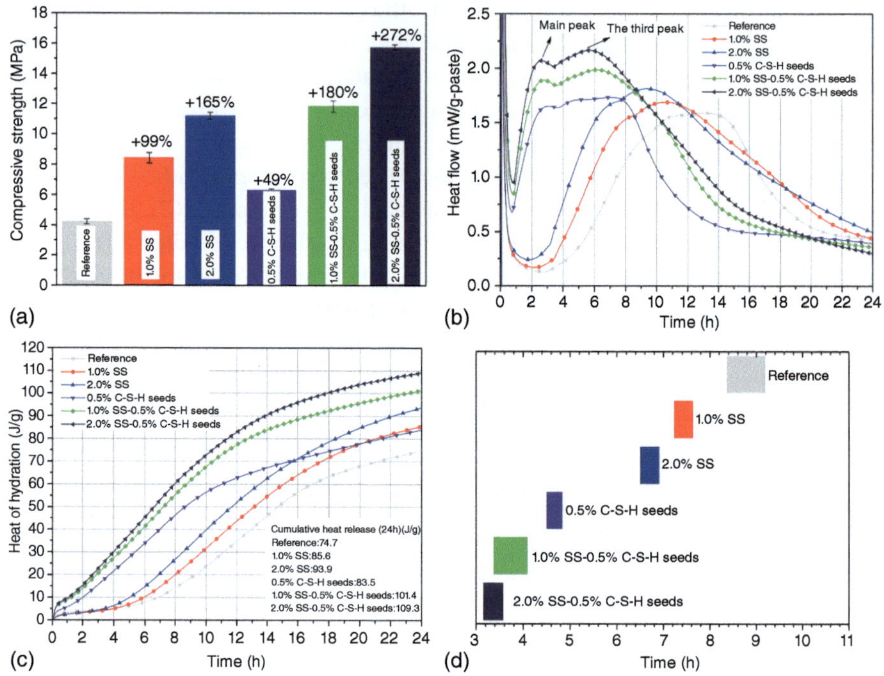

Figure 3.6 Effect of sodium sulfate and C-S-H seeding on the acceleration of hydration of fly ash cement paste: (a) 1d compressive strength, (b) hydration heat flow, (c) cumulative heat release, and (d) setting times.

Table 3.3 Degree of 1 d reaction between cement and fly ash in different systems.

Sample	Degree of reaction of cement (%)	Degree of reaction of fly ash (%)
Unhydrated slurry	—	—
Control the hydrated slurry	45.86	5.41
2.0% sodium sulfate	55.17	9.81
0.5% C-S-H seeding	51.57	6.79
2.0% sodium sulfate –0.5% C-S-H seeding	64.94	18.72

hydration of the two low-calcium fly ash cement pastes. The higher the content of the amorphous aluminum phase in low-calcium fly ash and the earlier the age of the slurry, the more obvious the synergistic promotion effect [25].

Although C-S-H seeding has been extensively researched and applied in cementitious materials, researchers have also found that C-S-H seeding has some shortcomings. First of all, the early strength effect of C-S-H seeding is significantly different when the dosage of C-S-H seeding is increased in cement with different mineral compositions. In addition, there exists a threshold value

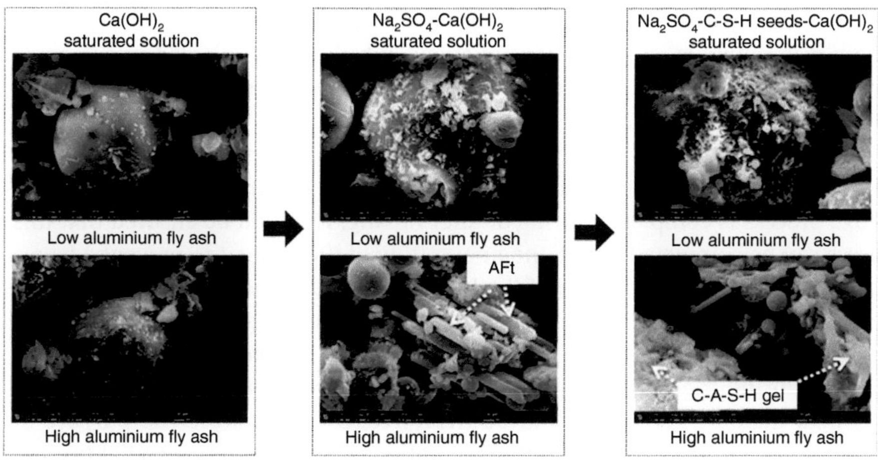

Figure 3.7 Mechanism diagram of the reaction process of sodium sulfate and C-S-H seeding with fly ash in the cement pore solution. *Source:* With permission of [25] / American Chemical Society.

for the effect of C-S-H seeding on hydration promotion by the dosage of C-S-H seeding. Excessive dosage leads to agglomeration in cementitious materials and a decrease in slurry compactness. In addition, threshold doping is related to the mineral phase composition in the cementitious materials. Finally, the effect of C-S-H seeding on cementitious materials with different water–cement ratios varied widely. The six hours compressive strengths of mortars with two water–cement ratios (0.30 and 0.35) with C-S-H seeding were significantly improved. 1d compressive strength of mortars with a water–cement ratio of 0.30 was basically at the same level as that of the blank group, while mortars with a water–cement ratio of 0.35 still showed an increase of more than 20%. 3d and 28d compressive strengths of mortars with a water–cement ratio of 0.30 had shown a strength regression of more than 15%, while those with a water–cement ratio of 0.35 showed a strength regression of more than 20%. The mortar with a 0.30 water/cement ratio showed more than 15% strength regression at 3d and 28d, while the mortar with a 0.35 water/cement ratio showed no strength regression.

3.6 Internal Curing Functional Materials

3.6.1 Overview of Internal Curing

Internal curing mainly uses lightweight internal curing functional materials as carriers to introduce water sources into low water–cement ratio cement-based materials and maintain the internal relative humidity of cement-based materials, thereby reducing the self-weight of concrete, compensating for shrinkage, and enhancing the matrix. Internal curing functional materials refer to a class of materials that adsorb moisture through chemical or physical action and have certain water storage and retention functions, which can be divided into organic superabsorbent polymer (SAP) materials and porous inorganic materials. Internal curing functional materials can either pre-absorb and store water before concrete mixing or complete water storage during the mixing process, depending on the material properties of the internal curing functional materials. When the water storage process is completed during or after mixing, it should be ensured that this process can be completed before the final setting of the cement to prevent adverse effects on the internal structure of the hardened cement paste. At the same time, the internal curing functional materials also need to have appropriate mechanical strength to ensure that they will not break during the concrete mixing process. In addition, during the cement hydration process, the water in the internal curing functional materials is required to be freely released to achieve the purpose of compensating for water and promoting hydration. The water in porous lightweight aggregates (LA) is generally released in turn from large pores in LA to small pores in hardened cement paste under the action of capillary pore pressure, while the water inside SAPs is mainly released in turn under the action of osmotic pressure and humidity difference. Finally, the internal curing water is required to be uniformly distributed in the cement matrix, so the uniform dispersion of internal curing functional materials is required. Uniformly dispersed internal curing water can more effectively increase the water content as a whole, promote cement hydration, and thus improve the overall structure of the hardened cement paste. For porous LA, their particle size is generally large, and there is no agglomeration phenomenon, which makes them easy to disperse, while SAPs are difficult to disperse due to their small particle size and the formation of hydrogels after water absorption. The uniform dispersion of internal curing functional materials is not only affected by the properties of the materials themselves but also by the mixing process.

The author's team has carried out systematic research on the absorption and release mechanism of organic and inorganic internal curing functional materials in cement paste and the influence of these materials on the properties of concrete. Based on this, high-strength LA concrete and rapid repair concrete for islands in the South China Sea have been prepared and successfully applied to the construction of urban high-rise buildings and islands and reefs in the South China Sea.

3.6.2 Organic Superabsorbent Polymer Materials

SAP is a polymer material with a three-dimensional crosslinked network structure that can absorb water hundreds or even thousands of times its weight. It was initially widely used in fields such as food, medicine and health, agriculture, and forestry. The use of SAP as an internal curing agent in concrete was first proposed by Jensen et al. in 2002 [26, 27]. The dosage of SAP in cement-based materials is low, generally 2%–1% of the mass of cementitious materials. According to the different sources of raw materials, SAP can be divided into three categories: starch-based, cellulose-based, and synthetic. Among them, the most commonly used SAPs for concrete internal curing are polyacrylic acid (PAA), polyacrylamide (PAM), or

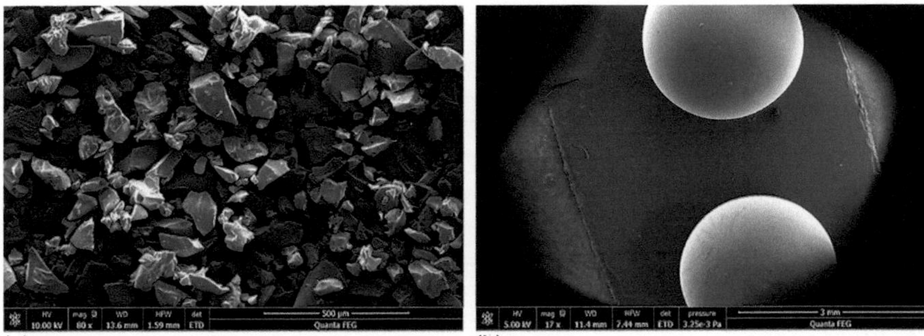

(a) (b)

Figure 3.8 SAP prepared by the sol–gel method (a) and the inverse suspension method (b). *Source:* With permission of [31] / Elsevier.

their copolymer (PAA-PAM), which all belong to the synthetic category. Currently, there are two main synthesis methods for SAP: the sol–gel method and the inverse suspension method. The sol–gel method involves dissolving polymer monomers, crosslinking agents, and initiators in water, and after a certain preparation process under certain conditions, the product is obtained after drying and powdering. Therefore, the SAP prepared by the sol–gel method is basically irregular broken particles with uneven particle size distribution. The inverse suspension method uses oil as the dispersing medium, and monomers as the aqueous phase, which are dispersed in the oil medium to form a suspension by relying on suspension stabilizers. Therefore, the SAP prepared by the inverse suspension method is basically spherical with uniform particle size distribution [28–30]. Figure 3.8 shows the different particle morphologies of SAP prepared by the sol–gel method and inverse suspension method, as photographed by the author's team.

3.6.2.1 Water Release Model of SAP in Cement Paste

The early state evolution of SAP inside the cement paste is the key to the influence of SAP on the microstructure and macroscopic properties of cement-based materials. At the same time, the internal hydration environment of cement-based materials is very complex, and there are many potential factors (such as ion type, ion concentration, pH value, water–cement ratio, and humidity) that affect the water release behavior of SAP. The research team explored the water release behavior and the main factors affecting the water release of SAP at different stages (early plastic stage and later hardening stage) by simulating the ion concentration, pH value, and internal relative humidity during the cement hydration process, and established a water release model and mechanism of SAP after water absorption in cement-based materials at different stages (including both the early plastic stage and the later hardening stage) [32].

(1) Water Release Law of SAP under Different Environments

(1) To investigate the water release behavior under different solution concentrations, which simulates the salt concentration environment inside concrete, salt solutions with the same pH value but varying total concentrations were prepared

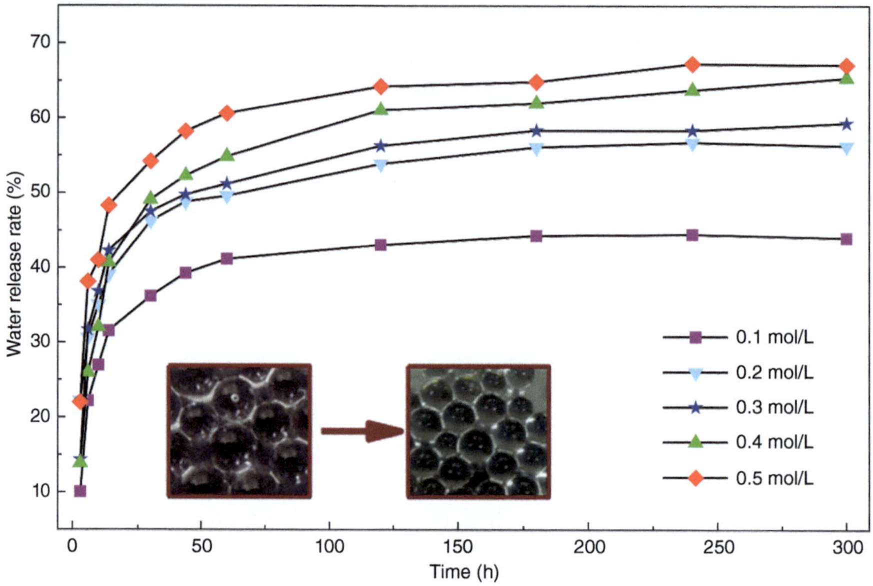

Figure 3.9 Water release law of spherical SAP in solutions with different concentrations.

using Na_2HPO_4-NaOH [31]. The water release law of saturated spherical SAP is shown in Figure 3.9. It can be seen from the figure that as the solution concentration increases, the final water release rate of the saturated spherical SAP also increases, and the water release range is as high as 40–70%. This phenomenon can be explained by the van't Hoff osmotic pressure formula:

$$\prod = c_B RT \tag{3.10}$$

where $\prod$ is the osmotic pressure, Pa; c_B is the concentration of solute in solution, mol/L; R is the ideal gas constant, 8.314 J/(mol·K); and T is the thermodynamic temperature, K.

It can be seen that the magnitude of the osmotic pressure of the solution is only determined by the concentration of solute in the solution. The outer wall of the saturated spherical SAP can be regarded as a semi-permeable membrane, and osmotic pressure is generated due to the concentration difference of solute inside and outside the membrane. The spherical SAP releases water under the action of the osmotic pressure of the solution, and both the water release speed and range are relatively large. In principle, the time when the saturated SAP starts to release water in the freshly mixed concrete can be adjusted by controlling the concentration of the solution absorbed by the SAP. The concentration of the freshly mixed cement paste will gradually increase as hydration progresses. When the concentration of the solution absorbed by the SAP is the same as or slightly lower than the initial concentration of the freshly mixed paste, the saturated SAP will always be in a state of water release in the paste; when the concentration of the solution absorbed by the SAP is higher than the initial concentration of the paste, the saturated

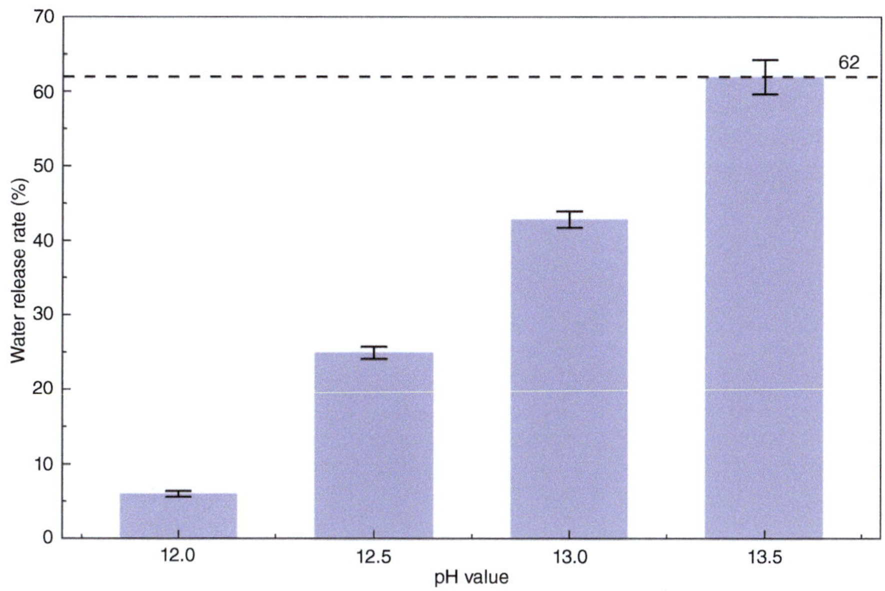

Figure 3.10 Water release behavior of spherical SAP at different pH values (NaOH).

SAP will have a process of first absorbing water and then releasing water in the paste.

(2) Water Release Behavior under Different Alkalinity Environments

To study the action law of spherical SAP in the alkaline environment inside concrete, two series of alkaline solutions were prepared using NaOH and Na_2HPO_4-NaOH buffer solutions, respectively, and the results are shown in Figures 3.10 and 3.11. It can be clearly seen from Figure 3.10 that as the pH value increases, the final water release rate of the saturated spherical SAP in the NaOH solution also increases. What is different is that in Figure 3.11, the water release rate of the saturated spherical SAP does not show a law of increasing with increasing pH value, but basically stabilizes at around 63%, which suggests that the pH value is not the essential reason for the water release of the saturated SAP.

By comparing the composition of these two solutions, it can be found that although the pH value of the Na_2HPO_4-NaOH buffer solution increases, its total solution concentration remains stable at 0.3 mol/L, while the concentration of the NaOH solution increases from 0.01 to 0.3 mol/L as the pH value increases. From this, it can be inferred that the action mechanism of saturated SAP in an alkaline environment has little to do with the pH value, and the real factor is the osmotic pressure caused by the concentration of the alkaline solution. It should be pointed out that the concentration of the solution with a theoretical pH value of 13.5 prepared using NaOH is also about 0.3 mol/L, and the water release rate of the saturated SAP in this environment is 62%, which is very close to the water release rate of SAP in the 0.3 mol/L Na_2HPO_4-NaOH buffer solution (63%), which is consistent with the above inference and proves the correctness of the above

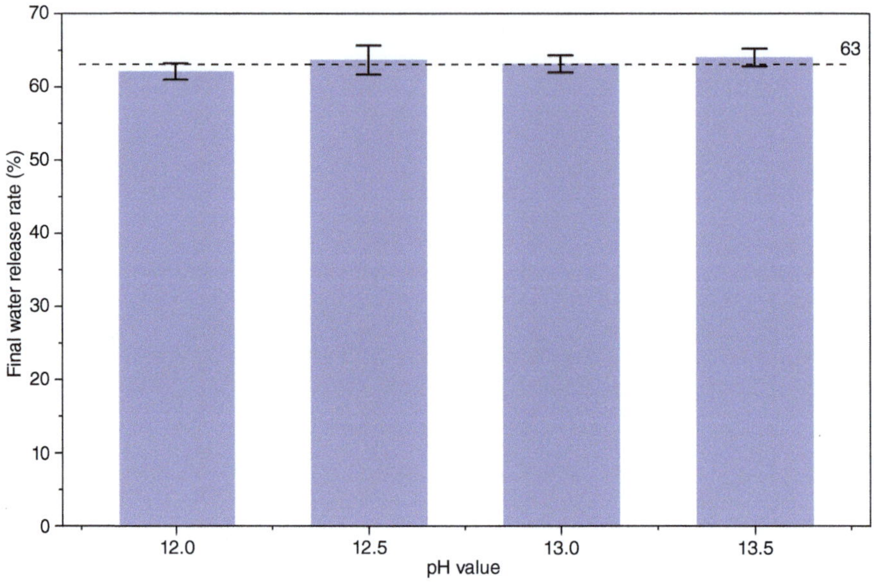

Figure 3.11 Water release behavior of spherical SAP at different pH values (Na_2HPO_4-NaOH).

inference, that is, the pH value is not the real reason affecting the water release of SAP and its intrinsic factor is the osmotic pressure caused by the solution concentration.

(3) Water Release Behavior under Different Humidity Environments

To study the behavior of saturated spherical SAP under different humidity environments inside concrete, four different environmental humidities were controlled, and the results are shown in Figure 3.12. Spherical SAP shows a good linear water release pattern in different humidity environments, and the lower the relative humidity, the faster the water release rate. It was found through fitting that there is a good logarithmic correlation between the water release rate of saturated spherical SAP and the relative humidity at different times, that is, the water release rate of spherical SAP at the same time shows a near-logarithmic growth trend as the relative humidity decreases (Figure 3.13). Taking 14 h as an example, the water release rates of saturated spherical SAP in 50–90% relative humidity (RH) environments are 60%, 34%, 20%, and 7%, respectively, showing an approximate logarithmic water release law. This indicates that the lower the relative humidity, the faster the growth of the water release rate of spherical SAP will be. Since the internal humidity of the saturated spherical SAP is higher than the internal humidity of the concrete, according to Fick's law, the water inside the saturated spherical SAP will diffuse into the hardened cement paste under the drive of the humidity difference. The greater the humidity difference, that is, the lower the internal relative humidity of the concrete, the greater the driving force of water diffusion, and the faster the release rate of water inside the SAP. It can be seen that the water release of saturated SAP can affect the internal

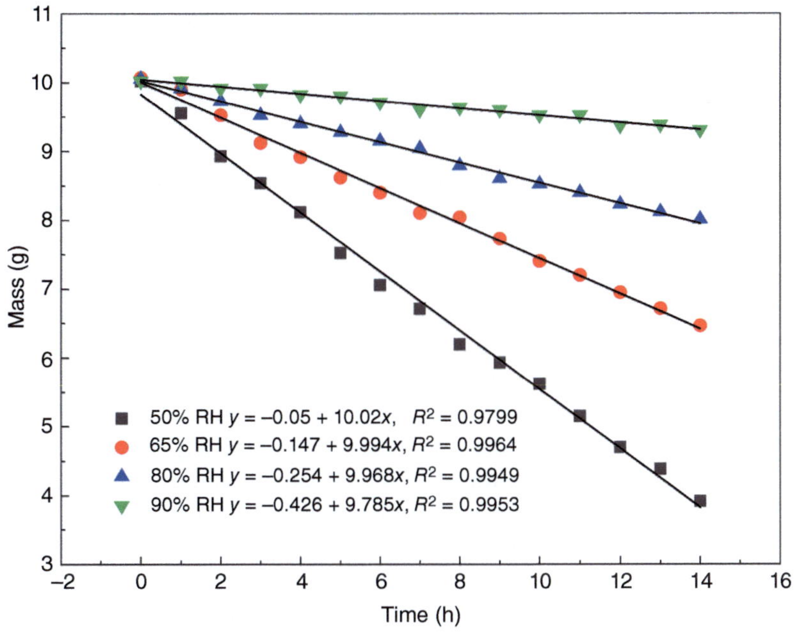

Figure 3.12 Mass change of spherical SAP with time.

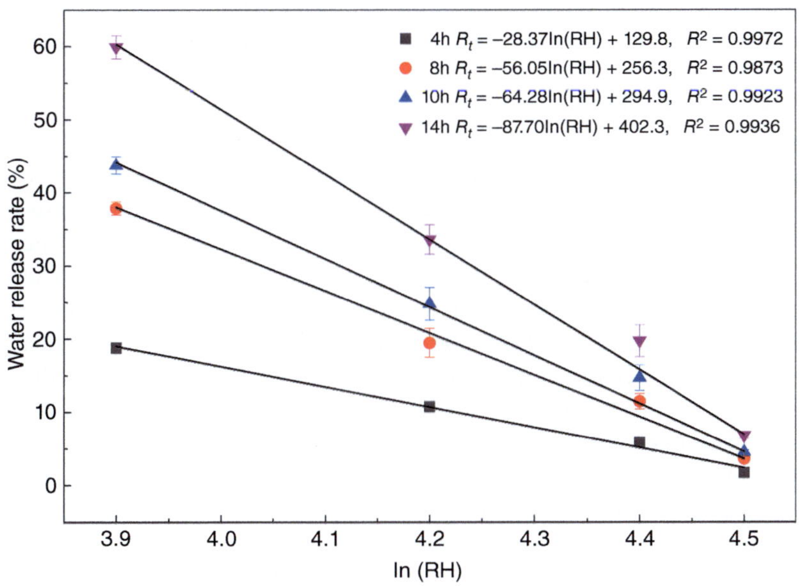

Figure 3.13 Change of water release rate of spherical SAP with relative humidity.

relative humidity of concrete, while the internal relative humidity of concrete in turn affects the water release of SAP.

(2) Full-Process Water Release Model of SAP in Concrete

In fact, the water release of saturated spherical SAP in concrete is not affected by a single factor, but by the combined action of multiple factors [33–35]. Comparing Figures 3.9 and 3.12, it can be found that at 45 minutes, the water release rate of saturated spherical SAP driven by osmotic pressure has reached 40–60%, while the water release rate of saturated spherical SAP driven by humidity difference is only up to 6%, indicating that osmotic pressure has a greater influence on the water release of spherical SAP. It can be assumed that osmotic pressure plays a major role in all stages. If the water release rate driven by osmotic pressure is estimated, as the water content inside the concrete decreases and the concentration of the internal pore solution increases, the saturated spherical SAP will quickly release water completely. In fact, the author's team found in the research that even after 28d, a considerable portion of large-particle spherical SAP still exists inside the concrete, and the water content is not low. The pores of the saturated superabsorbent polymer are not as numerous and large as those of LA. Generally, the water molecules inside are enclosed in a spatial grid with a side length of 1–10 nm. In addition, after the volume of SAP shrinks, it will lose contact with the cement matrix, and the capillary force between two solid porous materials needs to be realized through contact. Therefore, it is believed that there may not be a direct capillary pressure between the superabsorbent polymer and the cement matrix, and the capillary force mainly plays a role in transmitting water throughout the process. In general, the water release process of saturated SAP inside cement-based materials is significantly different from that of LA.

It is analyzed that with the progress of cement hydration, the water content of the freshly mixed concrete paste decreases, the ion concentration increases and the saturated spherical SAP is in direct contact with the matrix before the concrete sets, so the osmotic pressure plays a major role and the water release rate of the saturated spherical SAP is relatively fast (stage I). However, with the formation of the internal structure of the concrete and the shrinkage of the volume of the spherical SAP, it forms spherical pores inside the concrete, the contact area with the concrete is greatly reduced, and the effect of osmotic pressure will be gradually weakened (Figure 3.14). After seven hours of hydration, especially after the final setting, the total ion concentration inside the cement paste increases significantly, which leads to a gradual increase in osmotic pressure. Therefore, the duration of stage I should be around the final setting of concrete. In addition, the internal humidity of concrete is still at a high level (>90% RH) at this time, and the driving effect of humidity difference is still very weak, so the water release rate of spherical SAP turns to a lower level (stage II). The duration of this stage is related to factors such as the volume content of SAP inside the concrete and the environmental conditions, such as relative humidity. When the SAP content is low or the environmental humidity is low, the duration of this stage is relatively short. With the decrease of the internal humidity of concrete in the later stage, the humidity difference between the cement

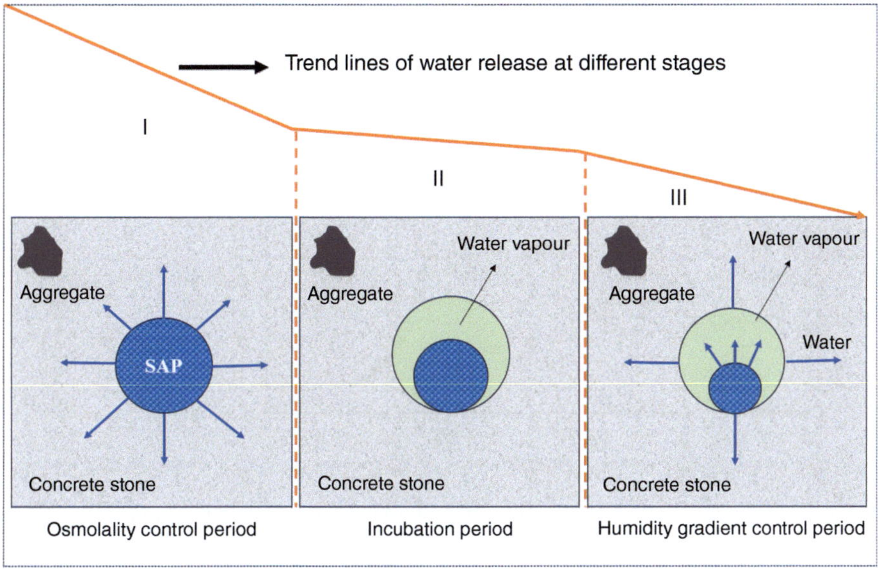

Figure 3.14 Ideal water release model of SAP in cement-based materials. *Source:* Modified from [31].

matrix and the water vapor in the pores begins to play a major role, and the water release rate of spherical SAP increases. The SAP itself also slows down the decrease of the internal humidity of concrete, so it can continue to release water evenly for a considerable period of time (stage III) [29–32], thus playing a role in compensating for water, which is of practical significance for compensating for autogenous shrinkage.

3.6.2.2 Influence of SAP on the Autogenous Shrinkage of Mortar

Cement hydration is a process of continuous water consumption, accompanied by a decrease in the relative humidity inside the mortar, an increase in the negative pressure of the meniscus, and a decrease in the critical radius. During the transition from the initial setting to the final setting, the skeleton structure of the cement paste forms rapidly, and this process is also the period of the most rapid development of self-desiccation, manifested as a rapid decrease in relative humidity (Figure 3.15), leading to a significant volumetric autogenous shrinkage of concrete [36–39].

It can be seen from Figure 3.15 that at the same hydration time, the time and extent of the decrease in relative humidity of mortar with SAP are lower than those of the control group, and this becomes more obvious with the increase of the pre-absorption ratio. Due to the curing effect of the water stored inside the SAP, the decrease in internal relative humidity is significantly delayed, which is very beneficial for limiting the early self-desiccation of hardened cement paste. As shown in Figure 3.16, the autogenous shrinkage of mortar with SAP is compensated to varying degrees, and the higher the pre-absorption ratio, the more obvious the compensation effect of autogenous shrinkage. For low pre-absorption ratios, the re-absorption behavior of SAP will lead to a decrease in the effective water–cement

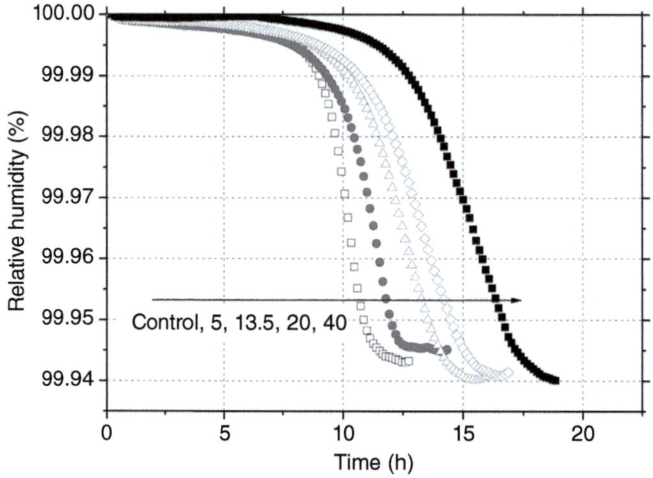

Figure 3.15 Influence law of SAP on the early internal relative humidity of mortar.

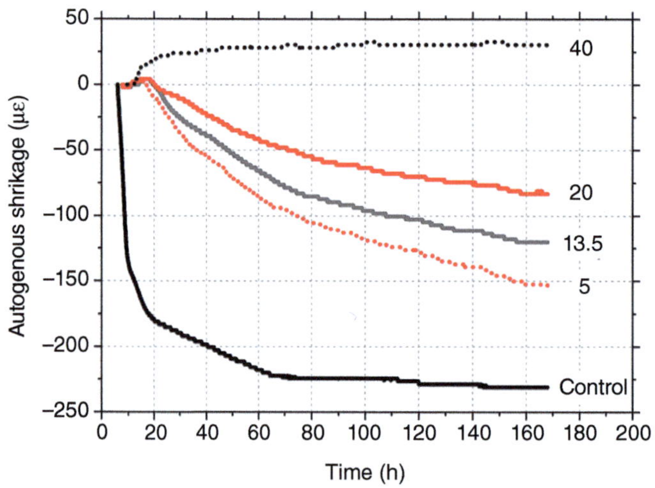

Figure 3.16 Influence law of SAP on the autogenous shrinkage of mortar.

ratio, and cement paste with a low water–cement ratio generally has a higher autogenous shrinkage value. It should be noted that although SAP absorbs a portion of the mixing water, this stored water will eventually return to the cement paste in the form of internal curing. From the perspective of the total water–cement ratio, the low water introduction ratio mortar still has a higher total water–cement ratio than the blank control group, which is also the reason why SAP-5 and SAP-13.5 still have a significant shrinkage reduction effect. For high pre-absorption ratios, although SAP will quickly release its internally stored water, the autogenous shrinkage of the mortar is still significantly compensated, especially in the SAP-40 group, where the autogenous shrinkage of the mortar is completely compensated and shows a micro-expansion state of 30×10^{-6}. This micro-expansion state can generate a certain amount of self-stress in the restrained mortar, which is very beneficial for reducing the cracking of tensile members.

3.6.3 Inorganic Water-releasing Materials

3.6.3.1 Functional Principle of Water-releasing Agent

According to the internal curing function, an inorganic material with a porous structure [40] is designed, which has a large number of tubular pore structures that can adjust and convert storage and water release functions through the external environment and its capillary action. This type of material is called a water-releasing agent. The water in the water-releasing agent does not participate in chemical reactions in the early hydration stage, but when the free water in the system is reduced to a certain level, it can release water to maintain the reaction in the system. When participating in cement hydration, the water-releasing material is in a closed, compressed state with saturated water in the tubular pores. At this time, the hydration reaction of the cement-based material is supplied by free water. As hydration continues, the free water in the system gradually decreases. At this time, the humidity of the cement-based material system decreases, and the capillary pores generate negative pressure, forming a pressure difference with the water pressure in the pores of the water-releasing material, causing the water in the pores of the water-releasing material to be released and enter the capillary pores of the hardened cement paste to participate in the reaction. By controlling the water release time, and release amount, and adjusting the composition of the expansive agent, the expansion value of the cement-based material can be controlled. The functional principle is shown in Figures 3.17 and Figure 3.18 is an actual microscopic structure photo of the water-releasing material.

According to the functional principle, the relationship between the structural parameters of the water-releasing agent, such as particle shape, size, surface state, internal pore diameter, pore shape, pore connectivity, and reaction characteristics,

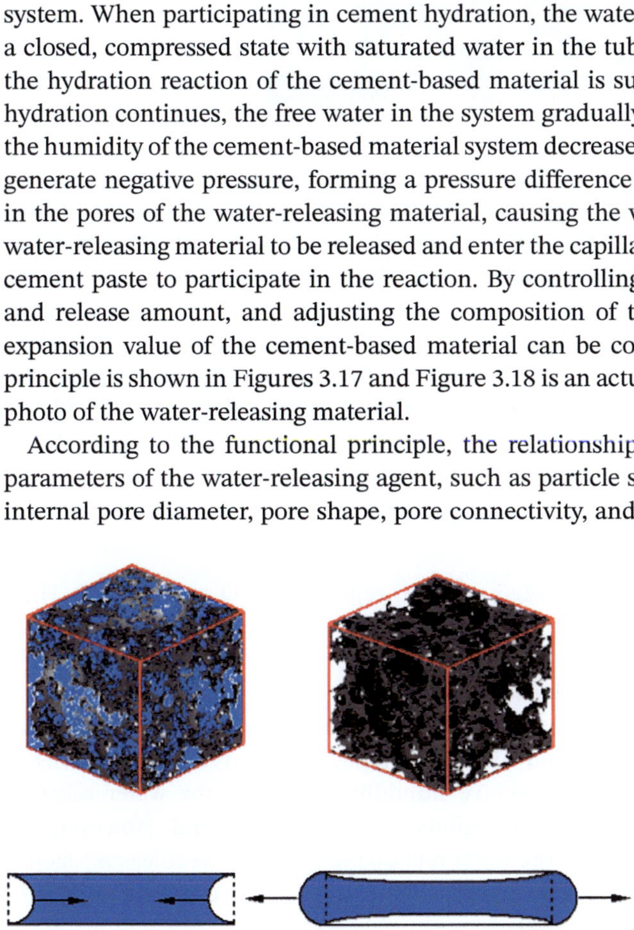

(a) (b)

Figure 3.17 Schematic diagram of the functional principle of water-releasing agent: (a) state of water storage in the bore of a tube and (b) state of water release from the bore of a tube.

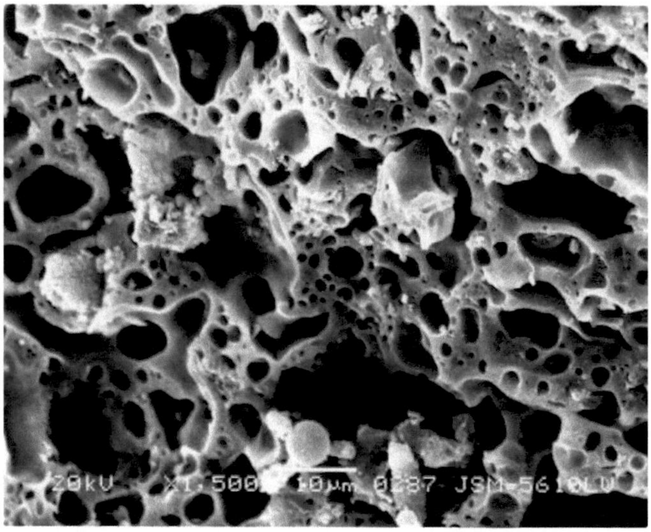

Figure 3.18 Microstructure of water-releasing material.

is studied. On this basis, the composition and structural parameters of the water-releasing material are designed. From the perspective of material composition and performance, the water-releasing agent itself should be one of the components of the cement-based material to maintain good compatibility and bonding with the cement-based material. A silicate material with a similar composition to the cement-based material is selected, and the chemical composition and range of its raw materials are: SiO_2 55–65%, Al_2O_3 14–25%, Fe_2O_3 4–6%, FeO 0.5%, CaO + MgO 6–8%, $K_2O + Na_2O$ 2.5–5.0%. Experimental research shows that the composition content should be controlled according to Eq. (3.11):

$$\frac{SiO_2 + Al_2O_3}{Fe_2O_3 + RO + R_2O + FeO} = 5.0 - 7.5 \tag{3.11}$$

where RO represents CaO and/or MgO and R_2O represents K_2O and/or Na_2O. The loss on ignition is controlled to be 3–5% of the total weight.

3.6.3.2 The Role of Water-releasing Agent

The rate of water release from the saturated water-releasing agent in environments with relative humidity of 60% and 85% was studied. The curves in Figure 3.19 show that as the environmental relative humidity decreases, the water release process of the water-releasing agent exhibits an accelerating trend. However, as time progresses, the difference in the water release rate of the water-releasing agent at different relative humidities is not significant and is close to the water absorption rate of the water-releasing agent [41].

3.6.3.3 Humidity Regulation by Water-releasing Agent

Figure 3.20 shows the trend of internal relative humidity of cement-based materials with four formulations over time [42]. The curves in the figure indicate that the

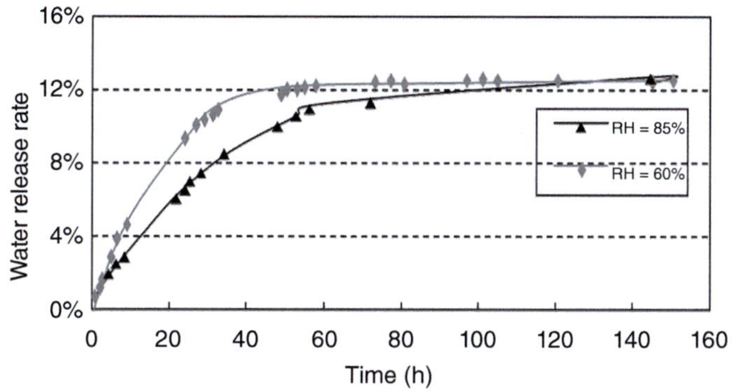

Figure 3.19 Water release rate of saturated water-releasing agent at RH = 60% and RH = 85%.

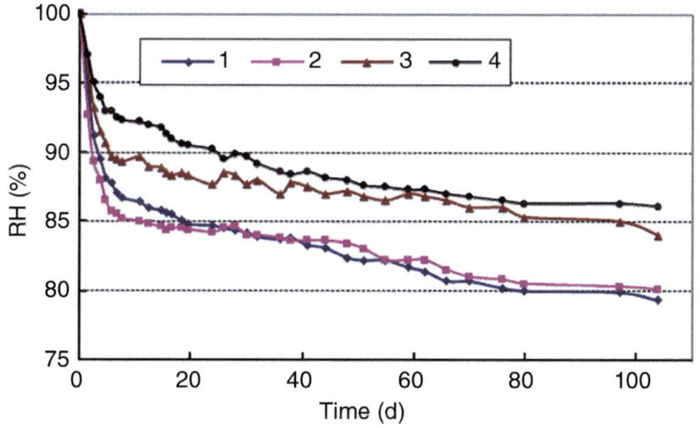

Figure 3.20 Correlation between internal relative humidity and time of cement-based materials.

internal relative humidity change trend of Groups 1 and 2 (with expansive agent) cement-based materials without saturated water-releasing agent is similar, with the internal relative humidity of Group 2 cement-based materials slightly lower than that of Group 1 at early ages; while the internal relative humidity of Group 3 cement-based materials with saturated water-releasing agent (volume ratio of 10%) and Group 4 with saturated water-releasing agent (volume ratio of 20%) is significantly higher than that of Groups 1 and 2. Moreover, as the volume ratio of the water-releasing agent increases, the internal relative humidity also increases. It can be seen that the addition of the water-releasing agent effectively regulates the internal relative humidity of cement-based materials, significantly reducing the rate and extent of its decline. Of course, the regulation effect is closely related to the particle characteristics, pore structure characteristics of the water-releasing agent itself, and the capillary pore distribution characteristics of the hardened cement paste. The more reasonable the design of the water-releasing agent, more significant the effect will be.

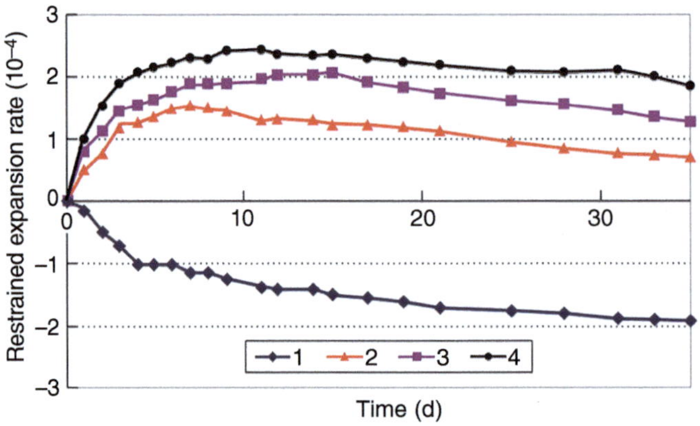

Figure 3.21 Restrained expansion rate of cement-based materials.

3.6.3.4 Influence of Water-releasing Agent on Volumetric Deformation Properties

The restrained expansion rate of cement-based materials under restricted conditions with four formulations was studied, as shown in Figure 3.21. Group 1 cement-based materials exhibited a shrinkage deformation trend from the beginning, reaching a shrinkage deformation rate of 2×10^{-4} at 35 d, while the other three groups of cement-based materials all showed an expansion deformation trend. The restrained expansion rate of Group 3 and Group 4 cement-based materials with saturated water-releasing agent and expansive agent was larger than that of Group 2 cement-based materials with only expansive agent; the restrained expansion rate of Group 4 with higher water-releasing agent content was higher than that of Group 3. This is mainly because when the internal relative humidity of the cement-based material decreases, the water in the water-releasing agent enters the hardened cement paste through the internal pores and capillary pores in the hardened cement paste. This water-releasing process is closely related to the change of relative humidity in the hardened cement paste; at the same time, the water-releasing behavior of the water-releasing agent in high-strength micro-expansive steel tube core cement-based materials can play three roles: first, it improves the self-desiccation phenomenon inside the cement-based material and reduces its shrinkage deformation; second, the released water is conducive to further hydration of the expansive agent to generate ettringite and produce volumetric expansion; third, some of the released water continues to react with unhydrated cement, increasing the solid-phase volume.

3.6.3.5 Influence of Water-releasing Agent on Compressive Strength of Materials

As shown in Figure 3.22, the addition of 10% by volume of saturated water-releasing factor has a slight impact on compressive strength, but the impact is not significant. When the volume ratio of the saturated water-releasing agent is 20%, the compressive strength decreases more significantly. This indicates that the addition of a large amount of saturated water-releasing agent has a certain impact on the compressive

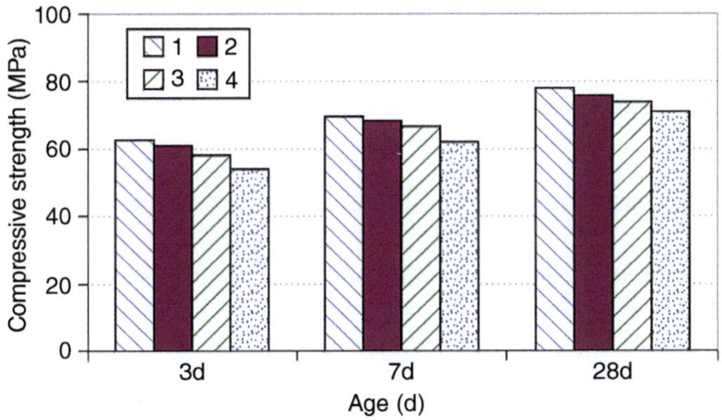

Figure 3.22 Compressive strength of cement-based materials.

strength of high-strength cement-based materials, which may be related to two reasons: first, when a large amount of water-releasing agent is added to high-fluidity cement-based materials, it is easy to cause the problem of segregation, which affects its homogeneity; second, for high-strength cement-based materials, the strength of the aggregate itself is very important. After adding a large amount of water-releasing agent, the water-releasing agent may become the source of crack propagation when the high-strength cement-based material is subjected to pressure. Therefore, when applying the water-releasing agent, its dosage should be reasonably selected.

3.6.3.6 Application Technology of Water-releasing Agent

Before use, the water-releasing agent needs to be saturated with water under vacuum for 5–30 minutes or saturated in water for 24–36 h; before adding, the water-releasing agent should be kept in a saturated surface-dry state, and the water-releasing agent should be added to the cement-based material by replacing the sand in equal volume. The water in the water-releasing agent is not included in the unit water consumption of the cement-based material; the water-releasing agent should be mixed uniformly with other sand and stone aggregates and cement before adding the water reducer.

(1) Dosage of Water-Releasing Agent

Let the amount of sand in the cement-based material be S, the water absorption rate of the water-releasing agent be $X\%$, and the volume replacement rate of sand by the water-releasing agent be $N\%$; then the dosage of the water-releasing agent Q is calculated according to Eq. (3.12):

$$Q = S/2650 \times N\% \times 1450 \times (1 + X\%) \tag{3.12}$$

(2) Size Selection of Water-Releasing Agent

The size of the water-releasing agent is determined according to the strength and construction conditions of the cement-based material. When the strength of the cement-based material is greater than 50 MPa, the water-releasing agent should be controlled at 2.5–5.0 mm; when the strength of the cement-based material is less than 50 MPa, a water-releasing agent of 2.5–7.5 mm can be used.

(3) Application and Actual Effect of Water-Releasing Agent

(1) Raw materials.
  Cement: Three Gorges brand 52.5 Portland cement from Gezhouba Cement Plant.
  Fly ash: Grade I ash from Wuhan Yanglu Power Plant, with a water requirement ratio of 94%.
  Expansive agent: UEA (United Expanding Agent) expansive agent from Wuhan Sanyuan Admixture Plant.
  Water reducer: FDN-9000 high-efficiency water reducer from Wuhan Haoyuan Admixture Plant.
  Sand: Bahe medium sand with a fineness modulus of 2.6.
  Stone: Yangxin 5–20 mm continuously graded crushed stone.
  Water-releasing agent: prepared by crushing specially fired 800-grade shale ceramsite into a particle size range of 4.75–9.5 mm and treating the particle surface to a certain extent.
  Mixing water: clean tap water.
(2) Experimental method. The water-releasing agent was placed in water to absorb water until saturation, and its water absorption rate was measured to be 13%. The saturated water-releasing agent was placed in different environmental humidities (controlled by an environmental chamber), and its weight was measured at regular intervals to examine its water release under different humidity conditions. After the water-releasing agent accounting for 10% and 20% of the unit volume of the cement-based material was absorbed to a saturated state and air-dried to a saturated surface-dry state, it was used to replace the same volume ratio of sand and stone (50% each) for preparing high-strength micro-expansive steel tube cement-based materials, and the specific mix proportions are shown in Table 3.4. The cement-based materials with the above mix proportions were formed into 150 × 150 × 150 mm test blocks with a small hole reserved in the middle for measuring the change of internal relative

Table 3.4 Mix proportions of high-strength micro-expansive cement-based materials with water-releasing agent (kg/m^3).

No.	Water	Cement	Fly ash	Expansive agent	Water-releasing agent	Sand	Stone	High-efficiency water reducer
1	180	510	60	0	0	680	1020	6.8
2	180	460	60	50	0	680	1020	6.8
3	180	460	60	50	140	550	890	6.8
4	180	460	60	50	280	420	760	6.8

Note: The water introduced by the water-releasing agent is not included in the mixing water. The weight of the water-releasing agent in the above table refers to the weight before water absorption.

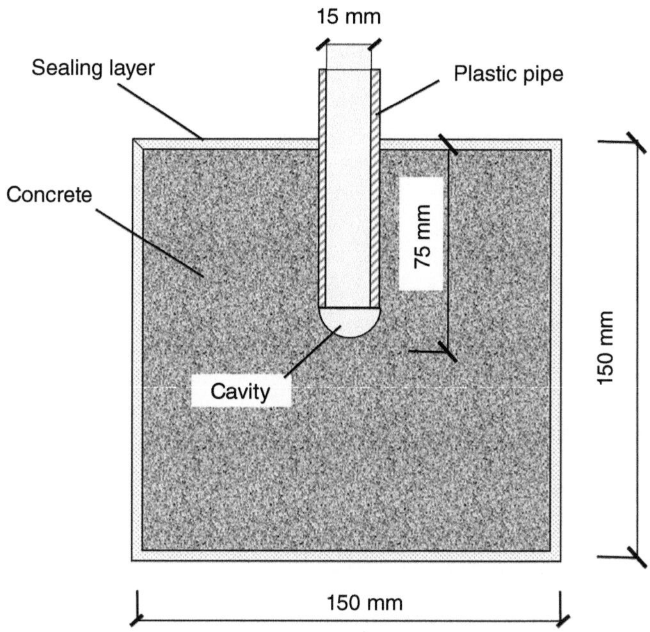

Figure 3.23 Measurement method of internal relative humidity of cement-based materials.

humidity [43] (see Figure 3.23). After demolding, the test blocks were wrapped with plastic film and then coated with paraffin to ensure sealing. A rod-type precision hygrometer was used to measure the internal relative humidity.

The cement-based materials with the above mix proportions were formed into 150 × 150 × 150 mm and 100 × 100 × 300 mm test blocks, cured in a moisture-free environment, and used for testing compressive strength and restrained linear deformation, respectively. The testing methods were carried out according to national standards. In Figure 3.24, the reddish-brown granular materials uniformly distributed in the cement-based material are the water-releasing agent. In Figure 3.25, the lower left part shows the microstructure of a water-releasing agent particle, and the upper left part shows the hydration products. It can be seen that these products grow in the area near the water-releasing agent, with a dense structure, and should have good expansive properties. Figure 3.26 shows the microstructure of an expansive cement-based material without the water-releasing agent. It can be seen that there are many voids and pores, and the structure is loose, which not only makes it difficult to guarantee expansive properties but may even cause shrinkage.

3.7 Functional Aggregates

Aggregates are an important component of concrete and also the part with the largest usage, generally accounting for about 70% of the concrete volume. Aggregates in

Figure 3.24 Distribution of water-releasing agent in cement-based materials.

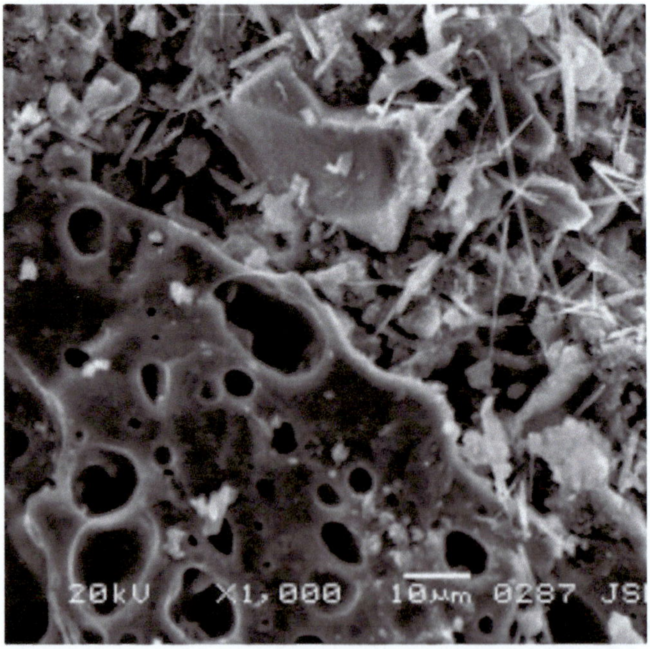

Figure 3.25 Effect of water-releasing agent (lower left corner) (×1000 magnification).

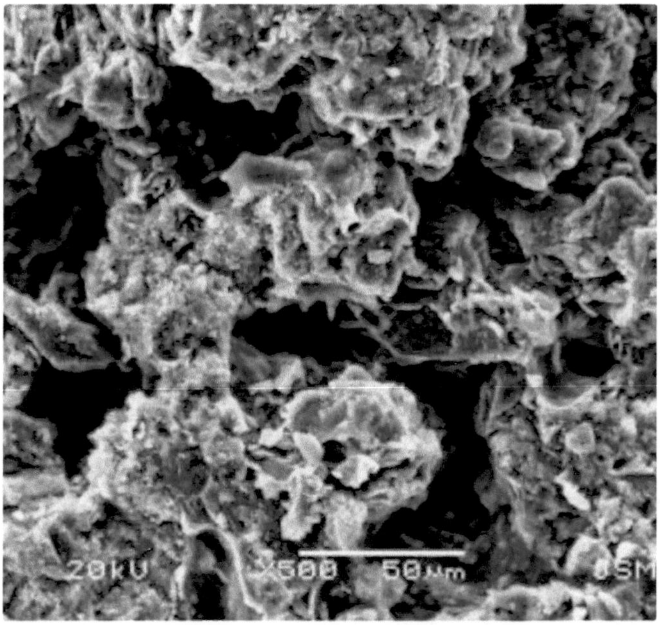

Figure 3.26 Late-stage structure without water-releasing agent (×500 magnification).

concrete mainly play the role of skeleton and reduce the volume change caused by the shrinkage and swelling of cementitious materials during the setting and hardening process, while also serving as an inexpensive filler to reduce the production cost of concrete. Aggregates have an important influence on the formation process and properties of concrete. Previous studies have shown that the weakening of the interfacial transition zone (ITZ) between concrete aggregates is an important reason for its durability problems. Traditional concrete preparation pays less attention to the function of aggregates and generally considers its economic feasibility. The research and application practice of the author's team show that high-tech concrete must select aggregates, and some special concrete cannot be separated from the function of aggregates. Applying the function of aggregates from the design of material structure and performance, and endowing concrete with the required performance through the function of aggregates is an important method for the design and preparation of high-tech concrete.

3.7.1 The Role of Functional Aggregates

3.7.1.1 The Influence of Aggregate Properties on Concrete Performance

In addition to the filling and skeleton effect in concrete, aggregates have varying degrees of influence on the mechanical properties, workability, volume stability, durability, and thermal properties of concrete. There has been a large amount of research work in this area. This section summarizes the influence of aggregate properties on concrete performance as shown in Table 3.5.

Table 3.5 The influence of aggregate properties on concrete performance.

Concrete properties		Corresponding aggregate properties
	Bulk density	Relative density, particle shape, maximum particle size, particle gradation
Mechanical properties	Strength	Strength, elastic modulus, particle shape, maximum particle size, particle gradation, surface characteristics, impurities
	Elastic modulus	Elastic modulus, Poisson's ratio
	Workability	Particle shape, surface characteristics, particle gradation, maximum particle size, moisture content, water absorption
Durability	Freeze-thaw resistance	Porosity, water absorption, permeability, pore structure
	Permeability resistance	Particle shape, surface condition, particle gradation, maximum particle size
	Alkali-aggregate reaction	Siliceous components in aggregates
	Abrasion resistance	Hardness
Shrinkage and creep		Elastic modulus, particle shape, particle gradation, maximum particle size, impurities
Thermal properties	Thermal conductivity	Thermal conductivity
	Thermal expansion coefficient	Thermal expansion coefficient, elastic modulus

3.7.1.2 The Mechanism of Functional Aggregates Improving Structure and Performance

In the composition of concrete, the ITZ between aggregates and hardened cement paste is the weakest region in concrete. Its main characteristics of ITZ are high water–cement ratio, high porosity, and oriented growth of crystalline hydration products (mainly $Ca(OH)_2$). The ITZ has an important influence on the mechanical and durability properties of concrete. The author's team utilizes the characteristics of artificial aggregates that can be designed in structure and performance, designing the outer shell of the aggregates as a hydration-active layer (mainly β-C_2S) and the interior as a porous structure with the water transport function. The hydration reaction of the aggregate shell is used to densify the ITZ; at the same time, the water storage and release function of the porous structure is used to increase the reaction degree between the aggregate surface layer and the hardened cement paste matrix, further improving the structure of the ITZ.

3.7.1.3 The Principle of Functional Aggregates Endowing Concrete with Special Properties

To endow concrete with certain special properties (such as air purification, conductivity, wave absorption, etc.), the author proposes a design idea of functional

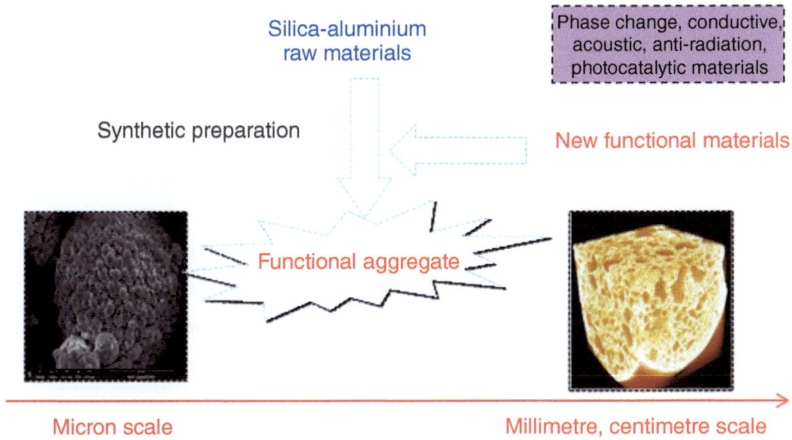

Figure 3.27 Design idea of special aggregates.

aggregates with special properties, as shown in Figure 3.27. Siliceous and aluminous raw materials are used as the matrix of functional aggregates, and corresponding functional phase materials are introduced to synthesize and prepare functional aggregates with different functions at different scales. This technical idea starts from aggregates and can avoid the adverse effects of cement hydration products and concrete service environment on functional phases and can also reduce the amount of functional phases in the concrete matrix, thus breaking through technical problems such as the difficulty of synergy between concrete function and structural performance and high cost.

3.7.2 Design and Preparation of Functional Aggregates

3.7.2.1 Selection of Natural Aggregates

In addition to widely used sand and gravel aggregates, natural aggregates also include some naturally formed functional aggregates. For example, pumice, pozzolanic slag, and pozzolanic tuff are natural LA with porous structures, which can be used for the preparation of lightweight and insulating concrete materials. Their properties are compared in Table 3.6.

Table 3.6 Comparison of properties of various natural lightweight aggregates.

Aggregate	Particle density (kg/m^3)	Packing density (kg/m^3)	24 h water absorption at normal pressure/%
pozzolanic tuff	1300–1900	Coarse aggregate: 700–1100 Fine aggregate: 200–500	7–30
Foam lava	1800–2800	800–1400	Around 10
Pumice	550-1650	350–650	Around 50

Table 3.7 Radiation shielding concrete and characteristics prepared with natural aggregates.

Aggregate type	Aggregate density (×10^3 kg/m^3)	Concrete bulk density (×10^3 kg/m^3)	Characteristics
Limonite	3.2–4.0	1.3–3.2	Used for neutron shielding
Hematite	5.0–5.3	3.2–4.0	Used for γ-ray shielding
Barite	4.3–4.7	3.2–4.0	Used for γ-ray shielding
Ilmenite	2.7–2.8	2.3–2.5	Used for γ-ray shielding
Serpentine	2.7–2.8	2.2–2.4	Used for neutron shielding
Basalt	2.6–2.8	2.3–2.5	Used for X-ray shielding

Some natural aggregates have high density and can promote the attenuation of radiation energy (such as γ rays and neutrons), which can be used as aggregates for radiation shielding concrete. Common natural high-density aggregates mainly include limonite ($2Fe_2O{\cdot}3H_2O$), hematite (Fe_2O_3), magnetite (Fe_3O_4), barite ($BaSO_4$), etc. The apparent density and characteristics of radiation shielding concrete prepared with natural heavy aggregates are shown in Table 3.7. In addition, high-density natural aggregates can also be used to prepare concrete that requires weighting and anti-floating functions, and some high-hardness aggregates can be used to prepare wear-resistant and abrasion-resistant concrete.

3.7.2.2 Preparation of Artificial Aggregates

Artificial aggregates refer to aggregates formed by artificial preparation. Due to the designability of composition and structure, artificial aggregates have great advantages in the preparation of functional aggregates. The preparation methods of different functional aggregates are slightly different. For example, the surface-activated aggregates that improve the structure and performance of concrete can be realized by the process of "two-step molding and one-step sintering," that is, first preparing the core material with strength function, and then coating a hydration-active layer on its outer layer through a certain process; for special aggregates, according to the characteristics of the introduced functional phase materials, an appropriate amount of functional phase can be mixed into the aggregate raw material and prepared by high-temperature calcination and molding; or through mechanical activation, surface treatment, and other means, active sites are constructed on the surface of siliceous and aluminous aggregates to make functional materials stably bond to the aggregate surface.

3.7.2.3 Ideal Optimization Model Design

The ideal optimization model of functional aggregates needs to be designed by comprehensively considering the functional characteristics and structural performance of the aggregates [44]. Figure 3.28(a) is an ideal model of surface-activated functional aggregates, which mainly consists of a surface-active layer and a

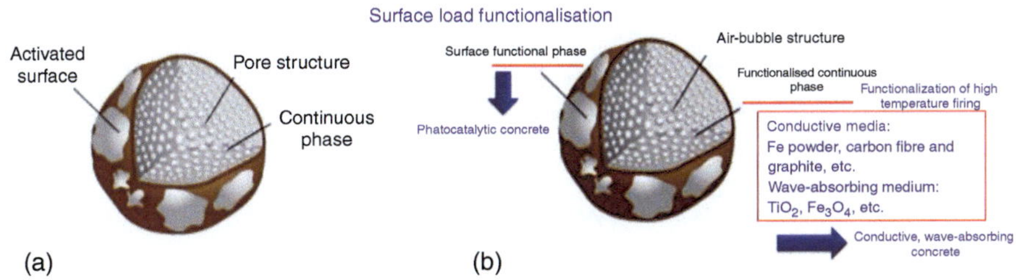

Figure 3.28 Ideal structural model of functional aggregates: (a) functional aggregates with surface activation layer and (b) characteristic functional aggregates.

high-strength porous core, and the two layers are tightly bonded through high-temperature solid-phase reaction. Figure 3.28(b) is an ideal optimization model of special functional aggregates, including two types: surface loading functionalization and high-temperature calcination functionalization.

3.7.3 Research and Development of Functional Aggregates

In recent years, the author's team has carried out some work on the research and development of artificial aggregates in conjunction with the research on relevant concrete materials, especially in the aspects of ultra-high-performance concrete and new functional concrete. This section provides a systematic brief introduction.

3.7.3.1 Aggregates for Optimizing Concrete Structure and Durability

Figure 3.29 shows the influence of surface-activated functional aggregates (FA) and ordinary LA on the microhardness of the ITZ. It can be seen that the microhardness of FA increases more than that of ordinary LA with age. The X-ray diffraction (XRD) analysis of the surface composition of FA with age (Figure 3.30) shows that the diffraction peak intensity of β-C_2S on the aggregate surface at 28d is significantly lower than that at 3d, and basically disappears at 90d, while the main peak of C-S-H

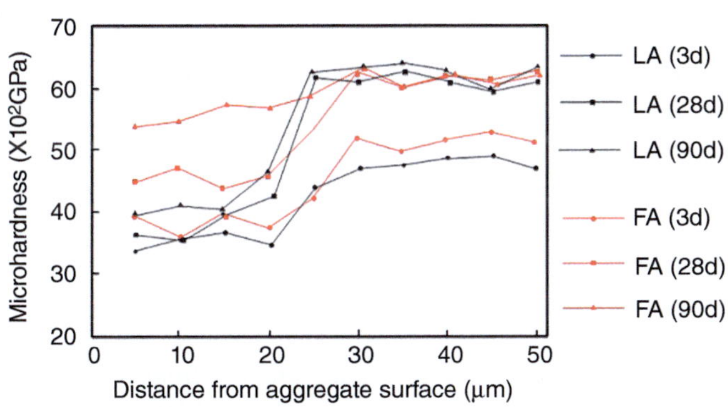

Figure 3.29 Microhardness of hardened cement paste-aggregate interfacial transition zone with different aggregates.

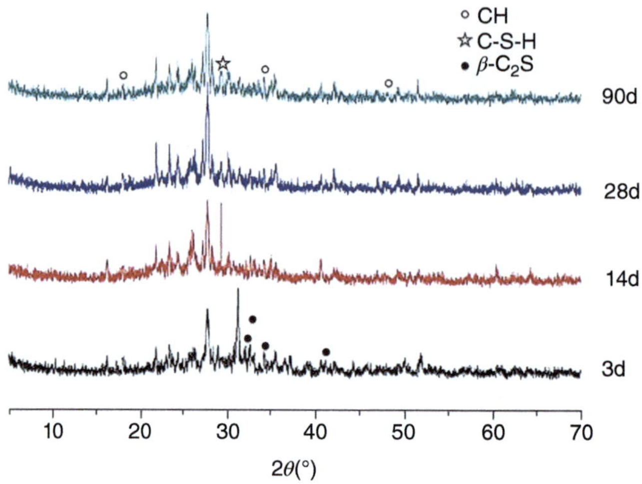

Figure 3.30 XRD analysis of change in aggregate surface composition with age.

appears at 28 d and the intensity increases significantly after 90 d, indicating that the surface layer of FA participates in hydration and improves the interfacial strength.

3.7.3.2 Lightweight Aggregates

Compared with other materials, the outstanding disadvantage of ordinary concrete is its low specific strength. In the commonly used strength range of concrete, reducing the self-weight of concrete is an important way to improve the specific strength. Reducing the weight of aggregates is the most effective method to reduce the weight of concrete. At the same time, by reducing the weight of concrete, the self-weight of buildings can be reduced, the cost of foundation treatment and maintenance can be lowered, and seismic safety can be increased. It has a wider application prospect in high-rise structures and prefabricated structures.

Aggregates can be divided into three categories according to density: lightweight, heavy, and ordinary aggregates, which can produce concrete with different apparent densities. LA are mainly obtained by selecting natural low-density lightweight materials or artificially preparing low-density lightweight porous aggregates, and the density range can be adjusted through the design of pore structures to obtain aggregates with different densities. The density of concrete prepared with different types of LA has a reasonable range, as shown in Figure 3.31.

3.7.3.3 Radiation Shielding Aggregates

High-density building shielding materials have a good attenuating effect on the energy of radiation rays. Therefore, natural high-density aggregates as described in Table 3.7 are generally selected to prepare radiation shielding concrete. However, natural high-density aggregates are limited in reserves, and the high density of aggregates leads to poor workability and homogeneity of concrete. The author's team has carried out research and development work on preparing radiation

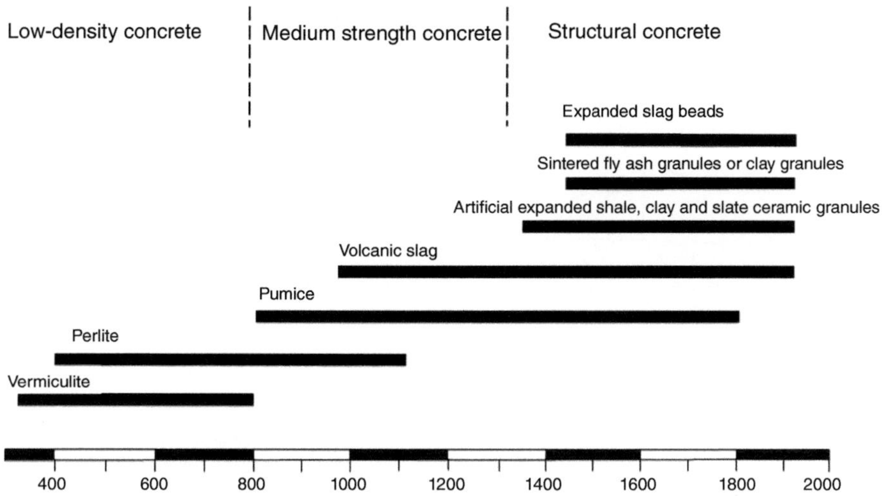

Figure 3.31 Relationship between lightweight aggregates used in concrete and their density.

shielding artificial aggregates using heavy metal-containing industrial sludge as raw materials, which not only utilizes the characteristics of heavy metals (Ba, Cr, Ni, Cu, Zn, etc.) in shielding high-energy rays (X-rays, γ-rays) but also solves the problem of weak radiation shielding areas caused by poor homogeneity of natural high-density aggregate radiation shielding concrete and achieves the ecological and environmental benefits of waste slag and sludge utilization. FA with a compressive strength of more than 50 MPa and shielding performance equivalent to iron ore (linear attenuation coefficient $\mu = 0.19\,\text{cm}^{-1}$) have been prepared [45].

3.7.3.4 Porous Functional Aggregates

The pore structure parameters of multi-FA mainly include porosity, pore size and distribution range, open or closed pores. Different pore structure parameters can be achieved by adjusting the chemical composition of raw materials and changing the sintering system and process, thereby obtaining porous aggregates with different functional properties. Porous functional materials have the effects of sound absorption, vibration reduction, damping, thermal insulation, and substance storage and release [46–48].

The sound absorption coefficient of sound-absorbing FA is directly related to open pores (Figure 3.32). The higher the open porosity, the better the sound absorption performance; under the condition of a certain connected porosity, the relatively smaller the pore size, the better the sound absorption performance. The thermal conductivity of thermal insulation aggregates is mainly related to porosity. As the porosity increases, the thermal conductivity of the material decreases. When the porosity is constant, the thermal conductivity of the material can be further reduced by refining the pore size distribution of the material and reducing the average pore size. The water release rate of water-releasing factors is mainly affected by

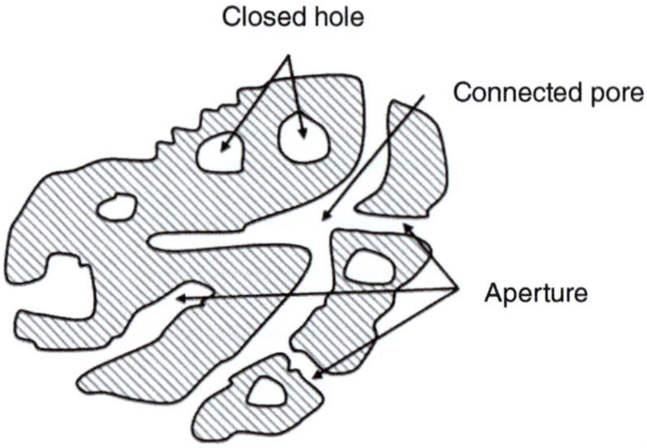

Figure 3.32 Schematic diagram of pore structure characteristics of porous FA.

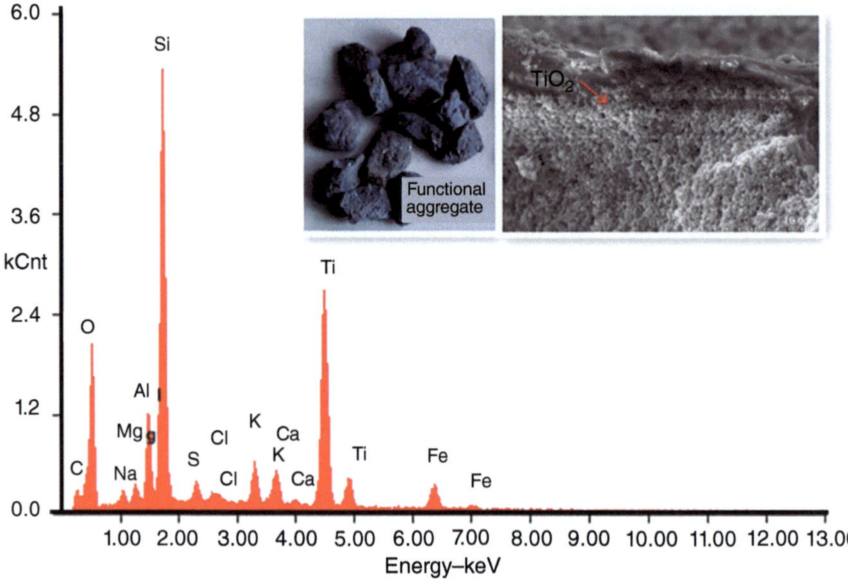

Figure 3.33 Loading of photocatalytic materials on the surface of functional aggregates.

the pore size distribution and pore connectivity. Under the same temperature and humidity conditions, the water in large pores and connected pores is more easily released.

3.7.3.5 Functional Aggregates for Purification

The author's team used porous aggregate shale ceramsite as the substrate and adopted special technical means to firmly load the photocatalytic functional material TiO_2 on the surface and interlayer pores of the ceramsite (Figure 3.33). Using this functional aggregate, concrete with purification function can be further prepared to effectively degrade organic pollutants in air and wastewater [49].

Based on ordinary LA (Figure 3.34), through the directional synthesis technology of calcium aluminate silicate hydrate gel (C-A-S-H gel), honeycomb structure

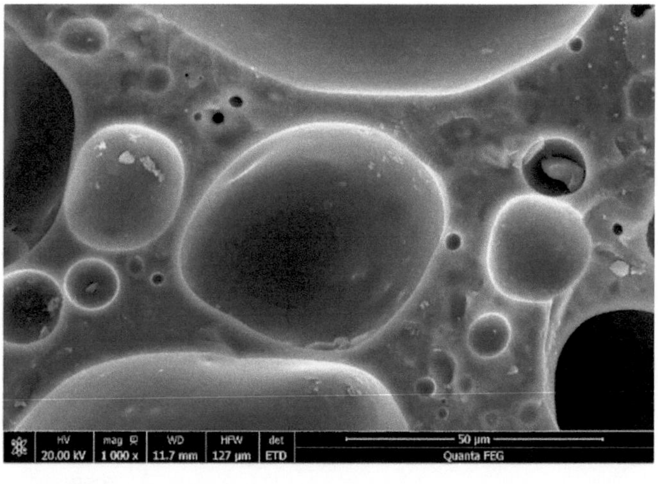

Figure 3.34 Ordinary porous lightweight aggregates.

Figure 3.35 Honeycomb structure functional aggregates.

FA were further synthesized and prepared (Figure 3.35), which have higher specific surface area and loading capacity than ordinary LA. Compared with unmodified FA, the photocatalytic ability of the modified FA is increased by more than 60%; the increase in durability is even more significant, still maintaining nearly 100% after six months.

At the same time, application of this honeycomb structure functional aggregate can be extended to heavy metal ion adsorption. By linking the chelating agent TCPS (3-thiocyanatopropyltriethoxysilane) on the honeycomb C-A-S-H surface, it is given the ability to chelate heavy metal ions (Figure 3.36). Experiments show that the linked FA have a strong adsorption capacity for heavy metals. The amount of Cd(II) ions adsorbed per surface area is 4.07 mg/m^2, and it also has the advantages of easy recovery and low cost.

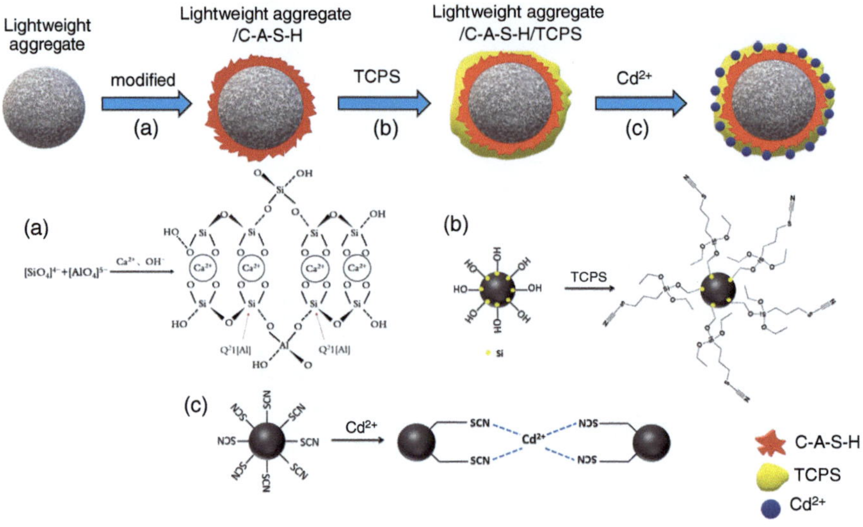

Figure 3.36 Technical route of heavy metal adsorption functional aggregates.

Table 3.8 Performance indicators of typical rock aggregates.

Type	Bulk density (kg/m^3)	Compressive strength (MPa)	Water absorption (%)
Limestone	2600	98.6	1.2
Granite	2710	118.9	0.6
Basalt	2800	145.6	0.6

3.7.3.6 High-strength and Wear-resistant Aggregates

(1) Natural wear-resistant aggregates

High-strength concrete preparation usually requires high-quality aggregates. Table 3.8 shows the performance indicators of three typical natural rock aggregates. Studies have shown that concrete prepared with basalt aggregates with higher parent rock strength can increase one strength grade compared to limestone aggregate concrete under the same conditions.

(2) Artificially Prepared Wear-Resistant Aggregates

(1) **Design of Aggregate Mineral Composition:** The aggregate matrix is strengthened through the design of raw material composition and calcination system. A certain amount of cordierite crystals (Figure 3.37) are introduced into LA. By controlling the relative proportion and spatial distribution of cordierite and mullite crystal phases, the thermal shock resistance of the aggregates is improved, the formation of microcracks in the aggregates during the cooling process is prevented, the strength of the LA is increased, and the water absorption is reduced.

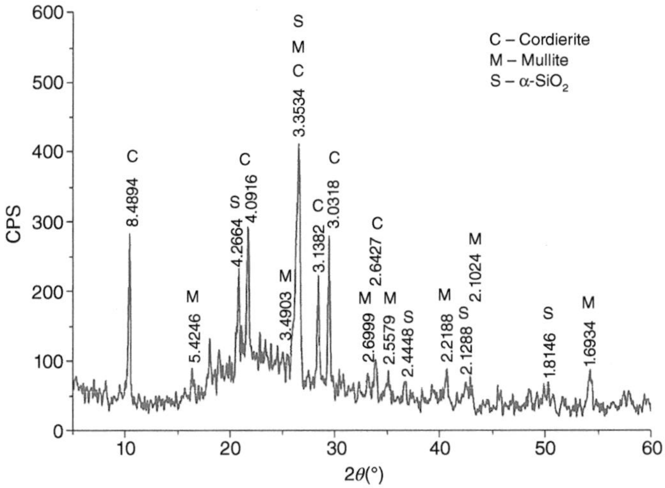

Figure 3.37 Mineral structure of high-strength and high-quality lightweight aggregates.

(2) **Bubble Structure Control:** Through the design of the chemical reaction Eq. (3.13), bubbles are generated during high-temperature chemical reactions to form a porous structure, reducing the density and water absorption of LA.

$$\frac{CaCO_3 \xrightarrow{600\text{–}1050\,^\circ C} CaO + CO_2}{Fe_2O_3 \xrightarrow{1000\text{–}1550\,^\circ C} Fe_3O_4 + O_2} \tag{3.13}$$

The optimal chemical composition range of raw materials for high-strength LA was obtained: SiO_2 55–65%, Al_2O_3 18–25%, Fe_2O_3 6–10%, RO 4–6%, R_2O 1.5–4.0%. The cylinder compressive strength of the prepared 800-grade LA reaches above 8.0 MPa, the water absorption is less than 4.0%, and the water absorption under 6 MPa pressure is less than 8.0%, reaching the international advanced level in technical indicators.

(3) **Surface Activation Technology:** Due to the influence of the glazed outer shell on the bonding effect between LA and cement paste, resulting in a decrease in the coating performance of LA in the mixture and problems such as strength and durability of hardened concrete, this technology improves the surface properties of LA through the design of the raw material composition, pelletizing and sintering process, making the surface of LA form clinker mineral components that have good compatibility and reactivity with cement paste; the surface activation slurry of aggregates is made of cement raw materials mixed with fluorite and gypsum powder, and the ratio value range is controlled: limestone saturation factor (KH) = 0.65–0.70, silica ratio (SM) = 2.0–2.6, alumina ratio (IM) = 1.4–2.0, and solid content 30–40%. XRD test results show that the surface layer minerals of LA after surface activation contain minerals such as C_2S, and the microhardness of the interface zone between aggregates and hardened cement paste is increased by more than 20% [50].

High-wear-resistant concrete is mainly used in hydraulic concrete, pavement concrete, or the surface layer of floor slabs of load-bearing industrial buildings that require wear resistance. The wear resistance of concrete largely depends on the hardness of the aggregates used and the bonding performance between the aggregates and cement paste. Therefore, aggregates with high strength and hard texture, such as calcined bauxite, diamond, corundum, or quartz, should be selected. The wear resistance of aggregates can be tested by the Los Angeles experiment, first grinding and crushing with steel balls for a certain period of time, then sieving with a 1.70 mm sieve, and measuring its mass loss.

3.7.3.7 Shock-absorbing Aggregates

The damping function of concrete is the ability to convert shock energy into other forms of energy dissipation, which is of great significance for improving the structural stability and safety of concrete subjected to long-term vibration load. However, ordinary concrete materials have a low damping function, and there is a contradiction between damping function and mechanical properties such as strength. The author's team designed and prepared shock-absorbing FA by using porous high-strength matrix materials (particle strength greater than 7.0 MPa) to load polymer damping materials, which can significantly improve the damping function of the material without reducing its mechanical properties, providing key components for the design and control of the damping function of high-strength concrete.

3.7.3.8 Solid Waste Used as Aggregates

The use of industrial solid waste in concrete has a certain history, and due to its large usage, it has become an effective way to consume solid waste. Typical solid wastes that can be used as concrete aggregates are shown in Table 3.9. The use of solid waste to prepare aggregates not only needs to consider economic efficiency but also must consider its impact on the service performance and environmental safety of concrete.

Table 3.9 Typical solid wastes that can be used as concrete aggregates.

Material	Components	Industry
Mineral waste	Natural rock	Mining and mineral processing
Blast furnace slag	Silicates, aluminosilicates, magnesium silicates	Steel and iron
Metallurgical slag	Silicates, aluminosilicates, glass	Metal refining
Bottom ash	Aluminosilicates	Power generation
Fly ash	Aluminosilicates	Power generation
Municipal waste slag	Paper, glass, plastic, metal	Commercial and household waste
Construction debris	Bricks, concrete, steel bars	Demolished buildings

References

1 GB/T 8075-2005. (2024). Definition, classification, nomenclature and terminology of concrete admixtures. 2024 (in Chinese).
2 Ji, C. (1997). *Principle and Application of Concrete Admixtures.* Beijing: Science Press (in Chinese).
3 Miao, C. (2008). *High Performance Concrete Admixtures.* Beijing: Chemical Industry Press (in Chinese).
4 Mehta, P. and Moneteiro, P. (2008). *Concrete Micrrostructure, Properties and Materials.* Beijing: China Electric Power Press (in Chinese).
5 Hu, S. (2009). *Advanced Cement-Based Composite Materials.* Beijing: Science Press (in Chinese).
6 Hu, S. (2007). *Steel Tube Confined Concrete.* Beijing: Science Press (in Chinese).
7 Hu, S. (1990). A study of XPS on the interface composition and structure of high strength polymer cement paste. Shanghai: 6th International Congress on Polymer in Concrete, 466.
8 Hu, S. (1993). Structural analysis of polymer-cement interface bond layer. *Journal of Wuhan University of Technology* 15 (4): 12–15. (in Chinese).
9 Hu, S. (1995). IPCN Structure and Its Formation Condition in MDF Cement. *New Development in Concrete Science & Technology* (1): 127–131.
10 Hu, S. (1992). *Study on Polymer Cement-Based Composites and Their Interface Strengthening Mechanism.* Wuhan: Wuhan University of Technology (in Chinese).
11 Hu, J. and Hu, S. (2004). Effect on anti-shrinkkage property of concrete with multiple anti-cracking additive. *Concrete* 1: 23–24+40. (in Chinese).
12 Ren, T., Hu, S., and Ding, Q. (2016). The study on preparation of anti-radiation aggregate used by electroplating sludge Eenriched Ni and Fe. *Journal of Wuhan University of Technology* 38 (08): 1–6. (in Chinese).
13 Lv, L., Chen, Y., Xiao, J. et al. (2017). Tensile performance of PVA fiber reinforced cementitious composites. *Engineering Journal of Wuhan University* 50 (1): 97–101+160. (in Chinese).
14 He, Y., Jin, S., Lü, L. et al. (2010). Effect of hydroxypropyl methyl cellulose on dispersibility of carbon fiber. *Journal of Functional Materials* 41 (06): 1034–1037+1041. (in Chinese).
15 Ding, Q., Li, Y., Hu, S. et al. (1998). Study on surface modified carbon fiber reinforced MDF cement and its mechanism. *Journal of Wuhan University of Technology* ((02)): 1–4. (in Chinese).
16 Ding, Q., Li, Y., Hu, S. et al. (1998). Study on carbon fiber reinforcement of MDF cement. *Journal of Wuhan University of Technology* (3): 14–17. (in Chinese).
17 Ding, Q., Hu, S., and Li, Y. (1997). Study on glass fiber reinforced MDF cement. *Journal of Wuhan University of Technology* (1): 24–27. (in Chinese).
18 Huang, X., Hu, S., Wang, F. et al. (2017). Properties of alkali-activated slag with addition of cation exchange material. *Construction and Building Materials* 146: 321–328.

19 Shen, P., Lu, L., Chen, W. et al. (2017). Efficiency of metakaolin in steam cured high strength concrete. *Construction and Building Materials* 152: 357–366.

20 Guan, X., Hu, S., and Yao, Y. (2004). Study on rheological properties of superfine cement grouting materials by subgroup theory. *Journal of Yangtze River Scientific Research Institute* (6): 27–30. (in Chinese).

21 Hu, S., Peng, Y., and Ding, Q. (2008). Strength and chloride ion permeability of reactive powder concrete containing steel slag powder and ultra-fine fly ash. Proceedings of the First International Conference on Microstructure Related Durability of Cementitious Composites, 437–444.

22 Peng, Y., Hu, S., and Ding, Q. (2009). Dense packing properties of mineral admixtures in cementitious material. *Particuology* 7 (5): 399–402.

23 John, E., Matschei, T., and Stephan, D. (2018). Nucleation seeding with calcium silicate hydrate – a review. *Cement and Concrete Research* 113: 74–85.

24 Zou, F., Hu, C., Wang, F. et al. (2020). Enhancement of early-age strength of the high content fly ash blended cement paste by sodium sulfate and C–S–H seeds towards a greener binder. *Journal of Cleaner Production* 244: 118566.

25 Zou, F., Shen, K., Hu, C. et al. (2020). Effect of sodium sulfate and C–S–H seeds on the reaction of fly ash with different amorphous alumina contents. *ACS Sustainable Chemistry & Engineering* 8 (3): 1659–1670.

26 Jensen, O.M. and Hansen, P.F. (2001). Water-entrained cement-based materials I. Principles and theoretical background. *Cement and Concrete Research* 31 (4): 647–654.

27 Jensen, O.M. and Hansen, P.F. (2002). Water-entrained cement-based materials II. Experimental observations. *Cement and Concrete Research* 32 (6): 973–978.

28 Zhou, Y. (2008). *Research on Internal Curing Mechanism and Control Technology of High Strength Concrete*. Wuhan: Wuhan University of Technology (in Chinese).

29 Yang, J., Wang, F., Huang, J. et al. (2015). Study on properties and application of internal curing commercial concrete. *Concrete* 7: 113–117. (in Chinese).

30 Yang, J., Wang, F., Huang, J. et al. (2016). Comparative analysis of reduction effects of different shrinkage reducing agents in concrete. *Journal of Building Materials* 19 (1): 53–58. (in Chinese).

31 Yang, J., Wang, F., Liu, Z. et al. (2019). Early-state water migration characteristics of superabsorbent polymers in cement pastes. *Cement and Concrete Research* 118: 25–37.

32 Yang, J. (2017). *Microstructure and Performance of Cementitious Materials Internally Cured by Superabsorbent Polymers*. Wuhan: Wuhan University of Technology (in Chinese).

33 Wang, F., Yang, J., Cheng, H. et al. (2015). Study on the mechanism of desorption behavior of saturated superabsorbent polymers in concrete. *ACI Materials Journal* 112 (3): 463–469.

34 Yang, J., Wang, F., He, X. et al. (2019). Pore structure of affected zone around saturated and large superabsorbent polymers in cement paste. *Cement and Concrete Composites* 97: 54–67.

35 Yang, J. and Wang, F. (2019). Influence of assumed absorption capacity of superabsorbent polymers on the microstructure and performance of cement mortars. *Construction and Building Materials* 204: 468–478.

36 Wang, F., Zhou, Y., Ding, Q. et al. (2008). Influence of pre-wetted lightweight aggregate on internal relative humidity of concrete. *Materials Science and Technology* (3): 366–369. (in Chinese).

37 Wang, F., Zhou, Y., Peng, B. et al. (2009). Autogenous shrinkage of concrete with super-absorbent polymer. *ACI Materials Journal* 106 (2): 123–127.

38 Hu, S., Zhou, Y., Wang, F. et al. (2008). Effect of superabsorbent resin particles on self-shrinkage and strength of concrete. *Journal of Huzhong University of Science and Technology (Urban Science Edition)* (1): 1–4+16. (in Chinese).

39 Wang, F., Yang, J., Hu, S. et al. (2016). Influence of superabsorbent polymers on the surrounding cement paste. *Cement and Concrete Research* 81: 112–121.

40 Hu, S., He, Y., and Lü, L. (2006). Water-releasing factor technology to modify the internal relative humidity of concrete and its application. *Journal of Railway Science and Engineering* (2): 11–14. (in Chinese).

41 Yang, W., Lü, L., Wu, J. et al. (2008). Influence of absorbing and releasing water performance of water-releasing factor on internal relative humidity of concrete. *Concrete* 3: 30–34. (in Chinese).

42 Hu, S., Yang, W., and Lü, L. (2006). Experimental study on influencing factors of water absorption and release process of light aggregate. *Highway* 10: 155–159. (in Chinese).

43 Lü, L., Yang, W., Wu, J. et al. (2008). Influence of water absorption and release of water-releasing factor on volume deformation of concrete. *New Building Materials* 3: 5–8. (in Chinese).

44 Yang, T. (2010). *The Research on Cement Paste-Aggregate Interface Based on Functional Design of Aggregate*. Wuhan: Wuhan University of Technology (in Chinese).

45 Huang, X. (2011). *Study on Preparation of Radiation-Proof Functional Aggregate and Concrete from Sludge Containing Heavy Metals*. Wuhan: Wuhan University of Technology (in Chinese).

46 Hu, S., Zhao, D., Wang, F. et al. (2020, January 14). In-situ synthesis of honeycomb C-A-S-H gel membrane composite porous aggregate method. CN 201710539135.9. (in Chinese).

47 Guo, K., Xu, M., and Ding, Q. (2010). Study on the preparation of cement-based traffic noise reduction material by sludge clay sand. *Highway* 3: 151–153. (in Chinese).

48 Tian, Y. (2008). *Research on Design and Performance of Damping and High Strength Concrete*. Wuhan: Wuhan University of Technology (in Chinese).

49 Wang, F., Dong, Y., Yang, L. et al. (2014, October 01). The invention relates to a concrete material with the function of purifying gas-solid pollutants and a preparation method thereof. CN 201210348398.9. (in Chinese).

50 Hu, S., Ding, Q., He, Y. et al. (2006, July 19). High strength light aggregate with surface reactivity and preparation method thereof. CN 200410012911.2. (in Chinese).

4

Ultra-high-strength Concrete

In the early 1980s, a wave of research on ultra-high-strength concrete materials emerged worldwide. Birchall and his collaborators from Imperial Chemical Industries and the University of Oxford developed a new type of high-strength cement material called macro-defect free cement (MDF), with a compressive strength exceeding 200 MPa, a tensile strength of over 50 MPa, and an elastic modulus of 25–50 GPa, exhibiting outstanding mechanical properties. The introduction of MDF cement immediately drew attention from researchers in cement materials worldwide. Countries such as the United States, the Soviet Union, Japan, Denmark, and China conducted research in this area, subsequently introducing a series of similar ultra-high-strength concrete materials, including densified systems of homogeneously arranged ultra-fine particles (DSP), compact reinforced composites (CRCs), reactive powder concrete (RPC), and others. This chapter will introduce several influential and representative categories based on the characteristics of raw materials and manufacturing processes.

Since the mid-1980s, the author's team has been conducting systematic research on MDF cement. They have achieved a series of results in the adaptation and optimization of raw materials [1–3], simplification of practical preparation methods [4, 5], design of material properties [6–8], improvement of material water damage resistance, and enhancement of volume stability [9, 10], as well as the development of new research methods [11–13] and exploration of reinforcement mechanisms, revealing the essential characteristics of the interface reinforcement mechanism of this low water–binder ratio polymer cementitious composite material [11, 14, 15]. This has provided theoretical guidance and experimental experience for addressing its inherent volume instability and late-stage strength retrogression issues and advancing the practical application of the material [16–20].

4.1 Macro-defect Free Cement

4.1.1 Formulation Principles and Preparation Process

In 1981, Birchall and others from Imperial Chemical Industries in the United Kingdom invented a cement-polymer composite material with unique properties

High-Tech Concrete Materials: Design, Preparation, and Applications, First Edition.
Fazhou Wang and Shuguang Hu.

using special manufacturing techniques and an extremely low water–cement ratio. Because the hardened paste of this material lacks macroscopic defects such as large pores and coarse crystals, it is called MDF cement [21, 22]. The main components of MDF cement are cement, water-soluble polymers, superplasticizers, and water, and the cement used is Portland cement or aluminate cement, with the water–cement ratio controlled within the range of 0.12–0.16. Due to the use of such a low water–cement ratio and thorough mixing of the components during preparation while preventing air ingress, pores or defects larger than 100 μm are eliminated from the paste, resulting in MDF cement hardened pastes with extremely high strength.

The key points of the MDF cement preparation process are as follows:

1) After mixing various components, a homogeneous slurry is prepared in a dedicated mixer.
2) The slurry is then molded using methods such as hot pressing, extrusion, roll pressing, or injection to obtain MDF cement products.
3) The molded products undergo appropriate curing, with products made from Portland cement typically cured in room temperature air for about three weeks, while products made from aluminate cement are preferably cured under dry heat.
4) Different types of cement should be matched with compatible polymers. Polyacrylamide (PAM) is generally used for Portland cement, while polyvinyl alcohol (PVA) is commonly used for aluminate cement.

The mechanical properties of the hardened paste of MDF cement far exceed those of ordinary concrete, with a compressive strength of 200–300 MPa, a tensile strength of 50–120 MPa, a flexural strength of 150–200 MPa, an elastic modulus of 25–50 GPa, and a critical stress intensity factor of $3\ N\cdot m^{3/2}$. MDF cement made with aluminate cement generally exhibits higher mechanical properties than those made with Portland cement. The hardened paste of MDF cement also possesses good sound absorption, antistatic properties, and crack resistance at low temperatures. Its main drawback is poor water stability, with a significant decrease in strength after moisture absorption, especially pronounced for those made with aluminate cement, primarily due to polymer swelling caused by water absorption. Therefore, the widespread application of this new composite material is somewhat limited. Significant progress has been made in improving its moisture absorption stability through targeted research, with several effective methods including

a) Substituting hydrophobic polymers for traditional water-soluble polymers in MDF material preparation. For example, using a methanol solution of phenol-formaldehyde resin precursor and a small amount of modifier to develop MDF materials with good water resistance [23]. Due to the hydrophobic nature of the resin component and the strong chemical bonding between cement and resin, this composite material exhibits no significant decrease in flexural strength even after soaking in room-temperature water for three months and can withstand temperatures up to 250 °C.
b) Modification of MDF materials using coupling agents. By adding trace amounts of silane coupling agents or titanate coupling agents during the preparation of

MDF materials, the flexural strength and fracture toughness of the materials are significantly increased, and their water resistance is also improved [24]. This modified MDF material retains approximately 85% of its initial flexural strength after soaking in room temperature water for three months.

c) Incorporating fibers to produce fiber reinforcement macro-defect free cement (FRMDF). However, this method has certain limitations regarding the selected fibers and manufacturing process. Intense mixing of fibers with cement and polymers, followed by extrusion or roll pressing of the mixture, can cause severe damage to the fibers due to shear or bending.

4.1.2 Performance, Development, and Application Potential

MDF cement exhibits excellent engineering performance. According to the research results of Birchall et al., besides significant improvements in mechanical properties, it also holds vast development prospects in electrical, magnetic, acoustic, and low-temperature performance [25]. Table 4.1 lists some engineering properties of MDF cement.

MDF possesses extremely high compressive and flexural strength; along with various other excellent properties, FRMDF combines high toughness and impact resistance. After overcoming the weakness of poor water resistance in the base material, applications can be found in several areas:

Table 4.1 Engineering properties of MDF cement.

Property	Value
Compressive strength	300 MPa
Flexural strength	150–200 MPa
Elastic modulus	50 GPa
Notched pendulum impact	$3\ kJ/m^2$
Critical stress intensity factor	$3\ N{\cdot}m^{3/2}$
Poisson's ratio	0.2
Density	$2500\ kg/m^3$
Porosity	1%
Permeability (Oxygen)	$-10^{-16}\ m/s$
Thermal expansion	$9.7 \times 10^{-6}/°C$
Volume resistivity	10^9–$10^{11}\ \Omega{\cdot}cm$
Initiation discharge voltage	10 kV
Electric field strength	9 kV/mm
Average dielectric constant (1–10 KHz)	9
Shielding 30–1000 MHz (with 30% Vol iron grains)	50 dB
Sound transmission loss angle tgδ	0.1

1) Armor and bulletproof materials for defense engineering.
2) Materials and containers for low-temperature engineering.
3) Electromagnetic radiation shielding materials for the electronics industry.
4) Processing and forming materials for the mechanical industry.
5) Sound damping materials.
6) Surface inlay materials for concrete.

4.2 Densified System of Homogeneously Arranged Ultra-fine Particles

4.2.1 Formulation Principles and Preparation Process

Shortly after the emergence of MDF materials in 1981, Bache et al. from Denmark developed a densified system containing homogeneously arranged ultra-fine particles (DSP) [26]. The formulation principles of DSP are mainly as follows:

1) Mix an appropriate amount of cement with an average particle size of around 10 μm and fine powder with volcanic ash activity with an average particle size of around 0.1 μm uniformly to form the densest packing state.
2) Substantially reduce the water–cement ratio (0.16 or lower) to further reduce the porosity, prevent the aggregation of cement particles and fine powder, and strive to make them evenly distributed.
3) This results in a microstructure that not only extremely small pore size and low porosity but also a non-interconnected pore structure. Therefore, an appropriate amount of high-efficiency water-reducing agent must be added during the material preparation process.

In the initial stages of research, ordinary DSP made from Portland cement and silica fume could achieve a compressive strength of over 120 MPa, with the optimum dosage of silica fume being 7–15% of the cement mass. It was later found that if suitable high-strength aggregates (such as sintered alumina, stainless steel aggregates, etc.) with an average particle size of about 1/7–1/10 of the cement particle size were added to the above system, the compressive strength could exceed 250 MPa, thus obtaining a higher strength DSP. This is not only because the aggregates used have higher strength but also because of the high volcanic ash activity and filling effect of silica fume, which enhances the interface area between the aggregates and the cementitious material. In the range of 1500–3000 nm from the aggregate surface, the region exhibited randomly oriented $Ca(OH)_2$ crystals and significantly reduced porosity, thereby preventing the interface area between the aggregates and the cementitious material from becoming the weak link of the curing system.

Figure 4.1 compares the stress–strain curves under compression of ordinary DSP, high-strength DSP, and ordinary high-strength concrete. It can be seen that although DSP has a relatively high compressive strength, the increase in its fracture toughness compared to ordinary concrete is not significant, and the flexural strength is only about 1/10 of the compressive strength. The stress–strain curve until final failure when subjected to tension or bending is a straight line, showing extremely obvious brittleness [7].

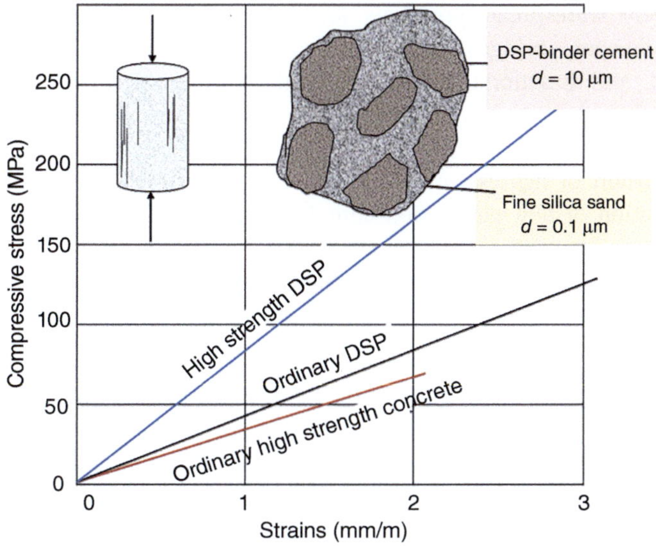

Figure 4.1 Stress–strain curves of DSP and ordinary high-strength concrete.

4.2.2 Performance and Prospects for Development and Application

The compressive strength of DSP materials can reach 5 to 10 times that of ordinary concrete, and the tensile strength is also significantly higher, often more than double. However, due to the limited increase in modulus of elasticity and fracture energy, DSP materials exhibit significant brittleness. They are prone to the generation of numerous microcracks during curing and drying shrinkage processes and may even experience "autogenous fracture," as observed in experiments where cylindrical specimens of pure DSP are fractured. The addition of fibers to DSP mortar can significantly reduce its brittleness, as shown in the experimental results presented in Table 4.2. Both the fracture mechanical properties of DSP mortar and neat paste are significantly higher than those of conventional cement paste and concrete, demonstrating characteristics of high strength and high ductility.

Among all ultra-high-strength cementitious materials, DSP stands out for its simplest forming process coupled with excellent performance. Therefore,

Table 4.2 Comparison of fracture mechanical data between DSP materials and concrete.

Material	Elastic modulus (GPa)	Tensile strength (MPa)	Fracture energy (J/m^2)	Crack tip opening displacement (CMOD) μm	Material ductility	Critical stress intensity factor (K_{IC}) (MN/m$^{2/3}$)
Cement paste	0.7	4	20	5	0.01	0.4
DSP paste	15	20	40	1	0.0008	0.5
Concrete	30	3	60	20	0.2	1.3
DSP mortar	50	20	100	5	0.0125	2.2

it has garnered relatively widespread prospects for applications, such as in high-rise buildings, long-span bridges, corrosion-resistant materials, highly wear-resistant materials, and functional materials. Furthermore, DSP material structures are dense, with extremely low porosity and excellent corrosion resistance. They not only prevent radioactive substances from leaking internally but also withstand the corrosion of aggressive external media, making them ideal materials for the preparation of new-generation containers for radioactive waste storage.

Based on the composition and preparation characteristics of DSP materials, several issues need to be considered for their development and performance improvement:

1) Compatibility between high-efficiency water-reducing agents and material systems. This is particularly critical when multiple ultra-fine powders are incorporated. If we polymers, coupling agents, etc. are introduced, based on the dense packing principle of MDF, the composition of the material system becomes more complex, and the compatibility issue becomes more complicated.
2) Coordinated application of material-related design theories. How to coordinate the principles of dense filling, fiber-reinforced concrete theory, and aggregate grading enveloping theory to achieve dense, high-performance cementitious materials.
3) Unhydrated cement issue. Considering replacing most of the unhydrated cement in the material with other blended materials can reduce cement consumption, improve material volume stability, and reduce material costs.
4) Ultra-fine particle size and packing density issue. According to particle packing principles, coarse and fine particles have significantly different effects on packing density. Therefore, particle selection needs to be optimized, and further research is needed on the particle characteristics of ultra-fine powders, their packing behavior, and their impact on material performance.
5) Autogenous shrinkage of DSP materials. Low water-binder ratios and silica fumes cause early autogenous shrinkage and microcracks in the material, which have a significant impact on its durability and become a bottleneck in improving material performance. Solutions need to be researched and developed to address this issue.

4.3 Compact Reinforced Composite

4.3.1 Formulation Principles and Preparation Process

In 1986, under Bache's leadership, Aalborg Portland Company developed an ultra-high-strength concrete material called CRC, which is based on DSP mortar and contains fiber reinforcement along with closely spaced steel reinforcement [27]. The base material of CRC consists of DSP material, with the main reinforcement similar to the steel reinforcement in reinforced concrete but in much larger quantities (at a volume ratio of 10–20%). Through the use of optimized aggregates and specific fibers, the base material tightly bonded to the reinforcement, resulting in high tensile strength, stiffness, and ductility.

Bache et al. conducted a comprehensive design of CRC materials and structures based on the fundamental principles of fracture mechanics. The key points are as follows: First, utilize the DSP binder composed of cement and ultra-fine particles (mainly silica fume) to achieve the densest particle packing and extremely low porosity. Second, add high-strength, high-modulus aggregate particles with a certain particle size range to the DSP binder to minimize the amount of binder material used. Additionally, incorporating a certain amount of uniformly dispersed high-strength and high-modulus fibers will enhance the rigidity and ductility of the composite material as much as possible. Third, to fully utilize the high compressive strength of DSP mortar, it is essential to incorporate closely spaced steel reinforcement with a high reinforcement ratio in the tensile zone of the flexural members, where they primarily bear tension and contribute to greater fracture toughness.

CRC is particularly suitable for the production of lightweight, slender precast components or components made on-site. According to research, the relevant parameters of CRC components can be summarized as follows:

1) The silica fume content in DSP mortar is equivalent to 7–15% of the weight of cement, with a water–binder ratio of 0.16–0.18. Quartz sand or sintered bauxite is used as aggregate, with a maximum particle size of 4 mm.
2) Steel fibers are generally used with a diameter of 0.15 mm and a length of 6 mm for stainless steel fibers. Alternatively, high-strength steel fibers with a diameter of 0.4 mm and a length of 12 mm can be used. The volume fraction of steel fibers is 6–9%.
3) For the tensile zone reinforcement, when the component thickness is 200–300 mm, the diameter of the reinforcement ranges from φ8 mm to φ25 mm, with a reinforcement ratio of 10–20%, a spacing of 10–15 mm, and a cover thickness of 10–15 mm. The spacing of the reinforcement and the cover thickness mainly depend on the size of the fibers and the maximum particle size of the aggregate in the DSP mortar.

The manufacturing process of CRC components is similar to that of conventional reinforced concrete components. However, due to the higher content of steel fibers, a high-efficiency mixer should be selected to ensure the uniform distribution of steel fibers in the DSP mortar. Generally, the process involves first preparing the DSP mortar and then adding the steel fibers. To ensure good workability of the mixed material under a low water–binder ratio and high fiber addition rate, an appropriate amount of high-efficiency water-reducing agent must be added to the mix. After pouring the mortar into the mold, it should be vibrated to achieve uniform distribution and compaction.

4.3.2 Performance and Prospects for Development and Application

The compressive strength of CRC typically reaches 200 MPa, with the potential to exceed 300 MPa, while its tensile strength is comparable to that of structural

Table 4.3 Fracture mechanical parameters of CRC and DSP mortar.

Material	Elastic modulus (GPa)	Tensile strength (MPa)	Fracture energy (J/m^2)	Crack tip opening displacement (CMOD) (μm)	Material ductility	Critical stress intensity factor (KIC) (MN/m$^{2/3}$)
CRC	100	120	1200	10,000	8	350
DSP mortar	50	20	100	5	0.0125	2.2

steel reinforcement and exhibits considerable ductility. Table 4.3 lists the fracture mechanical parameters of CRC compared to DSP mortar. From the data in the table, it can be seen that the tensile strength of CRC is six times that of DSP mortar, the fracture energy can reach 10,000 times that of DSP mortar, and the ductility is over 600 times that of DSP mortar. This indicates that, compared to DSP mortar, CRC not only has higher strength but also significantly greater deformability.

Bache et al. conducted bending tests on various combinations of CRC beams, with the results of one beam shown in Figure 4.2. The beam has a cross-sectional size of 100 × 150 mm and a length of 2100 mm. The DSP mortar used contains sintered bauxite aggregate with a maximum particle size of 4 mm and a volume fraction of 9%, along with steel fibers sized 0.15 × 6 mm. Six Φ15 mm steel prestressing wires are placed at the bottom of the beam, while four Φ16 mm variable-section steel bars are positioned at the top. The maximum load achieved in the bending test using the four-point method was 230 kN. Correspondingly, the calculated bending tensile stress was 323 MPa, and the corresponding strain was 0.008.

CRC exhibits excellent durability, surpassing that of typical high-performance concrete (HPC). This is primarily due to the extremely dense nature of its base material, which has a porosity of only around 1.5%. The majority of pores in the

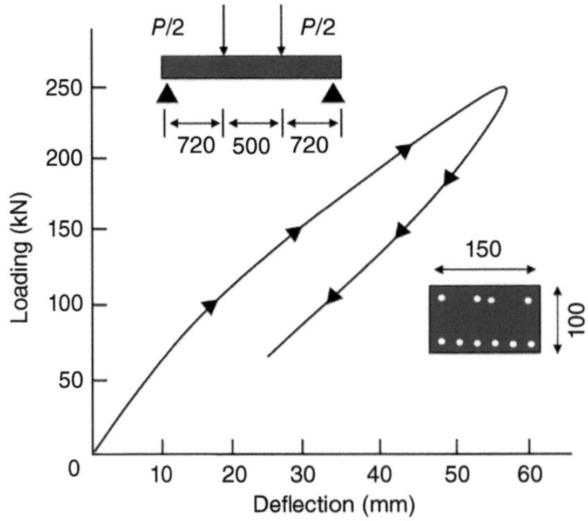

Figure 4.2 Load-deflection curve of a CRC beam tested in a bending test.

base material are gel pores, with capillary pores accounting for only about 14%. Therefore, CRC has extremely high resistance to freeze–thaw cycles, as there is no water freezing inside the pores, preventing structural damage from repeated freeze–thaw cycles. Additionally, CO_2 and chloride ions have difficulty penetrating CRC due to their refined pore network and discontinuous pore structure. Moreover, the base material contains a considerable amount of unhydrated cement particles, which can react with water in microcracks to generate hydration products that block the cracks. CRC also exhibits high resistance to projectile penetration. In one test case conducted by Aalborg Portland Company, two CRC panels, each 200 mm thick, were tested for ballistic resistance by combining them together. In the test, a 152 mm diameter, 500 mm long, 47 kg projectile was fired at the CRC panels using a cannon. The projectile was stopped by the CRC panels, preventing penetration, with only minimal cracks appearing on the backside of the second panel.

Since the 1990s, CRC has gradually entered the practical stage, being used for various applications such as tunnel lining blocks, cantilever balcony decks, support beams for stairs, manhole covers, and joints for precast slab-column systems poured on-site. Despite its practicality and effectiveness, CRC remains relatively expensive, making it impractical to replace conventional reinforced concrete components entirely. However, future development directions include applications in coastal structures, large-span bridges, buildings in earthquake-prone areas, and military defense projects.

4.4 Reactive Powder Concrete

4.4.1 Formulation Principles and Preparation Process

In 1993, Richard et al. from France developed an ultra-high-strength concrete known as RPC [28]. The principles and process for preparing RPC are as follows: First, based on the principle of closest packing, coarse aggregates are eliminated, and the particle size of fine aggregates is reduced to create a discontinuous particle size distribution curve, resulting in a highly uniform and dense particle system. Second, a large amount of high-efficiency water-reducing agent is used to significantly reduce the water content, decrease the porosity, and further enhance the density of the concrete. Some unhydrated cement particles can contribute to reinforcement. Third, fine steel fibers are added uniformly throughout the matrix to inhibit cracking, reduce brittleness, and improve toughness. Fourth, pressure is applied to evacuate air trapped during the setting and curing process, eliminating entrapped air bubbles and strengthening the bond between particles, thereby enhancing density. Fifth, heat treatment is applied after concrete curing to enhance chemical reactivity and internal structure density, thereby strengthening the microstructure of the hardened concrete.

The main ingredients used in preparing RPC are Portland cement, fine sand (with a maximum particle size generally not exceeding 400 μm), silica powder (with

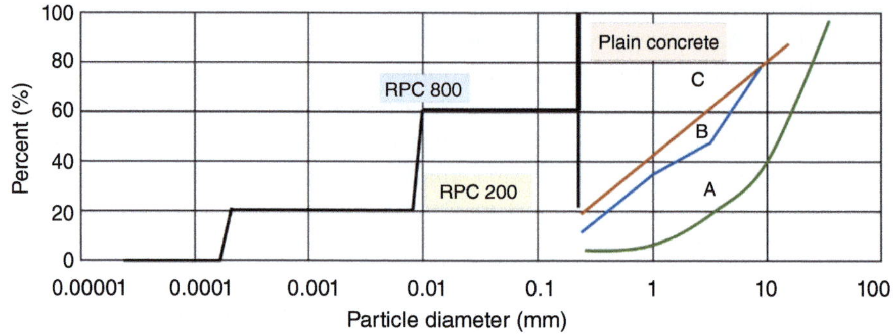

Figure 4.3 Particle size distribution curve of RPC.

Table 4.4 Formulations of two types of RPC materials.

Raw material	RPC200	RPC800
Portland cement type V	955	100
Fine sand	1051	500
Silica powder	—	390
Silica fume	229	230
Precipitated silica	10	—
High-efficiency water-reducing agent	13	18
Steel fibers (carbon steel, 12/0.18 mm)	191	—
Stainless steel fibers (3 mm length)	—	630
Water	153	180

a maximum particle size not exceeding 4 μm), silica fume (with a specific surface area of not less than 18,000 m^2/kg), precipitated silica (with a specific surface area of around 3000 m^2/kg), fine steel fibers, high-efficiency water-reducing agent, and water. The particle size distribution curves of various particles composing RPC, as shown in Figure 4.3, are discontinuous. The average particle size of each particle grade differs by a factor of 7 to 10, thereby achieving the minimum porosity particle packing density. Table 4.4 lists the formulations of two typical RPC grades: 200 and 800 MPa [29].

(1) Portland cement type V is similar to moderate-heat Portland cement clinker and sulfate-resistant Portland cement clinker.

Heating during the curing process promotes the reaction between cement and pozzolanic mineral powders, enhancing the density and early strength of the hardened material. Research findings indicate [30] the following:

a) Curing at 90 °C with hot water aids in reducing the porosity of RPC and increasing its strength by facilitating the reaction between $Ca(OH)_2$ generated from cement hydration and silica fume.

b) After curing in water at 20 °C for two days, subjecting RPC to heating at 200 °C in a furnace results in the formation of tobermorite crystals and abundant C-S-H gel, leading to very high compressive strength (around 250 MPa).
c) Curing at 300 °C under pressure after two days of curing in water at 20 °C results in a significant decrease in RPC strength due to the formation of xonotlite and α-C_2SH crystals.

4.4.2 Performance and Prospects for Development and Application

Based on different materials and processes, RPC is divided into two grades: RPC200 and RPC800. Their main mechanical properties and comparison with HPC are listed in Table 4.5 [11].

From Table 4.5, it is evident that

a) Compared to HPC, RPC exhibits more than double the compressive strength and approximately one order of magnitude higher flexural strength.
b) The fracture energy of RPC200 is two orders of magnitude higher than that of HPC. However, the fracture energy of RPC800 is significantly lower than that of RPC200, which is attributed to the stainless steel fibers in the former being pulled apart when RPC fails under tension.

Due to the extremely high ductility of RPC200, the ultimate elongation can reach $5000-7000 \times 10^{-6}$, making it unnecessary to use conventional steel reinforcement for structural components. Figure 4.4 presents a comparison of the cross sections of I-beams made from RPC, steel, prestressed concrete, and reinforced concrete, all with the same load-carrying capacity. Table 4.6 compares the dimensions, unit mass, and unit price of I-beams made from these four materials [29]. From Figure 4.4 and Table 4.6, it can be observed that the cross-sectional dimensions and unit mass of RPC beams are significantly lower than those of prestressed concrete beams and reinforced concrete beams but slightly larger than those of steel beams. Although the material unit cost of RPC beams is higher than that of prestressed concrete beams and reinforced concrete beams, the unit price is still lower due to the reduced total material consumption.

During the period from 1994 to 1998, RPC began to make its mark in civil engineering. For instance, in France, more than 2500 beams of varying sizes, totaling

Table 4.5 Comparison of mechanical properties between the two types of RPC and HPC.

Material properties	RPC200	RPC800	HPC
Compressive strength (MPa)	170–230	490–680	60–100
Flexural strength (MPa)	30–60	45–140	6–10
Fracture energy (J/m^2)	20–40	1.2–2	0.14
Elastic modulus (GPa)	50–60	65–75	35

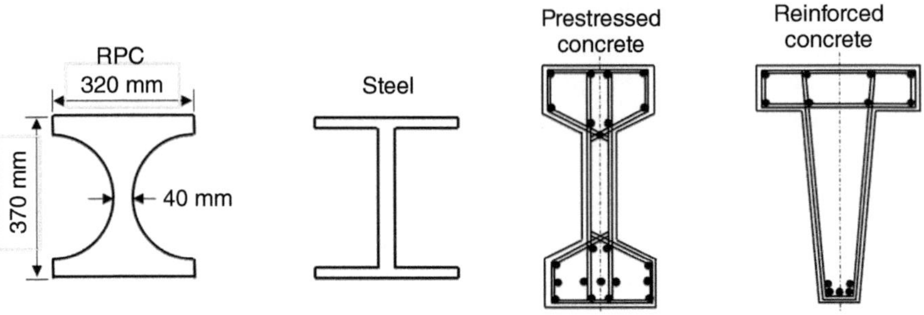

Figure 4.4 Cross sections of I-beams made from RPC, steel, prestressed concrete, and reinforced concrete.

Table 4.6 Cross-sectional dimensions, mass per unit length, and unit price of I-beams of four different materials.

Comparison items		**RPC beam**	**Steel beam**	**Prestressed concrete beam**	**Reinforced concrete beam**
Cross-sectional dimensions (mm)	Height	370	350	700	700
	Width	320	300	350	600
Mass (kg/m)		140	112	467	530
Unit price (yuan/m)		3328	3700	2885	2960

823 m^3 of RPC, were used in the cooling system of a nuclear power plant and extensively in the construction of numerous containers for nuclear waste storage [31]. In Sherbrooke, Canada, an RPC pedestrian/cycling bridge was constructed as part of the cycling network connecting Quebec with the United States [32]. This bridge, with a single span of 60 m and a deck width of 4.2 m, covering an area of 202 m^2, was erected in a harsh local climate characterized by high humidity and severe winter temperatures dropping to as low as −40 °C, requiring frequent de-icing with salt water. The RPC bridge deck panels, 30 mm thick with 70 mm high reinforcing ribs spaced at 1.7 m intervals, were prefabricated using conventional concrete techniques. Figure 4.5 depicts the actual scene of the bridge. A comparison of the material consumption estimates for RPC200, high-performance concrete HPC60, and normal concrete NC30 for the same project is presented in Table 4.7. It is evident from the table that using RPC significantly reduces the consumption of raw materials compared to HPC or normal concrete. With its high strength, toughness, density, and durability, RPC, in many cases, can be used as a structural material without the need for reinforcing steel, making it highly attractive to the engineering community.

Figure 4.5 RPC bridge in Sherbrooke, Canada.

Table 4.7 Comparison of raw material consumption for three different types of concrete used on the same project.

Parameter	NC30	HPC60	RPC200
Equivalent section thickness (mm)	500	400	150
Concrete volume (m^3)	126	100	33
Cement content (kg/m^3)	350	400	705
Cement usage (t)	44	40	27
Aggregate usage (t)	230	170	60

4.5 Reinforcement Mechanism of Ultra-high-strength Cementitious Composite

A large number of researchers have conducted studies on the high-strength mechanism of ultra-high-strength concrete materials, but the main work revolves around the principles of close packing of ultra-fine particles, fiber reinforcement mechanisms, and the analysis of the effect of ultra-high-pressure preparation on excluding material pores. However, the author contends that the essence of ultra-high-strength concrete materials is a type of low water–binder ratio organic–inorganic matrix composite material. From the range of successfully developed materials at present, the polymers used are all water-soluble polymers, which contain a large number of polar groups and have certain reactivity in the cement–water system. From the perspective of material preparation technology, using a smaller water–binder ratio, high-speed shear mixing, grinding, or extrusion molding greatly increases the interface contact opportunities between the two different matrices, providing favorable conditions for the chemical reactions between the organic and inorganic matrices. Therefore, the interface interaction in such materials becomes a very important factor. This is because the porosity in such materials is extremely low-unlike in high water–binder ratio slurries where pores govern strength. In such materials, the water–binder ratio is very low, with a large amount of cement remaining unhydrated. In addition, the high density under compression reduces the effect of gel volume and hydration phase composition factors.

The author believes that although previous studies have shown that the mechanical properties of hardened cementitious materials are a function of various factors, for materials with low water-binder ratios and special shaping processes, the physical and chemical interactions at the interface may become extremely important factors. The author's team, from the perspective of composite materials, employs modern microscopic testing techniques and research methods to investigate the physical and chemical reaction behaviors at the interface between polymers and cement minerals, two different matrices. By analyzing the types of polymers involved in chemical reactions and their dissolution and ionization processes in aqueous solutions, as well as the composition of liquid and solid phase products in reaction systems and hydration characteristics, the authors reveal the chemical bonding nature between the two matrices. Based on chemical reactions, the authors analyze the adhesive properties and chemical bonding characteristics of organic and inorganic phases in the interfacial layer and propose a hierarchical model of composite interfaces based on composition and structural information, providing a basis for practical material design and development.

Through a comprehensive analysis of factors influencing cement strength, the importance of interface behavior in low water–binder ratio polymer-cementitious composite materials has been elucidated. By integrating the physical, chemical, and physicochemical interactions of polymers, fibers, ultra-fine particles, and cement within the material, the excellent performance and reinforcement mechanisms of ultra-high-strength cementitious materials have been revealed.

4.5.1 Composition and Structural Characteristics of Ultra-High-strength Cementitious Material with a Low Water–binder Ratio

4.5.1.1 Material Composition Characteristics

Ultra-high-strength cementitious composite materials, such as MDF, are primarily composed of unreacted cement, the polymer phase, and their interface phase. The interface phase plays a crucial role in such polymer-cementitious composite materials, enabling both cement and polymers to exhibit their respective properties while endowing the composite material with superior comprehensive performance compared to individual raw materials.

Polyvinyl alcohol (PVA), as a significant ingredient in the preparation of polymer-cementitious composite materials, is a polymer compound containing a large number of active hydroxyl groups. The –OH groups in its molecule serve both as reactive groups and hydrophilic groups. In aluminate polymer-cementitious composite materials, the main role of water-soluble polymer PVA is to ensure the formation of plastic substances under high shear mixing and to form important microstructural interface phases through chemical reactions with cement hydration products. Although PVA is essential for obtaining high flexural strength in polymer-cementitious composite materials, its presence also results in the high moisture sensitivity of these materials. Swelling after moisture absorption negatively affects flexural strength, dielectric constant, and dimensional stability.

Unreacted cement particles, acting as active fillers in composite materials, serve as both the matrix for hydration phases and as receptors for water (moisture). Their late-stage hydration also significantly affects the microstructure and moisture sensitivity of composite materials.

4.5.1.2 Material Microstructure Characteristics

Popoola et al.'s observations and measurements using transmission electron microscopy (TEM), high-resolution episcopic microscopy (HREM), energy dispersive spectrometry (EDS), and parallel electron energy-loss spectroscopy (PEELS) revealed that the microstructure of aluminate/PVA polymer cementitious composite materials consists of unreacted cement particles, interface phases (layers) encapsulating individual cement particles, and the bulk polymer phase, as shown in Figure 4.6. The interface phase region includes the crystalline hydration product $Ca_2Al_2O_5 \cdot 8H_2O$, the amorphous mixed phase $Al(OH)_3$, and a mixture of PVA phases cross-linked by Al^{3+}. The PVA phase inhibits the transformation of the metastable $CaAl_2O_5 \cdot 8H_2O$ phase to the thermodynamically stable $Ca_3Al_2O_5 \cdot 6H_2O$ phase.

Based on the understanding mentioned above, the author proposes that water erosion in polymer-cement composites occurs in three regions: (1) the surface of unreacted cement particles, (2) the polymer directly adjacent to the cement particles, and (3) the polymer bulk away from the interface phase (layer) [5, 33]. The reason for distinguishing between the bulk polymer and the interface region is that the polymer near the cement particles and the polymer farther away have different

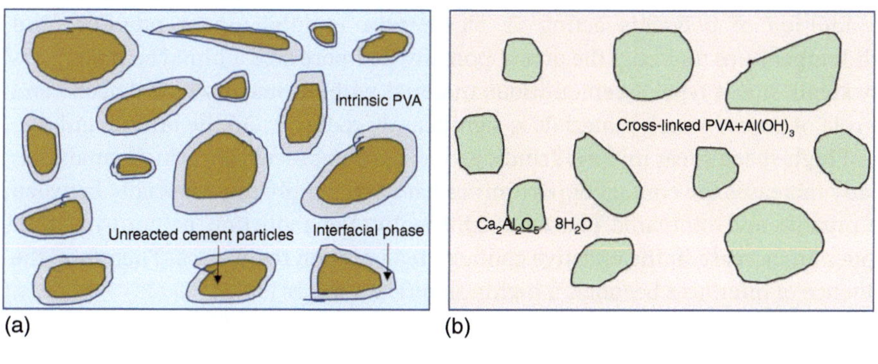

Figure 4.6 Schematic microstructure of aluminate/PVA polymer cementitious composites. (a) Schematic diagram of common microstructure and (b) magnified diagram of the interfacial phase.

chemical compositions and structures, and chemical reactions involving polymers and cement can occur at the interface [34, 35]. Since both the bulk polymer and the interface phase can bond unreacted cement particles, we collectively refer to them as the binding phase.

4.5.2 Factors Affecting the Strength of Ultra-high-strength Cementitious Material with a Low Water–binder Ratio

Through a comprehensive analysis of factors affecting cement paste strength, it has been recognized that they mainly include the total porosity, pore size, and distribution within the material; the composition of hydration products; gel volume; and the material's interface behavior. Among these factors, the influence of pores on material strength has been more extensively studied. This research approach is primarily based on Powers T.C.'s cement hydration cementitious model [36] and Griffith's fracture theory [37]. In typical cement slurries and fabricated materials, due to processing constraints, the actual water–binder ratio during hydration is generally higher than the theoretically critical water–binder ratio. Therefore, there are often many pores left in the material due to the evaporation of excess water, and some capillary water remains in the material. It is evident that typical cementitious materials are porous structural materials, and the pore structure plays a crucial role in material performance.

However, further in-depth research has revealed that in some cases [38], such as when the porosity is low, the established relationship between porosity and strength may no longer apply after reaching a certain threshold of porosity reduction. Materials composed of different raw materials may exhibit different macroscopic mechanical properties at the same porosity level. Particularly in the study of material interface behavior, it has been found that interface behavior has a significant impact on various physical properties of the material. Due to the complexity of cementitious materials, under current research conditions, it is still challenging to determine which factors are predominant for general materials. However, for some special materials, such analysis is feasible.

In low water–binder ratio polymer cementitious composite materials, due to the addition of polymers acting as plasticizers, molding under pressure, and high-temperature molding, the actual porosity and pore size within the material is very small. It is a type of cementitious material without macroscopic defects, and the role of pores in such materials is significantly reduced. On the other hand, the use of high-speed shear mixing, grinding, or extrusion processes during preparation greatly increases the contact opportunities between the phases, especially between the organic and inorganic phases. Additionally, the reduction in porosity itself implies an increase in the effective contact area between the phases. Therefore, the influence of interfaces becomes a highly significant factor [39].

4.5.3 Reinforcement Mechanism of Ultra-high-strength Cementitious Material

4.5.3.1 Enhancement of Functional Material in Ultra-high-strength Cementitious Material

1) Roles of polymers

The enhancement effects of polymers mainly include

1) **Plasticizing Effect**: Water-soluble polymers disperse cement particles in the cement slurry, wetting their surfaces, reducing slurry viscosity, facilitating easier relative sliding between particles, reducing inter-particle frictional resistance, and improving the flow properties of the slurry.
2) **Pore Reduction Effect**: Polymers added as plasticizers can decrease the amount of water used in the actual preparation process of cement slurry, effectively reducing the formation of voids and pores that may result from excess water. Polymers themselves possess good plasticity and can easily fill voids in some inorganic matrices, thereby filling and sealing pores.
3) **Toughening Effect**: Polymers bridge across cracks, impeding rapid crack propagation, absorbing energy, and thereby enhancing the material's toughness. Polymers not only initiate crack formation but also control crack propagation. The toughening effect of polymers can be generated through shear yielding in the matrix [39].

2) Roles of fibers

The reinforcement effects of fibers mainly include

1) **Crack Inhibition Effect**: Fibers integrate with the cement matrix and act as reinforcement, significantly reducing the occurrence of inherent cracks within the concrete and effectively preventing the initiation and propagation of cracks.
2) **Strength Enhancement Effect**: During initial loading (tension, bending), the cement matrix and fibers jointly bear external forces. After the matrix cracks, fibers spanning across the cracks become the primary load bearers. This can significantly improve the mechanical properties of concrete in tension, bending, and shear. Additionally the high modulus of elasticity of the fibers can greatly enhance the impact resistance of concrete.
3) **Toughening Effect**: After the matrix in the tensile region cracks, fibers bear the tensile forces and slow down the expansion of the matrix cracks. As the cracks expand, the residual stress between the matrix cracks gradually decreases. Fibers have significant deformability and can continue to bear the tensile forces on the cross section. With increasing loads, fibers transmit additional stress to the matrix through bonding stresses [39].

3) Roles of ultra-fine particles

The reinforcement effects of particles mainly include

1) **Filling Effect**: Since the particle size of ultra-fine powder is smaller than that of cement particles, it fills the voids between cement particles, thereby increasing the compactness of the cementitious material. After hydration with water, the compactness and strength of the cementitious material also increase.
2) **Strength Effect**: The active components SiO_2 and some Al_2O_3 in the powder undergo secondary hydration reactions with the hydration product $Ca(OH)_2$ of cement to generate a large amount of amorphous C-S-H gel, which fills the pores in the cementitious material, making it more compact and thereby improving its mechanical properties.
3) **Structural Optimization Effect**: Improving the interface structure between cement paste and coarse aggregates in concrete and improving the pore structure of cement paste in concrete. Eliminating the orientation arrangement of crystals at the interface and improving the interface structure between aggregates and cement paste [40, 41].

4.5.3.2 Interfacial Composite Enhancement Mechanism

1) The synthesis of water-soluble polymers is a prerequisite.
   As a surfactant, water-soluble polymers effectively act as plasticizers during material processing, increasing the opportunities for contact between organic and inorganic matrices and enhancing the degree of interface reaction. Simultaneously, water-soluble polymers can effectively fill some of the pores formed during material preparation that cannot be eliminated, thereby reducing the porosity within the material. Essentially, this increases the number of force bonds per unit area in the material. Moreover, polymers spanning across some cracks in the inorganic phase can absorb fracture energy, impede crack propagation, and enhance the limit of fracture energy. They also form a hydrostatic pressure ring around the inorganic phase, thereby improving the overall toughness of the material. From the perspective of composite materials, polymers, as one phase in the composite material, can also form their structural skeleton under certain conditions. They intertwine with the spatial skeleton formed by the inorganic cement slurry, forming an interpenetrating polymer cement network (IPCN) [39].
2) Interface-chemical interaction is crucial.
   In the presence of water, water-soluble polymers are prone to hydrolysis or ionization, forming unsaturated groups and ion-type polymers. Cement hydration generates new hydration product phases, resulting in various ions dissolved in the liquid phase of the system. Some of these ions can undergo chemical reactions with hydrolyzed polymers under alkaline conditions, primarily at the phase interface, which is constrained by reaction conditions. The chemical reactions increase the cross-linking of polymers in the system, facilitating the formation of a spatial network of polymer bodies. Additionally, these reactions bond organic

and inorganic matrices, significantly improving interfacial bonding properties. Furthermore, the addition of water-soluble polymers increases the concentration of hydrogen bonds, aiding in interphase connections. Moreover, the chemical reactions between polymers and inorganic ions generate new products, effectively enhancing their water resistance and increasing the stability of the system [42, 43].

3) Optimization of material composition and structure.
According to the analysis of factors affecting the strength of cementitious materials, in dense materials, the influence of pore defects will decrease, especially when large pores transition to smaller ones and the distribution of pores becomes uniform. Moreover, with the rise of interface factors, unlike ordinary cement materials or polymer cement materials, the interface behavior of low water–binder ratio polymer cementitious composite materials plays a crucial role in determining the performance characteristics of such materials. In low water–binder ratio polymer cementitious composite materials, the interface chemical interactions between polymers and cement effectively improve the interface structure and bonding properties of hydration product phases. This transformation converts weak connections into stronger chemical bonds, increasing the bonding strength between hydration products and significantly enhancing the mechanical properties of the material. Additionally, from a microstructural perspective, the interface chemical interactions facilitate the formation of an IPCN structure. This structure exhibits good compatibility and stability, thereby achieving composite overall performance of the composite material [40–44].

Based on the comprehensive research results, the author believes that in low water–binder ratio ultra-high-strength materials, physical effects, physicochemical effects, and chemical effects complement and alternate with each other. These interactions gradually transform the material from an initial plastic state with dispersed multiple phases containing water to a dense, high-strength composite material with no major defects at the macrolevel and interlacing networks at the microlevel. This process is the fundamental reason for the formation of properties in ultra-high-strength concrete materials.

References

1 Peng, Y., Chen, K., and Hu, S. (2011). Study on interfacial properties of ultra-high performance concrete containing steel slag powder and fly ash. *Advanced Materials Research* 194, 956–196, 960.

2 Peng, Y., Chen, K., and Hu, S. (2011). Influence of the characteristics of steel slag powder particle on compressive strength of reactive powder concrete. *Journal of Building Materials* 14 (4): 541–545. (in Chinese).

3 Geng, C., Ding, Q., Liu, Y. et al. (2018). Effects of granite powder on the performance of ultra high performance concrete under autoclaved condition. *Bulletin of the Chinese Ceramic Society* 37 (2): 533–540. (in Chinese).

4 Ding, Q., Liu, X., Liu, Y. et al. (2016). Effect of Curing Regimes on Microstructure of Ultra High Performance Concrete Cement Pastes. First International Conference on UHPC Materials and Structures. 371–379.

5 Shen, P., Lu, L., He, Y. et al. (2019). The effect of curing regimes on the mechanical properties, nano-mechanical properties and microstructure of ultra-high performance concrete. *Cement and Concrete Research* 118: 1–13.

6 Hu, S. and Yang, D. (1995). Optimizing design of high-tech concrete composites. Nanjing: Southeast University Press, New Development In Con. Sci. & Tec, (2): 960–966.

7 Guan, X., Hu, S., and Ding, Q. (2002). Research on Ultrafine High Performance Cement. The 5th International Symposium on Cement and Concrete. 227–232.

8 Hu, C., Yao, S., Zou, F. et al. (2019). Insights into the influencing factors on the micro-mechanical properties of calcium-silicate-hydrate gel. *Journal of the American Ceramic Society* 102 (4): 1942–1952.

9 Hu, S. and Yang, D. (1995). IPCN Structure and Its Formation Condition in MDF Cement. *New Development in Concrete Science & Technology* (1): 127–131.

10 Hu, S. and Li, Y. (1998). Studies on the formation characteristic and thermo-stability of hydrated calcium carboaluminate. 4th Beijing International Symposium on the Cement & Concrete. 2: 502.

11 Hu, S. (1990). A Study of XPS on the Interface Composition and Structure of High Strength Polymer Cement Paste. 6th International Congress on Polymer in Concrete, 466.

12 Hu, S. (1994). XPS nondestructive depth analysis method and its application in cement based composite materials. *Cement and Concrete Research* 24 (8): 1509–1514.

13 Hu, S., Wang, S., and Wang, S. (1994). Research on PAT hole technology and its application in MDF cement material. *Journal of the Chinese Ceramic Society* 22 (2): 188–194. (in Chinese).

14 Hu, S. (1992). *Study on Polymer Cement-Based Composites and Their Interface Strengthening Mechanism*. Wuhan: Wuhan University of Technology (in Chinese).

15 Hu, S., Yang, H., Liu, S. et al. (1994). *Thermochimica Acta* 24 (6): 129–140.

16 He, Y., Lv, L., Jie, M. et al. (2012). Preparation of ultra-high-strength self-compacting concrete and its high-drop casting construction in a pier structure. *Magazine of Concrete Research* 64 (12): 1049–1055.

17 Wang, F., Li, M., and Hu, S. (2014). Bond behavior of roughing FRP sheet bonded to concrete substrate. *Construction and Building Materials* 73: 145–152.

18 Huang, Q., Hu, S., Du, R. et al. (2015). Calculation method of cracking moment of prestressed RPC box girder. *China Civil Engineering Journal* 48 (S1): 15–21. (in Chinese).

19 Ding, Q., Liu, Y., Liu, X. et al. (2018). Effects of sulfate attack on microstructure of UHPC hydration products under different curing regimes. *Bulletin of the Chinese Ceramic Society* 37 (3): 772–780. (in Chinese).

20 Hu, S., Chen, Y., and Ma, J. (2004). *Research and Engineering Application of Cement Base Treatment Materials*. Wuhan: Wuhan University of Technology Press (in Chinese).

21 Birchall, J., Howard, A., and Kendall, K. (1981). Flexural strength and porosity of cements. *Nature* 289 (5796): 388–390.

22 Birchall, J., Kendall, K., and Howard, A. (1981, January 07). Cementitious product. EP0021682.

23 Hasegawa, M., Kobayashi, T., and Pushpalal, G.K.D. (1995). A new class of high strength, water and heat resistant polymer – cement composite solidified by an essentially anhydrous phenol resin precursor. *Cement and Concrete Research* 25 (6): 1191–1198.

24 Li, B. (1998). *Study on Humidity Sensitivity and Modification of Cement-Based Composites Without Macroscopic Defects*. Wuhan: Wuhan University of Technology (in Chinese).

25 Alford, N. and Birchall, J. (1985). Fiber toughening of MDF cement. *Journal of Materials Science* 20: 37–45.

26 Bache, H. (1981). Densified Cement/Ultrafine Particle-Based Materials. The Second International Conference on Superplasticizers in Concrete, 5–35.

27 Bache, H. (1989). Fracture Mechanics in Integrated Design of New, Ultra-Strong Materials and Structures. Fracture Mechanics of Concrete Structures: From Theory to Applications, 382–398.

28 Richard, P. and Cheyrezy, M. (1995). Composition of reactive powder concrexes. *Cement and Concrete Research* 25 (7): 1501–1511.

29 Richard, P. and Cheyrezy, M. (1994). Reactive powder concretes with high ductility and 200-800 MPa compressive strength. *Concrete Technology: Past, Present and Future* 144: 507–518.

30 Cheyrezy, M., Maret, V., and Frouin, L. (1995). Microstructural analysis of RPC (reactive powder concrete). *Cement and Concrete Research* 25 (7): 1491–1500.

31 Cheyrezy, M. and, Daniel, J. (1996). Specific production and manufacturing issues. High Performance Fiber Reinforced Composites. 25–42.

32 Adeline, R., Lacheme, M., and Blaif, P. (1998). Design and behavior of the sherbrooke footbridge. Proceedings of International Symposium on High Performance and Reactive Particle Concrete.

33 Feng, X. and Hu, S. (1991). XPS study on the composition and structure of polymer-aluminate-calcium interface. *Journal of the Chinese Ceramic Society* 19 (6): 481–487. (in Chinese).

34 Hu, S. (1996). Studies of MDF cement reinforced with glass fiber. MAETA Workshop on High Flexural Polymer-Cement Composite, 131–135.

35 Shen, P., Lu, L., Wang, F. et al. (2020). Water desorption characteristics of saturated lightweight fine aggregate in ultra-high performance concrete. *Cement and Concrete Composites* 106: 103456.

36 Powers, T.C. (1958). Structure and physical properties of hardened portland cement paste. *Journal of the American Ceramic Society* 41 (1): 1–6.

37 Griffith, A. (1921). The phenomena of rupture and flow in solids. *Philosophical Transactions of the Royal Society of London. Series A, Containing Papers of a Mathematical or Physical Character* 221 (582–593): 163–198.

38 Hu, S. and L Y. (1999). Research on the hydration, hardening mechanism, and microstructure of high performance expansive concrete. *Cement and Concrete Research* 29 (7): 1013–1017.

39 Hu, S. (2009). *Advanced Cement-Based Composite Materials.* Beijing: Science Press (in Chinese).

40 Zhu, M. and Hu, S. (2007). Study on the character and physicochemical property of cement-based material's grain by fractal theory. *Cement Engineering* 6 (4): 7–10. (in Chinese).

41 Peng, Y., Hu, S., and Ding, Q. (2010). Preparation of reactive powder concrete using fly ash and steel slag powder. *Journal of Wuhan University of Technology-Materials Science Edition* 25 (2): 349–354.

42 Hu, S. (1992). The increasing strength mechanism of the role of interfacial bond in MDF cement. 9th International Congress on the Chemistry of Cement. (2): 393.

43 Hu, S. (1993). An investigation of bonding behavior on the interface of polyacrylamide-aluminous composites. *Material Science Edition* 8 (2): 19–25.

44 Shen, P., Lu, L., He, Y. et al. (2020). Investigation on expansion effect of the expansive agents in ultra-high performance concrete. *Cement and Concrete Composites* 105: 103425.

5

Ultra-high-performance Concrete

5.1 Overview of Ultra-high-performance Concrete

Ultra-high-performance concrete (UHPC) is a new generation of cement-based building material, first proposed by Larrard and Sedran in 1994 [1]. The mix design of UHPC is based on the principle of maximum packing density to optimize the compatibility of superplasticizers and cementitious components, thereby enhancing the compactness and fluidity of the matrix. Short-cut steel fibers of 1–3% by volume are added to increase its toughness. It is prepared through on-site pouring and conventional curing methods and possesses high workability, ultra-high durability, ultra-high toughness, and ultra-long service performance. UHPC has now become an important development direction for research and application both domestically and internationally, attracting widespread attention from countries such as France, Japan, and the United States, which have invested a large amount of funds in researching materials and engineering applications. In the 1990s, China began to study UHPC, but there are still very few actual engineering applications [2]. This chapter will briefly describe the research and development status of UHPC from four aspects: material composition, design and preparation methods, mechanical properties, and durability.

Currently, there is no consensus on the definition of UHPC among countries worldwide. As shown in Table 5.1, the definition of UHPC in European and American countries mainly includes two aspects: mechanical properties and durability. Generally, it is specified that the 28 d compressive strength of UHPC should be at least 120 MPa, with excellent tensile strength and toughness, while also meeting the durability requirements.

The material composition of UHPC mainly includes cement, densified supplementary cementitious materials (silica fume, quartz powder, etc.), viscosity modifying admixtures (fly ash, ground granulated blast furnace slag, etc.), chemical functional materials (superplasticizers, early strength agents, etc.), fine aggregates (quartz sand), water, and short-cut steel fibers. Compared with traditional concrete, UHPC has the following significant characteristics [1–5]:

(1) Extremely low water-to-binder ratio. As shown in Figure 5.1, the water-to-binder ratio of high-strength and high-performance concrete is generally around 0.3,

High-Tech Concrete Materials: Design, Preparation, and Applications, First Edition.
Fazhou Wang and Shuguang Hu.

Table 5.1 Standards and definitions of ultra-high-performance concrete in various countries around the world.

Institution	Standard	Definition
American Concrete Institute (ACI)	ACI 239 "Structural Design of Ultra-High-Performance Concrete"	Ultra-high-performance concrete: minimum compressive strength of 150 MPa (22,000 psi), and meets specific requirements for durability, tensile ductility, and toughness, and usually incorporates fibers to achieve the above properties
American Society for Testing and Materials (ASTM)	ASTM C1856 "Standard Practice for Fabricating and Testing Specimens of Ultra-High Performance Concrete"	Ultra-high-performance concrete: a cement-based material with a minimum compressive strength of 120 MPa (17,000 psi), usually incorporating fibers, and other properties measured using standard test methods meet the requirements for durability, ductility, and toughness
Swiss Society of Engineers and Architects (SIA)	SIA 2052 "Ultra-High Performance Fiber Reinforced Cement-based composites (UHPFRC): Construction material, dimensioning, and application"	Ultra-high-performance concrete: a composite material containing cement, supplementary cementitious materials, fine aggregates, short-cut fibers, water, and admixtures, with extremely high impermeability due to the high compactness of the matrix, and the 28 d cube compressive strength is generally above 120 MPa
French Standardization Association (AFNOR)	NF P18-470 "Ultra-High Performance Fiber-Reinforced Concrete Specifications, Performance, Production and Conformity"	Ultra-high-performance concrete should meet the following indicators: Compressive strength: 130–250 MPa 90 d saturated porosity $\leq$9.0% 90 d Cl-diffusion coefficient $\leq 0.5 \times 10^{-12}$ m²/s

but the water-to-binder ratio of UHPC is below 0.25, generally between 0.16 and 0.20.

(2) High usage of cementitious materials. The cement content of UHPC is between 700 and 1000 kg/m^3, almost twice that of ordinary concrete. By comparing the volume components of typical high-performance concrete and UHPC, it can be seen from Figure 5.2 that the volume content of cement paste in UHPC is usually higher than 50%, while high-performance concrete is only about 30%.

(3) Incorporation of silica fume. The secondary hydration reaction between silica fume and the hydration product $Ca(OH)_2$ makes the hydration microstructure more compact and also plays a key role in the maximum packing density effect.

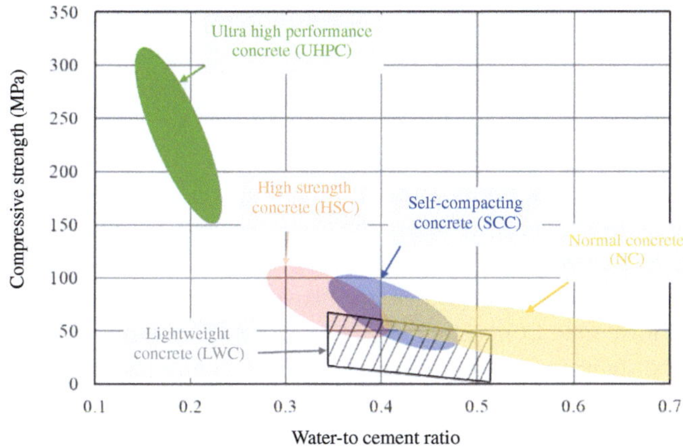

Figure 5.1 Relationship between compressive strength and water-to-binder ratio of typical concrete materials.

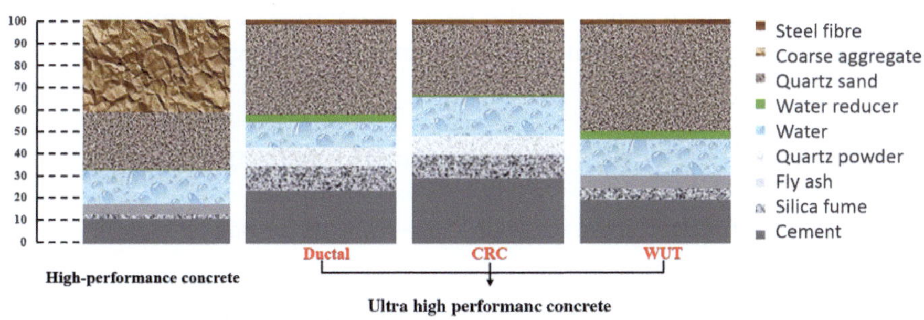

Figure 5.2 Volume components of typical high-performance and ultra-high-performance concrete materials.

The physical and chemical synergistic compaction effect of silica fume is also one of the important factors for the high strength and high durability of UHPC.

(4) High dosage of superplasticizers. In view of the high content of fine powders and the low water-to-binder ratio of UHPC, the dosage of superplasticizers accounts for 1–3% of the weight of cementitious components, several times higher than the dosage used in traditional concrete.

(5) Low degree of hydration. The extremely low water-to-binder ratio of UHPC limits the degree of hydration of cementitious materials. The degree of reaction after 28 d of hydration is less than 40%, resulting in a large number of unhydrated particles in the matrix. This characteristic also facilitates the resource utilization of industrial waste in UHPC [4–7].

To promote the research and standardization of UHPC, countries around the world have successively formulated and promulgated construction application standards for UHPC. In 2002, the French Association of Civil Engineering (AFGC) issued the earliest relatively complete UHPC structural design guide "Ultra-high performance Fiber Reinforced Concrete - Interim Recommendations," and based on subsequent research results and actual engineering application experience, issued a revised version of the guide in 2013. In 2004, the Japan Society of Civil Engineers (JSCE) issued the "Design and Construction Guidelines for Ultra High Strength Fiber Reinforced Concrete (UFC) Structures (Draft)." In 2013, the Federal Highway Administration (FHWA) of the United States issued the design guide "Design Guide for Precast UHPC Waffle Deck Panel System, Including Connections" for precast UHPC waffle bridge deck panels. In 2014, the Swiss Society of Engineers and Architects (SIA) released the SIA 2052 standard "UHPFRC: Building materials, design, and application." In 2019, China issued the "Basic Properties and Test Methods of Ultra-High Performance Concrete" (T/CCPA 7-2018). The formulation of the above standards not only promotes the research of UHPC materials but also provides a basis for the application of UHPC in construction engineering.

In recent years, the author's team has conducted in-depth research on the material system design, correlation between microstructure evolution and macroscopic properties, volume stability, and durability of UHPC, and has prepared green, lightweight, and high volume stability UHPC materials, which have been applied to important projects such as municipal solid waste pre-treatment plants and steel bridge deck paving, achieving significant social and economic benefits.

5.2 Design of Ultra-high-performance Concrete

5.2.1 Guidelines for the Preparation of Ultra-high-performance Concrete

The preparation of UHPC follows the following technical guidelines [2–4]:

(1) Optimization of the particle gradation to improve the compactness of the material.
The packing compactness of the concrete particle system has an important effect on its performance. The increase of concrete particle packing compactness can accelerate the hydration reaction process of the system, enhance the microstructure of the system, and improve its mechanical properties. Thus, optimizing the particle gradation of the constituent materials and improving their packing compactness is one of the keys to the preparation of UHPC.

(2) Adoption of highly efficient water-reducing agents with good compatibility with the active components to improve the fluidity of freshly mixed materials.
UHPC must select efficient water-reducing agents with good compatibility with the active components to reduce the water–cement ratio while allowing the paste to flow well, even with minimum water consumption. The compatibility of the cementitious material system with the agent can be characterized by the flow spread of the freshly mixed slurry.

(3) Minimization of weak areas in the matrix to improve the structural homogeneity of the material.
Concrete is a multiphase, porous, non-homogeneous material at the fine microscopic scale consisting of aggregates, cement pastes, and interfacial transition

zones between the two [8, 9]. The non-homogeneity of concrete is mainly reflected in mechanical (differences in the modulus of elasticity of aggregates and cement pastes), physical (differences in the coefficients of expansion of the phases), and chemical (differences like chemical shrinkage) aspects. Under the coupling effect of the service environment and external load, the internal deformation of concrete is easy to produce shear stress and tensile stress at the interface due to the inconsistency of deformation, which leads to the formation of microcracks, and ultimately reduces the mechanical and durability performance of concrete. UHPC eliminates defects and improves homogeneity mainly through the following ways:

① Reducing the aggregate particle size reduces the thickness and extent of the interfacial transition zone and at the same time reduces the chance of defects in the aggregate, thus improving the overall homogeneity of the system.

② Enhance the mechanical properties of cementite and reduce the variability of its modulus of elasticity with aggregates. UHPC contains more active components, and its secondary hydration reaction consumes a large number of $Ca(OH)_2$ crystals, while generating more hydrated calcium silicate gels and improving the microstructure of the interfacial zone and the compactness of the paste. The ratio of the modulus of elasticity of the aggregate to that of the hardened cement paste in UHPC is between 1.0 and 1.4, and the effect of the inhomogeneity of the two is almost eliminated.

③ The interfacial transition zone was strengthened by internal curing techniques. The interfacial transition zone is a weak region of concrete, and the denseness and strength of the interfacial transition zone are improved by introducing lightweight aggregates [10–14] and super-absorbent resins [15–18] to continuously replenish the water required for the hydration of the cementitious components.

④ Blended with short-cut steel fibers to improve toughness and ductility

The cement pastes and aggregate of concrete are typical quasi-brittle materials, which are prone to cracking under shear and tensile stresses. The extremely high tensile strength of short-cut steel fibers is used to give full play to their crack-bridging effect under load to improve the tensile strength and toughness of concrete. Usually, the diameter of steel fibers used in UHPC is 0.15–0.20 mm, the length is about 13 mm, and the volume mixing amount is 1.5–3%.

5.2.2 Design Method for Ultra-high-performance Concrete Material System Based on Response Surface Methodology

Currently, UHPC mainly adopts the tightest packing design method based on the Dinger–Funk equation, but there are problems in that the distribution modulus is difficult to determine, and the packing state of the particles deviates greatly from the calculation results. Response surface methodology is used to obtain experimental data by designing a series of experiments and testing [19, 20], and then the multiple quadratic regression equation is used to fit the functional relationship between

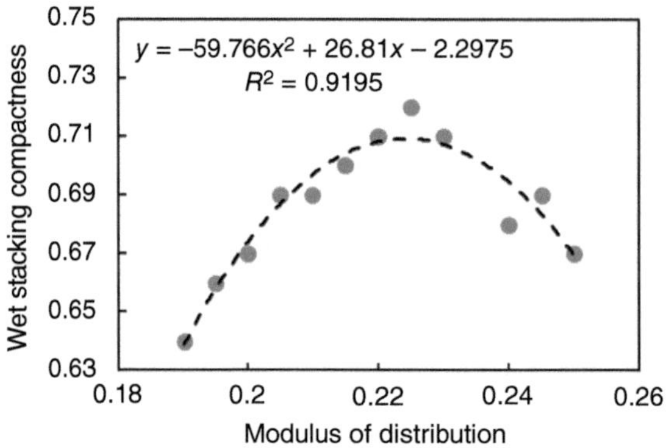

Figure 5.3 Wet packing densities of ultra-high-performance concrete with different distribution moduli.

the influencing factors and the response values, and the optimal parameters are sought through surface modeling and analysis, which can be used to achieve the optimization of multiple variables. In this section, it is proposed to use the response surface method to study the influence of three kinds of particle components, namely, cement, silica fume, and fly ash, on the packing density of ultra-high performance cementitious materials and the working performance of UHPC, and to establish the quadratic regression equations of the three variables, namely, cement, silica fume, and fly ash, with respect to the packing density and the working performance. The optimal design of the components was carried out for the two response values of pile compactness and workability, respectively, to determine the optimal data range and the optimal region of compactness and to obtain the optimal values of the variables affecting workability in this region so as to prepare a kind of UHPC with both a high degree of compactness and excellent workability.

5.2.2.1 Determination of the Optimal Distribution Modulus

First, cement–fly ash–silica fume ternary cementitious material systems with distribution modulus ranging from 0.19 to 0.25 were calculated using the Dinger–Funk equation, respectively, and then their maximum wet packing densities were determined using the minimum spread test method (with a fixed water-reducing agent of 2% of the cementitious material). As shown in Figure 5.3, the wet packing density of cementitious material first increases with the increase of distribution modulus and reaches the maximum value at 0.225, after which the maximum wet packing density continues to decrease with the increase of distribution modulus. The relationship between maximum wet packing density and distribution modulus is shown in Eq. (5.1):

$$\Phi = -59.776n^2 + 26.81 - 2.2975\ R^2 = 0.92 \tag{5.1}$$

where Φ is the maximum wet pile compactness and n is the modulus of distribution.

Therefore, the optimum distribution modulus of the cement–fly ash–silica fume ternary cementitious material system is between 0.22 and 0.23. Considering the large amount of cementitious material in UHPC, the distribution modulus was determined to be 0.225 for the Dinger–Funk equation for the closest packing

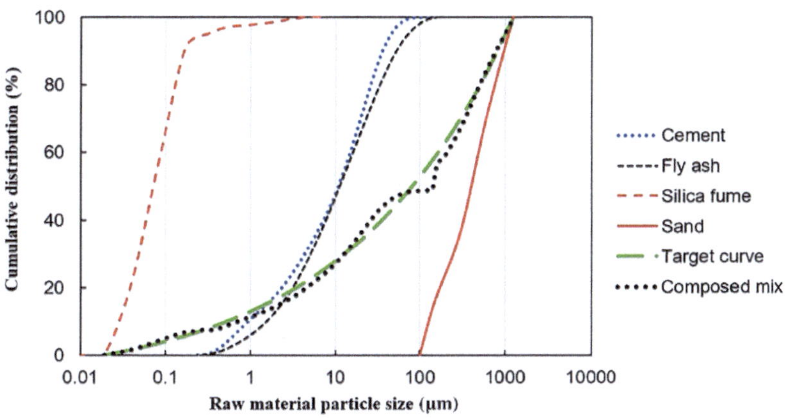

Figure 5.4 Criteria for optimum pile compactness and actual particle size distribution curves for ultra-high-performance concrete.

calculation. The particle size distribution of the raw material is shown in Figure 5.4. The standard curve for the tightest packing of the ternary cementitious system is calculated by Eq. (5.2). Then the mean square differences are compared between the cumulative sieve residue in the actual curve and the value of the target curve under each particle size distribution calculated, and the least squares method is used to minimize the mean square difference, i.e., to minimize the deviation between the actual gradation curve and the target curve. The UHPC particle fits obtained after iterative calculations are shown in Table 5.2:

Table 5.2 Ultra-high-performance concrete solid particle mix ratios based on Dinger–Funk equation calculations.

Raw materials	Cement	Fly ash	Silica fume	Sand
Mass ratio	0.80	0.20	0.15	1.25

$$U(D_p) = 100\frac{D_p^n - D_{min}^n}{D_{max}^n - D_{min}^n} \tag{5.2}$$

where D_p is the current particle size, D_{max} is the maximum particle size in the powder, D_{min} is the smallest particle size in the powder, n is the modulus of distribution, and $U(D_p)$ is the cumulative percent sieved for particle size D_p.

5.2.2.2 Central Composite Design

In response surface methodology, for the relationship between response values and variables, a central combination design is used for the design of the experimental process, and the design schematic is shown in Figure 5.5. The corner points of the cube in the figure represent the possibilities of +1 and −1 in the spatial distribution

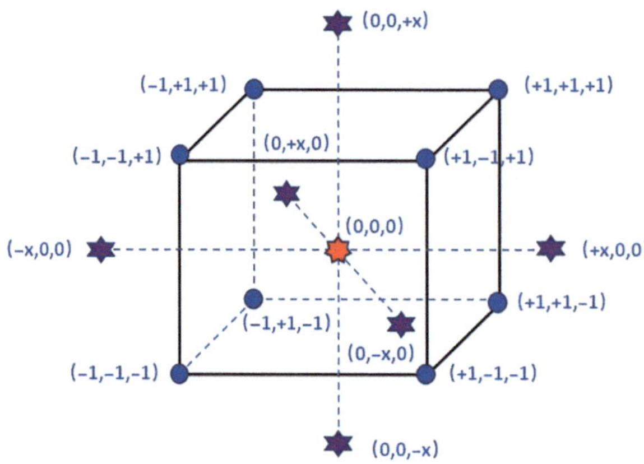

Figure 5.5 Schematic representation of the distribution of multiplication steps, axes, and centroids in the design of central combinations.

Table 5.3 Response surface analysis factor size values.

	Minimum and maximum values (g)		
Types	**−1**	**0**	**+1**
Cement	320	360	400
Fly ash	40	80	120
Silica fume	40	60	80

of the combinatorial design, and the axis center points represent the distance of each face of the cube from the center, where +1 and −1 are taken within the scale range of Table 5.3.

Based on the design points and test response values for the above center combination, the experimental data were fitted using a second-order response surface function with a quadratic polynomial as shown in Eq. (5.3):

$$y = \beta_0 + \Sigma\beta_i x_i + \Sigma\beta_{ii} x^2 + \Sigma\beta_{ij} x_i x_j \tag{5.3}$$

where β_0, β_i, β_{ii}, and β_{ij} are quadratic terms to be determined; x is variable; and y is the response value.

The least squares method was used to solve the regression equation between the variables and the response values accurately. First, cement, silica fume, and fly ash dosage were identified as the key variables, wet packing compactness and spread were the response values, and the relationship between the three-factor variables and the response values was investigated. Based on the mass proportions of cement, silica fume, and fly ash in Table 5.2, the ranges of the three variable factors were determined by floating within the range of the calculated proportions, as shown in Table 5.3, and the results of the wet packing compactness and spread tests are shown in Tables 5.4 and 5.5.

Table 5.4 Experimental results on central combination design variables and wet packing compactness.

No.	Cement (g)	Fly ash (g)	Silica fume (g)	Wet packing density
1	400	120	80	0.710
2	400	120	40	0.684
3	400	40	80	0.659
4	400	40	40	0.679
5	320	120	80	0.706
6	320	120	40	0.723
7	320	40	80	0.691
8	320	40	40	0.701
9	430	80	60	0.703
10	360	159.6	60	0.720
11	360	80	96.6	0.691
12	290	80	60	0.704
13	360	10	60	0.679
14	360	80	25.4	0.723
15	360	80	60	0.709

5.2.2.3 Fitting the Model and Validation

Quadratic multinomial equations were fitted to the data results in Tables 5.4 and 5.5 using Design Expert design software to optimize the three variables, cement, fly ash, and silica fume, to obtain a computational model with wet packing compactness and spread as the response, and the three variables as the independent variables as shown in Eqs. (5.4) and (5.5):

$$y_1 = 3.71 + 5.32x_1 + 16.05x_2 + 3.42x_3 + 0.43x_1x_2 + 1.08x_1x_3 + 1.7x_2x_3 + 1.67x_1^2 + 3.66x_2^2 + 0.93x^2 \quad (5.4)$$

$$y_2 = 50.37 + 0.47x_1 + 155.59x_2 + 157.32x_3 + 10.94x_1x_2 + 8.22x_1x_3 + 5.88x_2x_3 + 14.2x_2^2 \quad (5.5)$$

where y_1 and y_2 are the wet pile compactness and the degree of expansion, respectively; x_1, x_2 and x_3 are the amounts of cement, fly ash, and silica fume.

To verify the applicability and reliability of the model, analysis of variance (ANOVA) was used to determine the significance of the variables and the reliability of the fitted results, which are shown in Tables 5.6. The P-value of the misfit of the model is $0.0226 < 0.0500$, which indicates that the developed model has a significant level. The P-value of the misfit in the table characterizes the degree of fit of the model to the experimental results, and the P-value of the misfit of the model fitted to the wet pile compactness is less than 0.05, which indicates that the model correlates

Table 5.5 Experimental results on central combination design variables and spread.

No.	Cement (g)	Fly ash (g)	Silica fume (g)	Water-to-binder ratio	Binder-to-sand ratio	Spread (mm)
1	400	120	80	0.18	1.25	140
2	400	120	40	0.18	1.25	225
3	400	40	80	0.18	1.25	120
4	400	40	40	0.18	1.25	165
5	320	120	80	0.18	1.25	180
6	320	120	40	0.18	1.25	220
7	320	40	80	0.18	1.25	110
8	320	40	40	0.18	1.25	130
9	430	80	60	0.18	1.25	165
10	360	159.6	60	0.18	1.25	190
11	360	80	96.6	0.18	1.25	130
12	290	80	60	0.18	1.25	180
13	360	10	60	0.18	1.25	115
14	360	80	25.4	0.18	1.25	235
15	360	80	60	0.18	1.25	170

Table 5.6 Analysis of variance (ANOVA) of response surface models for wet density and spread.

Types	Sum of squares		*P*-value	
	Compactness	Spread	Compactness	Spread
Cement (A)	89.4	29.9	0.0438	0.04977
Fly ash (B)	7986.3	9999.5	0.0025	0.0001
Silica fume (C)	1023.6	10110.5	0.0942	0.0001
A × B	536.2	703.1	0.0525	0.0063
A × C	624.5	528.1	0.3230	0.0142
B × C	465.3	378.1	0.0221	0.0320
A2	126.3	–	0.0226	–
B2	953.2	912.4	0.0849	0.0027
C2	759.2	–	0.0356	–

well with the experimental data and the quadratic fitting equations are well fitted. From the *P*-values of the three variables, cement, fly ash, and silica fume in the table, it can be seen that the three factors have a significant effect on wet packing compactness and their interaction.

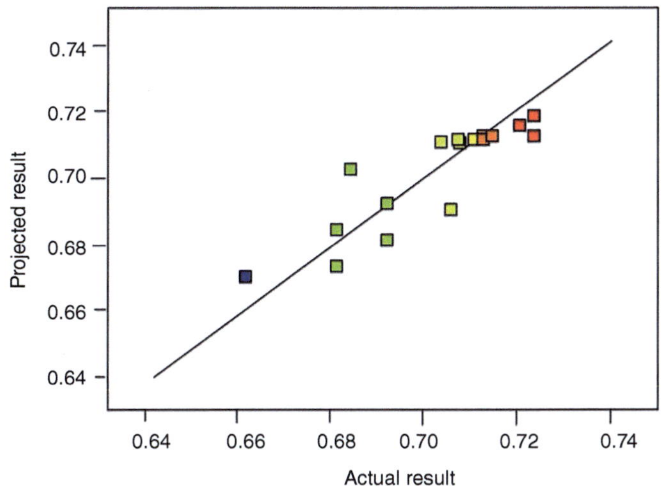

Figure 5.6 Relationship between predicted and measured values of wet packing compactness.

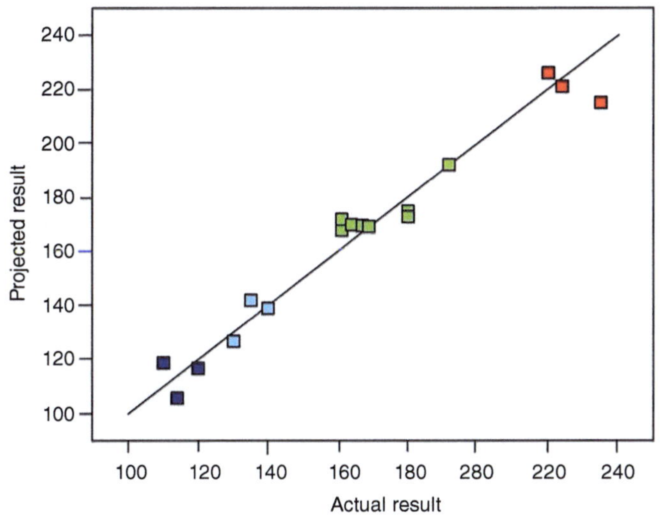

Figure 5.7 Relationship between predicted and measured values of spread.

From Figures 5.6 and 5.7, it can be seen that the predicted values of wet packing densities of UHPC are close to the experimental test values, and the correlation is calculated by Eq. (5.6), and the wet packing densities of the experimental tests are distributed in the predicted straight line and its vicinity, and the correlation coefficient $R^2 = 0.89$, which indicates that the regression model is accurate and reliable. The predicted extensions in regression Eq. (5.5) correlate well with the experimental test values with a correlation coefficient of $R^2 = 0.97$. In summary, the accuracy of the wet packing compactness model and spread model of the materials obtained

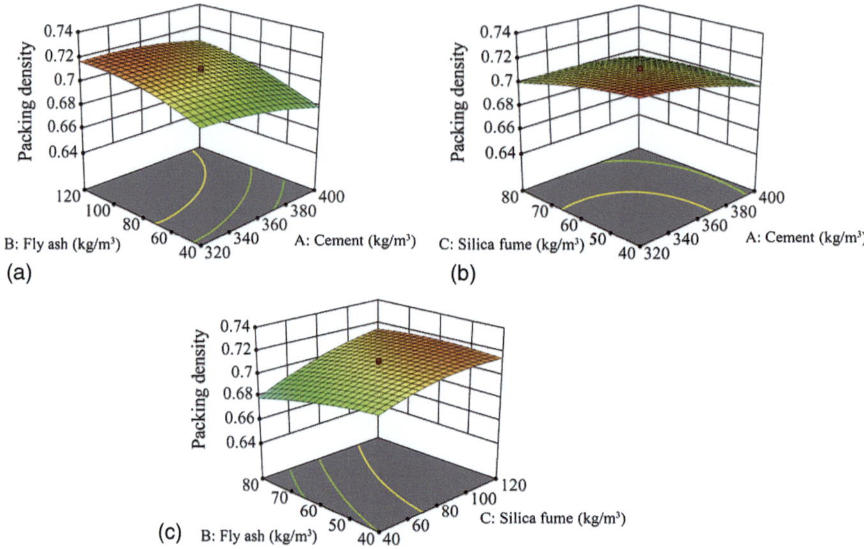

Figure 5.8 Response surface three-dimensional plots of the effect of ultra-high-performance concrete component variables on wet packing compactness. (a) Cement and fly ash, (b) cement and silica fume, and (c) silica fume and fly ash.

from Eqs. (5.4) and (5.5) is good, and they can predict and analyze their packing compactness and spread more accurately:

$$R^2 = y_{ij} - \overline{y} \tag{5.6}$$

where R is the variance; y_{ij} is the value of each response; and $\overline{y}$ is the average of the response values.

5.2.2.4 Response Surface Analysis of Wet Packing Compactness and Spread

Figure 5.8 shows the spatial surfaces and contour plots of the materials corresponding to the variables cement, fly ash, and silica fume for the wet packing compactness model, which visualizes the effect of the variables on the response values of the stacking compactness. The degree of variation of the response surface in the figure can reflect the degree of influence of the variables on the response, and the shape of its contour lines can reflect the influence of the interaction of the two variables, where the circular contour lines indicate that the interaction is not significant, while the elliptical contour lines indicate that the interaction is significant.

As can be seen from Figure 5.8(a), the wet packing compactness increases with the increase of fly ash dosage, indicating that fly ash with lower fineness is favorable for stacking compactness, and the maximum wet packing compactness is reached at the fly ash dosage of 120 g. From Figure 5.8(b), it can be seen that at a fixed amount of fly ash, the densities showed a tendency to increase with decreasing amounts of silica fume and cement, but the distortion of the three-dimensional response surface was not obvious, which indicated that the interaction of the two components was weak, and the maximum wet packing densities were obtained at a silica fume

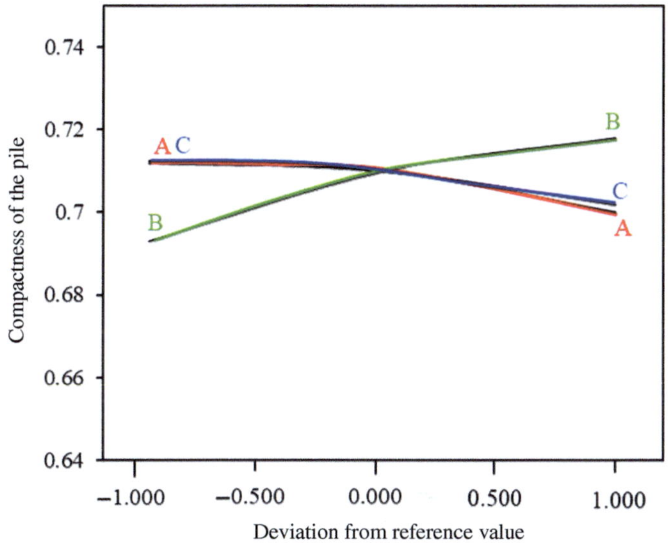

Figure 5.9 Influence of variables of compositional changes on wet pile compactness.

dosage of 40 g and a cement dosage of 320 g. The maximum wet packing densities were obtained at a silica fume dosage of 40 g and a cement dosage of 320 g, respectively. From Figure 5.8(c), it can be seen that when the cement dosage is fixed, the wet pile compactness varies with the ratio of fly ash and silica fume, and their interaction has a significant effect on the pile compactness. Figure 5.9 shows the sensitivities of the three independent variables, cement (a), fly ash (b), and silica fume (c), in the computational model, from which it can be seen that the pile compactness of UHPC is mainly affected by the amount of fly ash admixture, and the selection of a suitable ratio of fly ash to silica fume is favorable for the degree of compactness.

Figures 5.10 show the spatial surfaces and contour plots formed by the corresponding variables cement, fly ash, and silica fume for the UHPC spread regression model. From Figure 5.10(a), it can be seen that the spread increases with the increase of fly ash dosage, which is reflected by the rapid change of the surface and dense contour lines, indicating that the effect of fly ash on its spread is more significant. From Figure 5.10(b), it can be seen that when the amount of cement or silica fume is kept constant, the spread increases with the increase of silica fume or cement, and the three-dimensional response surface is distorted, indicating that the interaction between the two is significant. From the 3D surface variation and contour variation, it can be seen that the silica fume dosage has a large effect on the degree of extension, which decreases rapidly with the increase of the dosage. To be able to prepare UHPC with excellent workability, the amount of silica fume needs to be minimized while ensuring the degree of packing compactness. From Figure 5.10(c), it can be seen that the three-dimensional response surfaces of the silica fume and fly ash variables are basically undistorted, which laterally indicates that the interaction between the two is small. From the contour distribution density, the degree of influence of silica fume

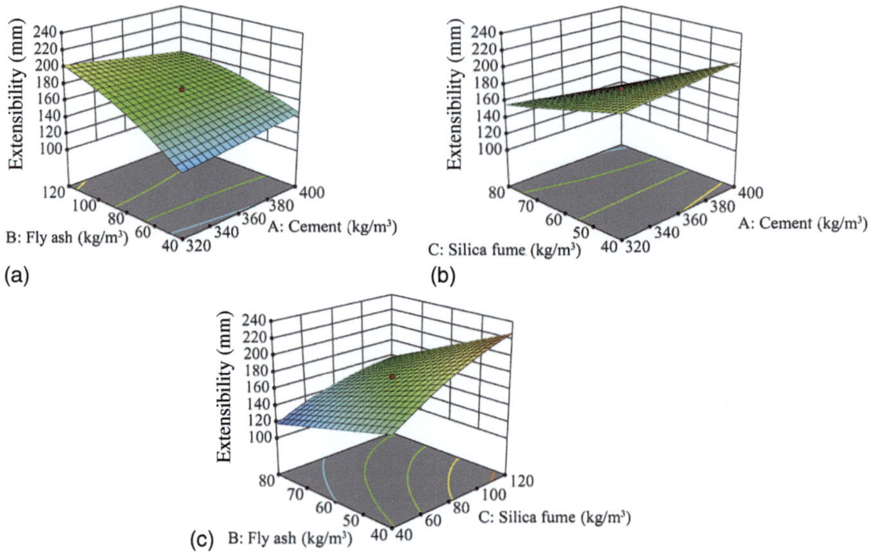

Figure 5.10 Response surface three-dimensional plots of the effect of ultra-high-performance concrete component variables on spread. (a) Cement and fly ash, (b) cement and silica fume, and (c) silica fume and fly ash.

and fly ash on the extension degree is similar, but their effects are opposite. It can be seen that the degree of significance of the influence on the degree of extension is in the following order: fly ash > silica fume > cement, and at the same time, the absolute value of the coefficients of the three variables in the regression equation also follow this order, which indicates that the results of the surface analysis are reliable. Figure 5.11 shows the sensitivity of the three independent variables of (a) cement, (b) fly ash, and (c) silica fume in the computational model. The workability of UHPC is mainly affected by the dosage of fly ash and silica fume, and an increase in the dosage of fly ash and a decrease in the dosage of silica fume can significantly improve workability.

The optimal mix ratio obtained from the above response surface method analysis is based on taking into account the two properties of wet packing compactness and spread, therefore, the response law of the above components to the two main properties is synthesized, and the expectation function analysis is carried out by using numerical optimization method, which is given in the following formulas:

$$D = \left(d_1^{r_1} \times d_2^{r_2} \times d_3^{r_3} \times \cdots \times d_n^{r_n}\right)^{1/\sum r_1} = \left[\prod_i^n d_i^{r_1}\right]^{1/\sum r_i} \tag{5.7}$$

where D is the value of the expectation function; n is the number of response parameters, five in total, which are cement, silica fume, fly ash, pile compactness, and spread; r_i is the level of importance of each response parameter; and r_i is the comparison value for each response parameter.

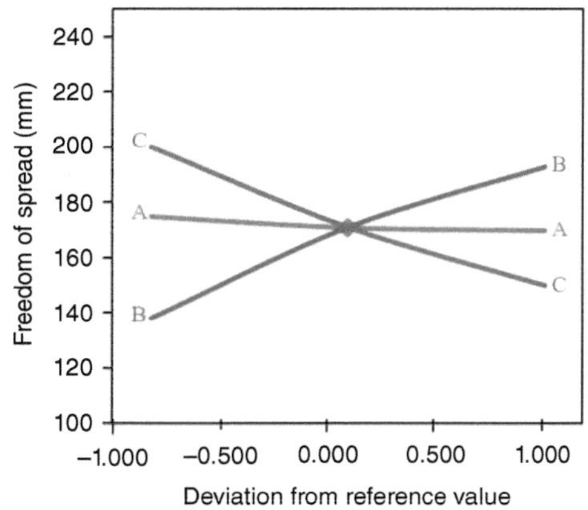

Figure 5.11 The effect of variables with compositional changes on the degree of expansion.

Table 5.7 Proportion of solid powder for ultra-high-performance concrete.

Raw materials	Cement	Fly ash	Silica fume	Sand
Mass ratio	0.80	0.30	0.10	1.25

In practice, while ensuring the flow properties, the pile compactness is increased as much as possible, which is obtained by using the numerical calculation program in Optimization in Design Export software. Cement: fly ash: silica fume is 320:120:40. To verify the accuracy of the optimization results of the UHPC ratios, six repetitions of the test were carried out, and the average values of pile compactness and spread obtained from the test were 0.72 and 220 mm, respectively, which indicated that the model optimization predicted the data with small errors. By the response surface method, UHPC with an extension of 220 mm can be prepared at a water–cement ratio of 0.18, with the proportion of solid powder as shown in Tables 5.7, and the parameters such as steel fiber admixture, water–cement ratio, and water-reducing agent dosage are further determined by the optimized design of the mix ratio.

5.3 Physical and Mechanical Properties

5.3.1 Workability

In the early stages of research, the ultra-high performance of UHPC was mainly reflected in two indicators: strength and durability. As UHPC is applied in steel bridge decking and high-rise construction, the control of workability has gradually become more important to facilitate on-site pouring or long-distance pumping operations in these applications. The ultra-low water-to-binder ratio and the complex and diverse binder components of UHPC give it unique characteristics in the mixing

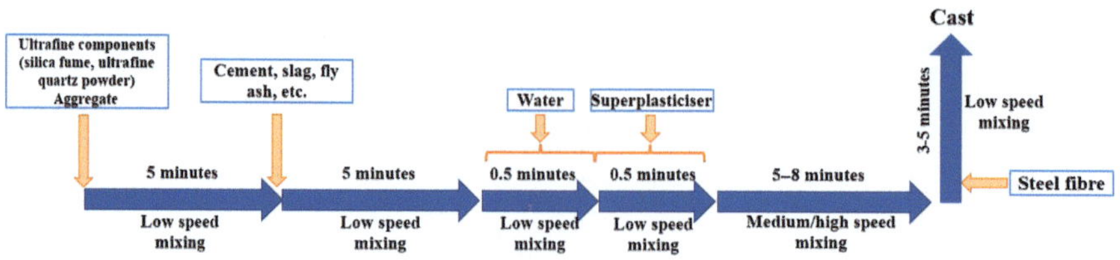

Figure 5.12 Typical mixing process for UHPC.

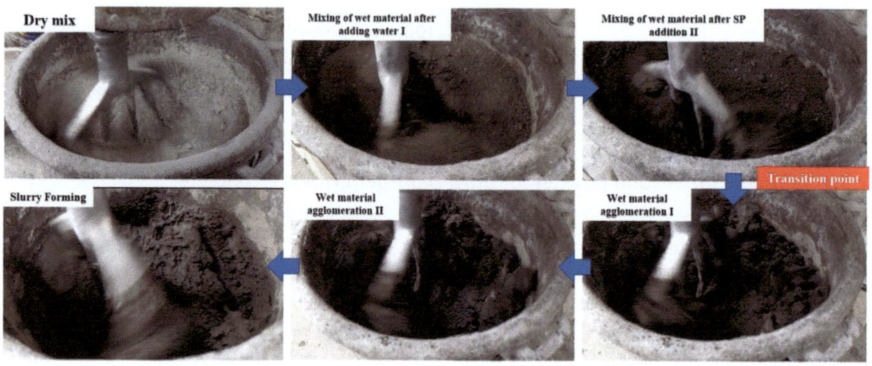

Figure 5.13 State transition of UHPC during mixing.

process, different from those of ordinary or high-performance concrete. First, UHPC requires longer mixing times and higher mixing power to achieve full dispersion of the densest packed powder system with limited water content. Figure 5.12 shows a typical mixing process for UHPC. It can be seen that UHPC needs a longer mixing time to transition from dry materials to a paste state. Second, in terms of the order of material addition, the ultra-fine components (such as silica fume or ultra-fine quartz powder) are generally added together with quartz sand, and under slow stirring, the stirring blades lift the quartz sand to disperse the agglomerated powder particles. Finally, UHPC powder undergoes a unique state change point during the mixing process, as shown in Figure 5.13. After adding water, the surface of the dry mix is gradually enveloped by a water film, changing color from light gray to gray. After the addition of a superplasticizer, the powder particles are further dispersed, releasing more water, and the color deepens further. With continued stirring, the wet mix gradually transitions from a powdery to a granular state, forming clumps, and this point in time is referred to as the state change time (turnover time), with the wet mix eventually becoming a uniform paste.

The main factors affecting the workability of UHPC can be divided into material properties and preparation properties. Material properties include the type and dosage of superplasticizers and the dosage and fineness of admixtures, while preparation properties are reflected in the degree of dispersion of the powder during the mixing process. As shown in Figure 5.14(a), the free expansion of UHPC prepared with three commercial superplasticizers at the same dosage varies significantly, mainly related to the compatibility of the superplasticizer with the composite binder components. With the increase of the superplasticizer dosage, the expansion of UHPC gradually increases until the dosage reaches saturation (Figure 5.14(b)). The SiO_2 content and production method (densified or not) of silica fume also affect the workability of UHPC. Generally speaking, the lower the SiO_2 content and the higher the content of impurities such as carbon, the more severe the adsorption of the superplasticizer, resulting in a lower expansion of the fresh mixture. Densified silica fume increases its density by compression during production for easier transportation, but it is more difficult to disperse during mixing, leading to a lower expansion of the paste (Figure 5.14(c)). The particles of slag powder present a smoother surface at the microscopic scale, which is conducive to the sliding of particles in a wet state, so replacing cement with slag powder increases the expansion of the mixture (Figure 5.14(d)). The slender geometric characteristics and random distribution state of short-cut steel fibers increase the flow resistance

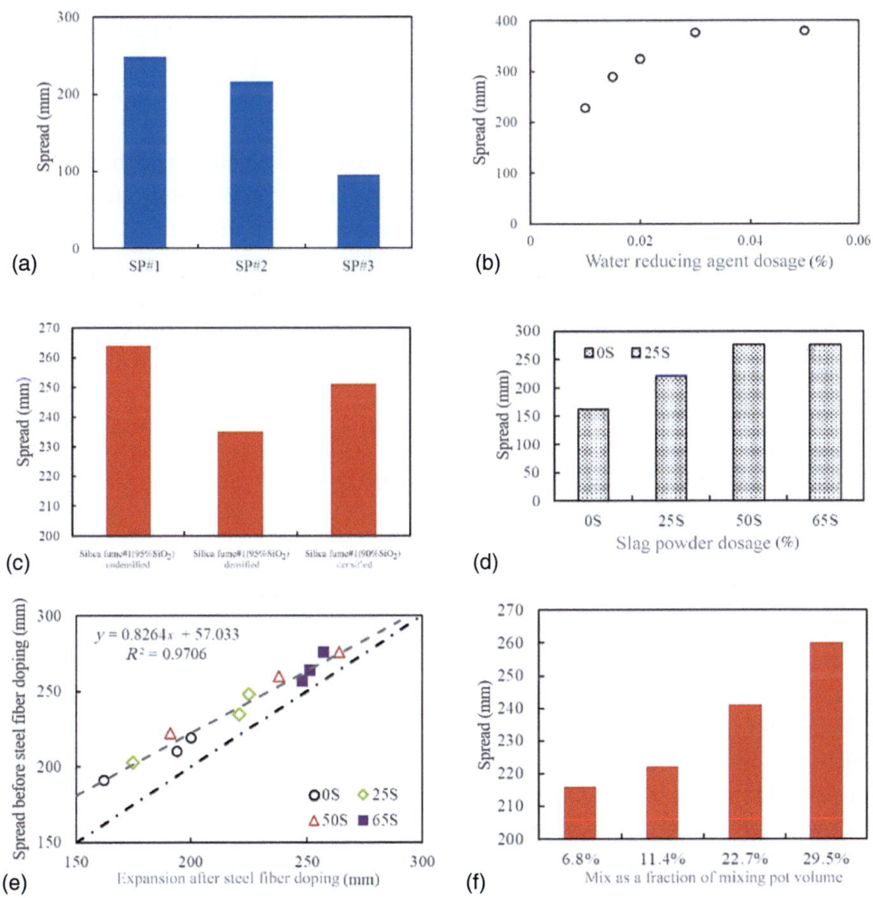

Figure 5.14 Factors affecting the flowability of UHPC. (a) Influence of water reducing agent type, (b) effect of water reducing agent dosage, (c) effect of silica fume type, (d) effect of slag powder mixing, (e) effect of steel fibers, and (f) effect of mixing capacity.

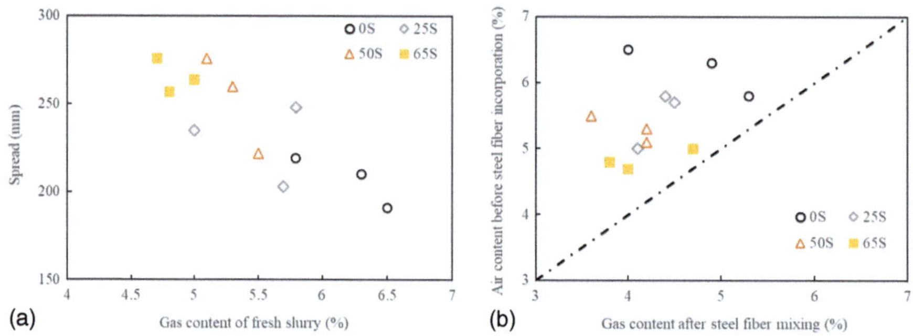

Figure 5.15 Factors affecting the air content of fresh UHPC mixture. (a) Impact of spread and (b) effect of steel fibers.

of the mixture, so after adding steel fibers, the fluidity of the mixture decreases, and the free expansion is only 82.6% of that of the plain mixture (Figure 5.14(e)). Regarding preparation properties, as the mixing volume increases, the stirring force transmitted to the mixture by the blades during the mixing process is greater, which is more conducive to the dispersion of the wet mixture, so the free expansion of the mixture is higher (Figure 5.14(f)).

At the same time, the viscosity of the UHPC mixture is higher than that of ordinary concrete, and with a higher dosage of superplasticizers, different amounts of air bubbles are introduced during mixing. As shown in Figure 5.15(a), the lower the expansion of the mixture, the more viscous the mixture, and the more difficult it is for the introduced air bubbles to escape during mixing, resulting in a higher air content. The addition of steel fibers, due to their sharp geometric characteristics, will puncture some bubbles during mixing, leading to a reduction in air content (Figure 5.15(b)).

5.3.2 Hydration and Microstructure Evolution

5.3.2.1 Hydration Behavior

The material composition characteristics of UHPC directly determine its special laws of hydration and microstructure evolution. As can be seen from Figure 5.16, UHPC is in a dormant period of hydration reaction within 10 h after mixing, with no significant heat release. After UHPC enters the acceleration phase of the reaction, it generates a large amount of heat, and a peak of exothermic reaction occurs. Around 22 h later, the rate of hydration reaction begins to decline, and it gradually enters a stable period of around 60 h. In comparison, the hydration reaction process of high-performance concrete is faster after mixing, with a clear peak of heat release from hydration occurring 3 h after mixing, and the intense hydration reaction lasts longer, approaching 25 h. As shown in Figure 5.17, the degree of reaction of UHPC after hydration for 28 d is only 33%. This is due to the extremely limited water content in UHPC, which severely restricts the extent of the reaction between cement and water, and at the same time, its highly dense matrix structure limits the growth space of the hydration products. The combined effect of these two factors results in a

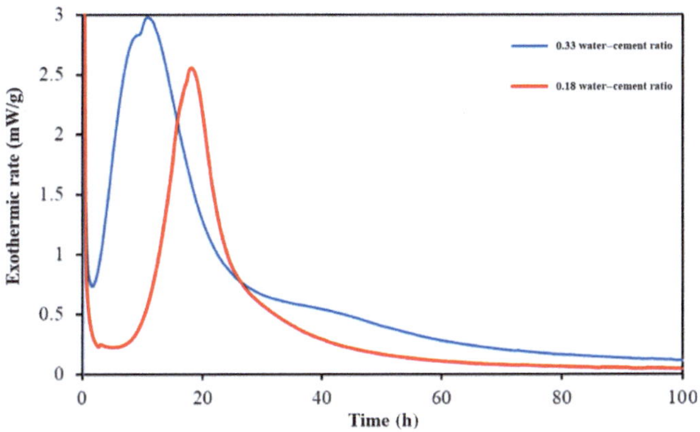

Figure 5.16 Hydration heat release of ultra-high performance and high-performance pastes.

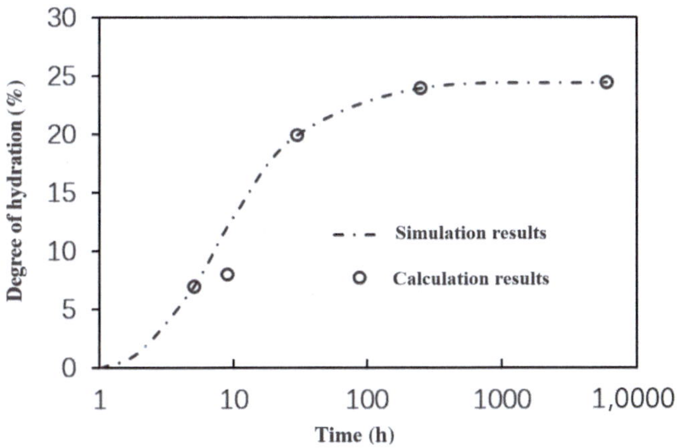

Figure 5.17 Degree of hydration development curve for ultra-high-performance concrete.

hydration degree of the binder components in the UHPC system that is much lower than that of ordinary concrete. The three-dimensional image of UHPC obtained through X-ray computed tomography clearly shows the presence of a large number of unhydrated cement particles (Figure 5.18). It is also noted that the induction period of UHPC is much longer than that of traditional concrete, generally between 7 and 15 h, mainly due to the use of a large amount of superplasticizers, which causes cement and other binder component particles to be wrapped in a protective film formed by the long-chain molecules of the superplasticizers for a long time, thus delaying the contact between cement and water. Figure 5.19 shows the hydration heat release of UHPC with different dosages of superplasticizers (1%, 2%, 4%, and 6%). It can be seen that when the dosage of the superplasticizer is 1%, the dormant period of the hydration reaction is 3 h, and when the dosages are 4% and 6%, the dormancy periods last for 8 and 11 h, respectively.

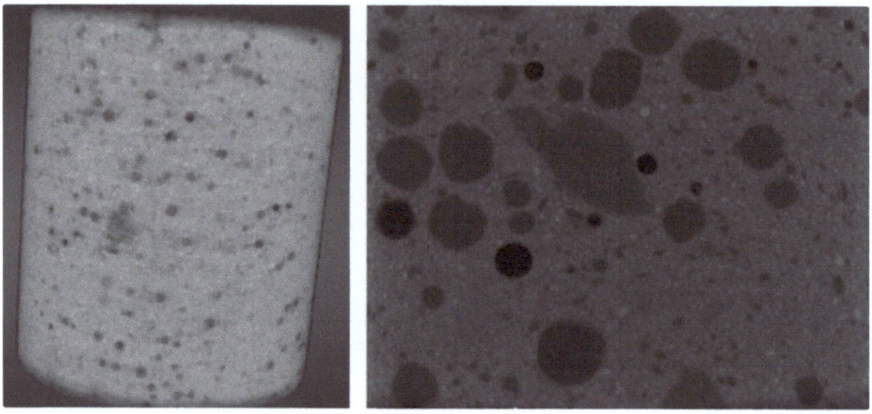

Figure 5.18 X-ray computed tomography image of ultra-high-performance concrete.

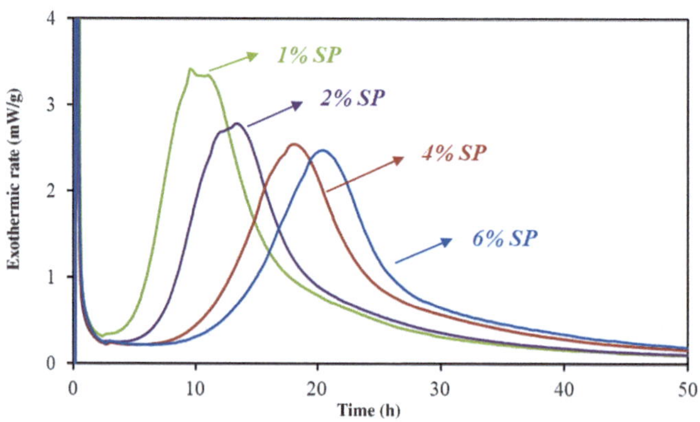

Figure 5.19 The effect of superplasticizers on the hydration heat release of ultra-high-performance concrete.

5.3.2.2 Pore Structure

The pore structure distribution of UHPC differs significantly from that of traditional cement-based materials. As shown in Figure 5.20, both types of samples exhibit a single-peak distribution characteristic in their pore size distribution, but the shapes of the peaks differ greatly. Ordinary cement paste has a higher porosity, with pores mainly being capillary pores larger than 100 nm. The primary pore size distribution of UHPC lies between 5 and 30 nm, and due to the lower degree of hydration, there are fewer gel pores below 10 nm. The good internal particle packing and the presence of ultra-fine silica fume result in fewer pores larger than 30 nm within UHPC.

Figure 5.21 and Table 5.9 show the change in the matrix pore structure distribution of UHPC with the curing age. As hydration time extends, the pore size distribution noticeably shifts toward smaller sizes, and the porosity significantly decreases. After 12 h of hydration, the internal pore sizes of the paste are mainly concentrated

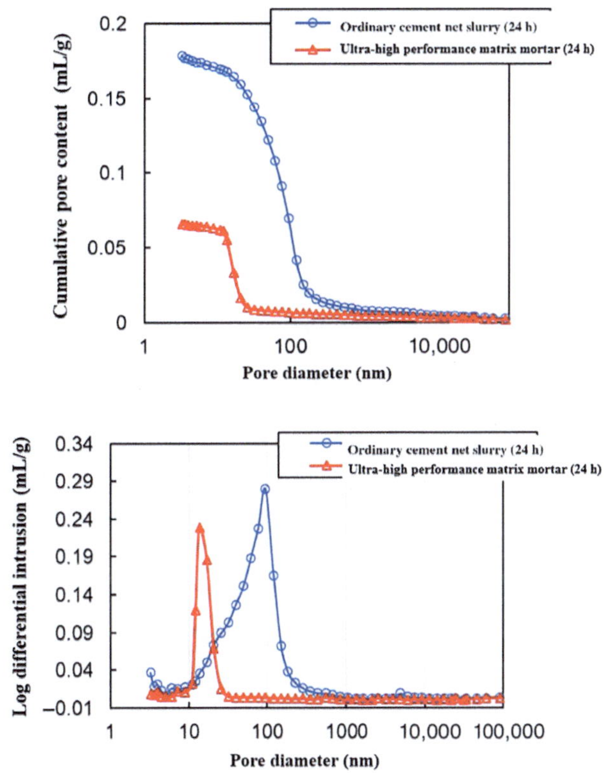

Figure 5.20 Comparison of pore sizes in UHPC matrix and ordinary cement paste at 1-day curing age.

at 100 nm, with a portion of larger pores, and the porosity is close to 17%. After 24 h of hydration, the pore sizes are noticeably refined, primarily distributed between 20 and 30 nm, with the porosity dropping to 14%. Continuing hydration for 7 d shows no significant refinement in pore size, with the porosity reduced to 11%. When hydrated for 28 d, the pore sizes are mainly distributed between 10 and 20 nm, with the porosity reduced to 6%. Based on the impact of pore size on durability, the pore size distribution of UHPC is statistically analyzed (Table 5.8). The low porosity and small pore sizes are also important reasons for the ultra-high strength and high durability of UHPC.

5.3.2.3 Characteristics of Hydration Products

Figure 5.22 shows the evolution of hydration products in UHPC over time. In the early stages of hydration, UHPC remains in a dormant period for a relatively long time, with no production of $Ca(OH)_2$ observed in the X-ray diffraction (XRD) spectrum. As hydration progresses, after 10.6 h, the hydration enters an acceleration phase, and a large amount of hydration products are generated, with a distinct $Ca(OH)_2$ peak visible in the XRD spectrum. As hydration continues, the content of $Ca(OH)_2$ keeps increasing. Later on, the formation of ettringite and the pozzolanic

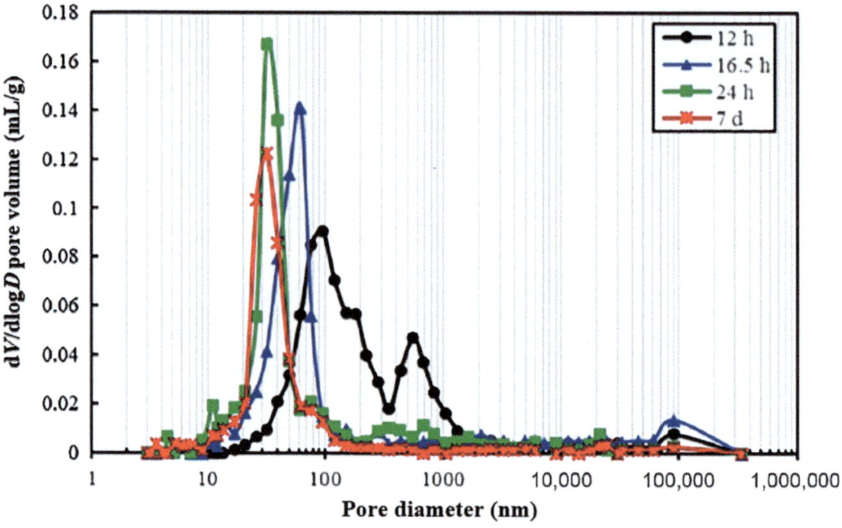

Figure 5.21 The change in the pore size distribution of UHPC with curing age.

Table 5.8 Statistical table of pore distribution in UHPC.

	<10 nm	10–100 nm	100–1000 nm	>1000 nm
12 h	0	4.26%	10.21%	2.71%
24 h	0.12%	10.52%	2.05%	1.53%
7 d	0.26%	9.19%	0.76%	0.80%
28 d	3.5%	7.70%	0.60%	0.30%

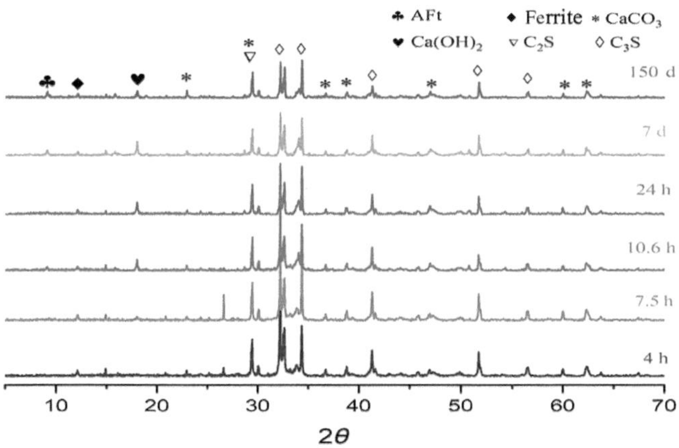

Figure 5.22 XRD of hydration products of ultra-high-performance concrete (UHPC) at different ages.

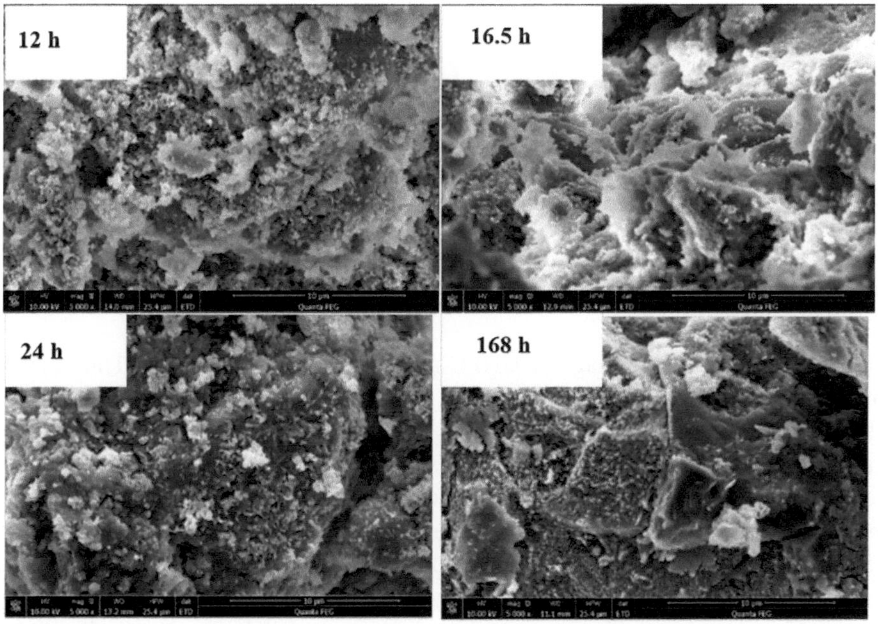

Figure 5.23 Morphology of hydration products of UHPC at different ages.

reaction of silica fume consume a portion of the produced $Ca(OH)_2$, leading to a reduction in the $Ca(OH)_2$ peak in the later XRD spectra. Figure 5.23 presents scanning electron microscope (SEM) images of UHPC at different ages. In the early stages of hydration, no significant hydration products are observed in the SEM images, and the overall appearance is relatively loose. As the hydration age increases, noticeable deposits of hydration products on the surfaces of raw material particles can be seen, and the overall appearance becomes denser.

(a) (b)

Figure 5.24 FE-SEM photo of hydration products in UHPC paste (magnification ×5000). (a) $Ca(OH)_2$ and AFt and (b) C-S-H.

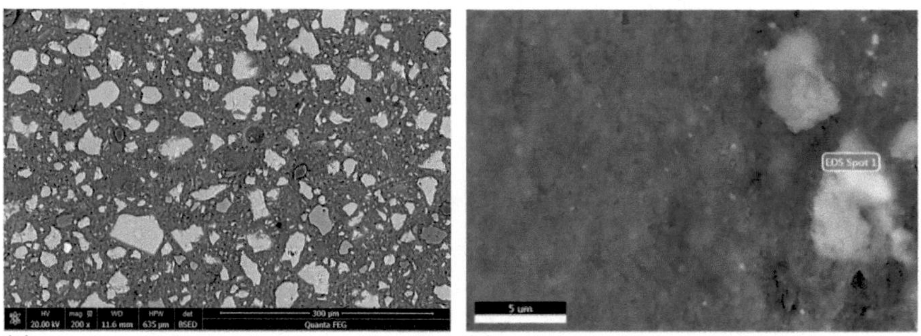

(a) (b)

Figure 5.25 (a) BSE image and (b) EDS analysis of UHPC paste after 28 d of curing.

Figure 5.24 is the SEM image of UHPC after 28 d of hydration [21]. The main hydration products are C-S-H gel, $Ca(OH)_2$, and AFt crystals, with AFt often accumulating in a needle-like form. The $Ca(OH)_2$ crystals are well-formed and appear as hexagonal plates [21–23]. Figure 5.25(a) presents the back scattered electron (BSE) image and energy dispersive spectroscopy (EDS) analysis of UHPC. Unhydrated cement particles, which have a higher average atomic number, appear brighter in the image; the gray areas correspond to hydration products, and the darkest black regions are the pores filled with epoxy resin during sample preparation. The presence of numerous and larger unhydrated cement particles suggests a limited degree of hydration under extremely low water-to-binder ratios. Figure 5.25(b) is a local magnification of the area within the rectangle in Figure 5.25(a), where the white particles indicate the alite phase, $Ca(OH)_2$ appears slightly less bright, and the gray C-S-H gel extensively fills the matrix.

5.3.3 Mechanical Properties

Concrete is a typical multi-scale and heterogeneous material in terms of phases, exhibiting specific properties from the macro to the micro-scale. Materials at smaller

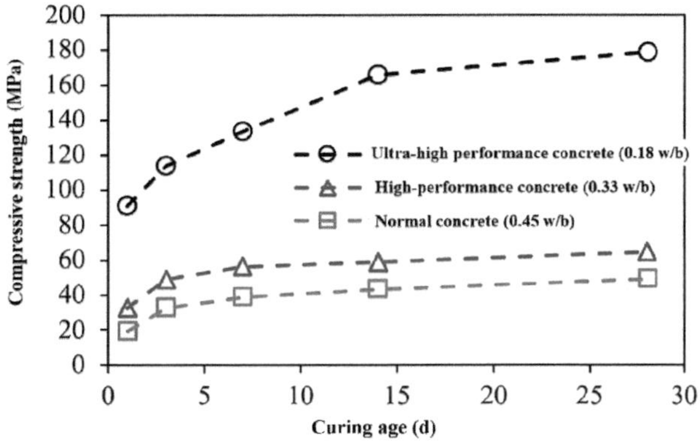

Figure 5.26 Strength comparison between ultra-high-performance concrete, high-performance concrete, and ordinary concrete.

scales are combined in various ways to form larger-scale materials, and there is usually a close connection between the properties at different scales. For mechanical properties, the bearing capacity at the macro scale is represented by compressive and flexural tensile strength, at the micrometer scale by micro-hardness values and at the nano/micro-scale by nano-mechanical properties.

5.3.3.1 Macro Mechanical Properties

The most prominent manifestation of the ultra-high performance of UHPC is its high strength and toughness. As shown in Figure 5.26, both the early and later strengths of UHPC are far higher than those of ordinary and high-performance concretes. The compressive strength after conventional curing for 1 day reaches 90 MPa, and the 28-day compressive strength is around 170 MPa. Figure 5.27(a) compares the 28-day mechanical strength results of UHPC specimens with different fiber contents. The increase in steel fiber content enhances the mechanical strength of UHPC. However, when the fiber content is high, further increasing the fiber content has a smaller effect on compressive strength. This is because the addition of a small amount of steel fibers can disperse stress and prevent stress concentration that leads to the formation of positive and negative corner pyramidal fracture forms in the matrix, but when the fiber content increases, more fibers are introduced, increasing the weak bonding interfaces of the matrix, thus reducing the increase in compressive strength. The addition of steel fibers also significantly improves the flexural strength of concrete (Figure 5.27(b)). Without steel fibers, the flexural strength of concrete is 19 MPa, and after adding 1% and 3% volume fraction of steel fibers, the flexural strength increases to 23 and 35 MPa, respectively.

The mechanical behavior of UHPC under tensile stress typically exhibits the characteristic of "strain hardening," which is reflected in a certain increase in strength after initial cracking. Figure 5.28 presents the ideal stress-strain curve for UHPC under uniaxial tension proposed by Graybeal [24], which mainly includes the following four stages. Stage 1 is linear elastic deformation; in Stage 2, the tensile stress exceeds the tensile strength of the UHPC matrix, beginning to show a plastic multi-crack deformation state; Stage 3 is the plastic strain hardening stage;

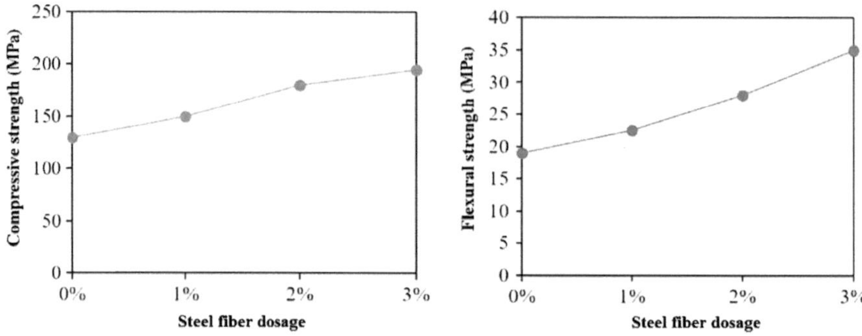

Figure 5.27 The effect of steel fiber volume fraction on the (a) compressive strength and (b) flexural strength of ultra-high-performance concrete at 28 d.

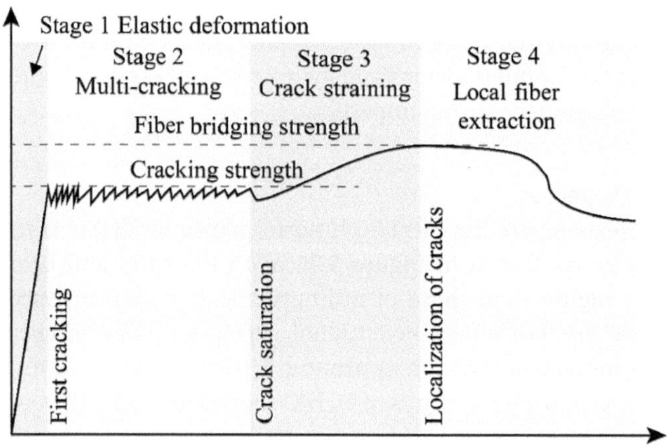

Figure 5.28 Ideal stress-strain development curve for ultra-high-performance concrete under uniaxial tension.

in Stage 4, individual cracks reach their strain limit, causing the fibers that bridge the cracks to begin being pulled out from the matrix.

5.3.3.2 Micro-Mechanical Properties

The aggregate-cement paste interfacial zone is a weak area in concrete, significantly affecting the mechanical and durability performance of concrete. UHPC, due to the reduction of coarse aggregate particle size and the incorporation of highly reactive mineral admixtures during its preparation process, has an inhibiting effect on the formation of the interfacial transition zone. Additionally, internal curing materials can significantly improve the microstructure of the interfacial transition zone. Figure 5.29 illustrates the pattern of micro-hardness variation in UHPC with varying water-to-binder ratios. As the water-to-binder ratio increases, both the interfacial zone and the cement-based matrix exhibit a noticeable decrease in

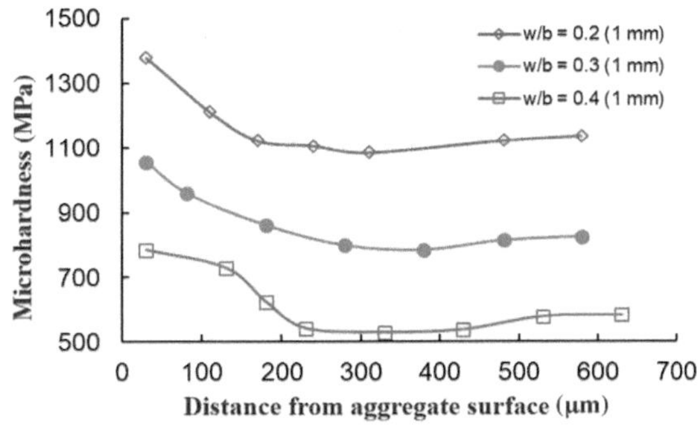

Figure 5.29 Micro-hardness of the quartz sand interfacial transition zone in ultra-high-performance concrete.

micro-hardness values, and the thickness of the interfacial zone also increases. The thickness of the interfacial enhancement zone increases from about 100 μm at a water-to-binder ratio of 0.2 to around 150 μm at a water-to-binder ratio of 0.4. However, looking at the distribution of micro-hardness in the interfacial enhancement area, the decrease in micro-hardness values over a smaller thickness range is greater for samples with a low water-to-binder ratio. The micro-hardness difference in the 100 μm interfacial enhancement zone is 256 MPa, which is a 60% increase compared to samples with a water-to-binder ratio of 0.4, indicating a smaller enhancement amplitude at high water-to-binder ratios. The dense layer formed in UHPC has a more pronounced effect on improving micro-mechanical properties, and the structure of this dense layer plays an important role in enhancing concrete performance. It should be noted that due to the large indenter used in micro-hardness testing and the larger testing distance, it is not conducive to judging the thickness of the enhancement layer. The next section will further test this using nanoindentation technology.

5.3.3.3 Nano-Mechanical Properties

Nanoindentation technology is a method for testing nano-mechanical properties, which offers higher precision and scale than micro-hardness testing. It can accurately characterize the elastic modulus and hardness of multiphase and multi-component materials in concrete, particularly obtaining the nano-mechanical properties of the main hydration product, C-S-H gel. Figure 5.30 presents the nanoindentation load-displacement curves for hydration products, quartz sand, and cement clinker minerals in UHPC. Under the same indentation loading, holding, and unloading conditions, different phases exhibit significantly different load-displacement curves. Unhydrated particles and quartz sand show higher load-bearing capacity with lower indentation depths, among which the indentation depth of cement clinker is only about 100 nm, while hydration products exhibit a more porous structure with larger indentation depths, and the tested indentation depths range from 300 to 500 nm. Through the load-displacement curves, the nano-mechanical properties of each phase can be well reflected and distinguished [23, 25]. At the same time, according to the classification of the load-displacement curves, three types of hydration products with different structures are found in

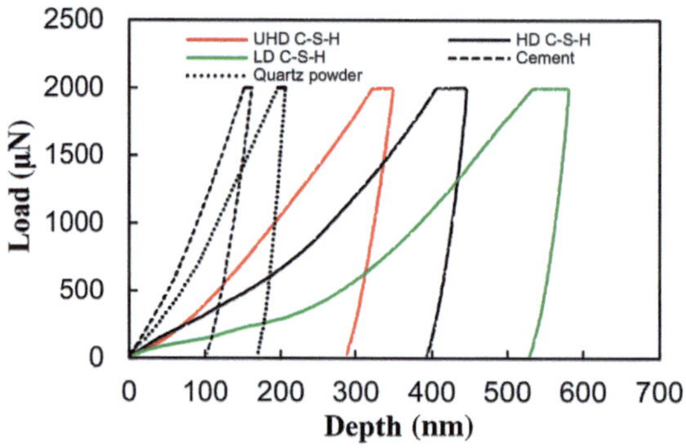

Figure 5.30 Load and displacement curves of hydration products, quartz sand, and cement clinker.

UHPC, namely high-density (HD) hydration products HD C-S-H gel, low-density (LD) hydration products LD C-S-H gel, and ultra-HD (UHD) hydration products UHD C-S-H gel. This indicates that UHPC is different from ordinary concrete, with the emergence of new UHD hydration products, which are of significant help for the enhancement of performance.

According to the load-displacement curves, the indentation hardness and indentation elastic modulus of each phase can be calculated using Eq. (5.8) [26]:

$$M = \frac{1}{2}\left(\frac{\mathrm{d}p}{\mathrm{d}h}\sqrt{\frac{\pi}{A}}\right) h = h_{\max} \tag{5.8}$$

where M represents the indentation modulus, GPa; p represents the applied load, N; h represents the maximum depth, m; and A represents the contact area calculated using the indentation depth and the Oliver and Pharr method, m^2.

A nanoindentation tester is used to test a 10×10 array of indentation points within the area shown in Figure 5.31, with a test point distance of 20 μm. A Gaussian function is used to perform a frequency distribution analysis of the elastic modulus of the cutting test point data obtained from the test, as shown in Figure 5.32. In ultra-high-performance cement-based materials, the hydration products are mainly UHD hydration products with an elastic modulus higher than 35 GPa. Referring to the method in reference [27], the hydration products' elastic modulus and hardness are classified according to the frequency distribution histogram obtained from the fitting, which are capillary pores (0–1.3 GPa, H = 0–0.4 GPa), LD C-S-H gel (13–22 GPa, H = 0.4–0.9 GPa), HD C-S-H gel (23–33 GPa, H = 0.9–1.2 GPa), $Ca(OH)_2$ (33–50 GPa, H = 1.2–2.0 GPa), and ultra-HD C-S-H gel. Due to the large amount of highly active admixtures in UHPC, which can significantly consume the $Ca(OH)_2$ produced by hydration, and the dense pore structure that limits the crystallization of $Ca(OH)_2$ crystals, resulting in a lower content of $Ca(OH)_2$, the phases with higher elastic modulus and hardness in the hydration products are mainly UHD hydration products. According to statistical calculations, the volume fractions of LD, HD, and UHD C-S-H, three types of hydration products, are 2%, 9%, and 56%, respectively. The content of LD and HD hydration products in UHPC is low, and the hydration products are mainly composed of UHD C-S-H components. The UHD C-S-H gel has an ultra-high elastic modulus (3.3–6.0 GPa) and hardness

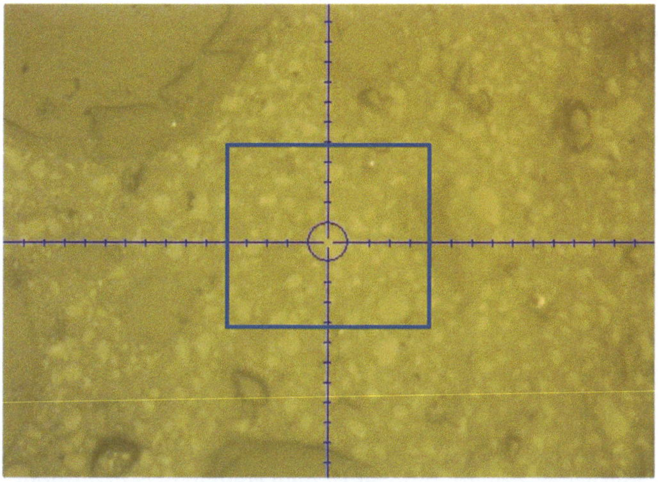

Figure 5.31 Test area of 10 × 10 indentation points in ultra-high-performance concrete.

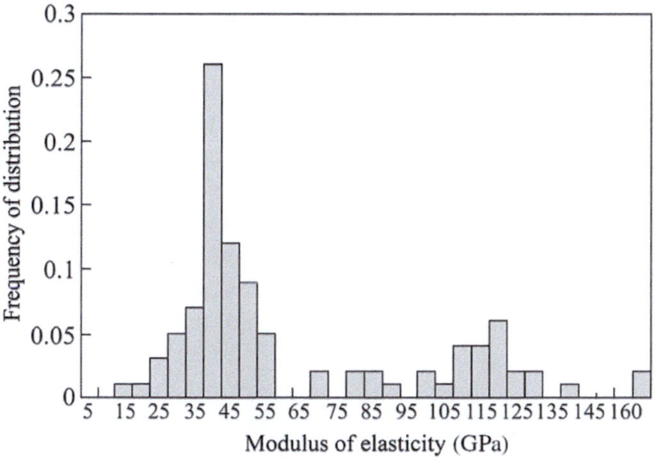

Figure 5.32 Frequency distribution histogram of elastic modulus in ultra-high-performance concrete.

(1.2–3 GPa); therefore, the hydration products in UHPC have a higher elastic modulus and hardness than those in ordinary and high-performance concrete, thereby exhibiting superior mechanical properties.

5.4 Volumetric Stability

The compositional characteristics of UHPC give it excellent performance while also causing the phenomenon of serious autogenous shrinkage. First of all, the

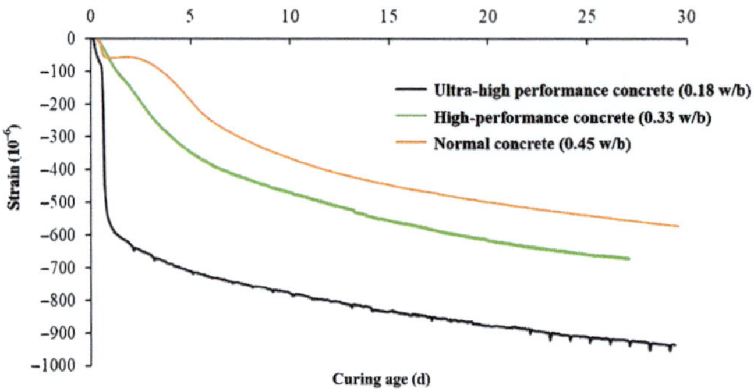

Figure 5.33 Autogenous shrinkage of UHPC netting vs. high-performance concrete and normal concrete.

UHPC has an extremely low water–cement ratio and a high content of cementitious materials, which leads to an intense hydration reaction, consuming a large amount of water in the capillary pores in a short period of time, resulting in significant self-drying effect and negative capillary pressure, which manifests itself as obvious volumetric autogenous shrinkage on the macroscopic scale. As can be seen from Figure 5.33, compared with ordinary concrete and high-performance concrete, the autogenous shrinkage of UHPC presents the significant characteristics of concentrated development trend and high final shrinkage rate, and its 28 d autogenous shrinkage rate is as high as 930×10^{-6}, while ordinary concrete and high-performance concrete are only 670×10^{-6} and 550×10^{-6}. This autogenous shrinkage phenomenon can cause large tensile stress and microcracks in UHPC members due to restricted shrinkage, thus affecting their mechanical and durability performance. In this section, the volumetric deformation characteristics and mechanisms of UHPC under conventional curing will be introduced as a guide for the preparation of UPC with low shrinkage.

5.4.1 Shrinkage Characteristics

Figure 5.34 shows the curves of linear shrinkage evolution with time for unadulterated steel fiber UHPC. The shrinkage development process of the UHPC matrix reflects the characteristics of concentrated shrinkage deformation, an obvious stage, and significant total shrinkage. The total shrinkage value of UHPC after 7 d of curing was as high as 512×10^{-6}, of which more than 70% of the shrinkage occurred centrally between 13 and 21 h. The shrinkage stage of UHPC can be seen from its autogenous shrinkage process in the early 48 h. As shown in Figure 5.35, its shrinkage development presents four obvious stages: there will be a short period of small linear shrinkage after halving the pouring, which lasts for about 9 h and is recorded as Stage 1; after that, the shrinkage development enters into a relatively stable stage, which lasts for nearly 6 h without any obvious volume deformation, and is recorded as Stage II; then, rapid and significant shrinkage deformation occurs, and the shrinkage value increases from 40×10^{-6} to 400×10^{-6}

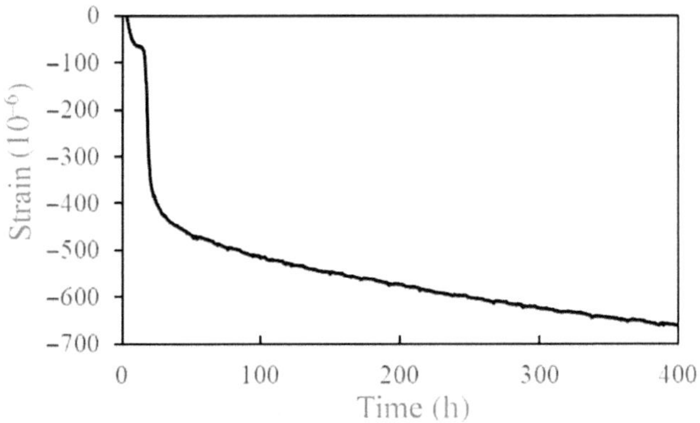

Figure 5.34 Shrinkage curve for UHPC.

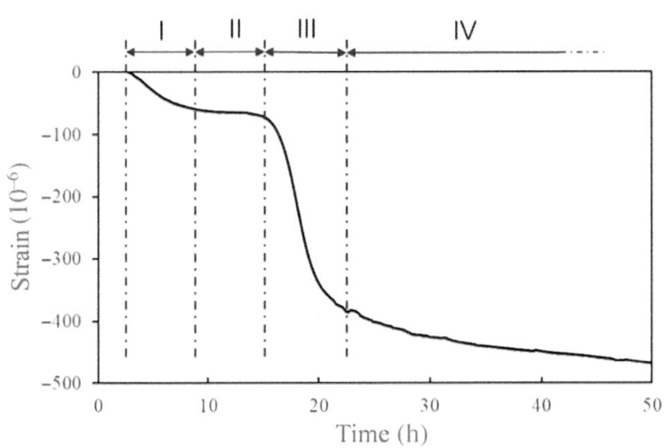

Figure 5.35 Early shrinkage curve of UHPC with classification of development stages.

within 8 h, and this stage is recorded as Stage II. This stage is recorded as Stage 3; then the shrinkage deformation gradually slows, relatively smooth, and this stage is recorded as Stage 4. Figure 5.35 for the linear shrinkage relative to the time to take the derivative of the rate of change of shrinkage (Figure 5.36), which further confirms that the division of its shrinkage development stage is closely related to the shrinkage rate. The larger changes in linear shrinkage rate are in the first stage as well as in the third stage, while the remaining two stages of shrinkage development proceed smoothly, and the rate of change is relatively stable.

5.4.2 Shrinkage Mechanism

Figure 5.37 shows the linear shrinkage vs. temperature development curves of UHPC specimens. As can be seen from the figure, the temperature of the specimens slowly decreased from the initial 28.2 °C to room temperature during the first stage of shrinkage, and then the temperature was kept basically constant during the second stage of shrinkage. With the rapid development of shrinkage in stage III,

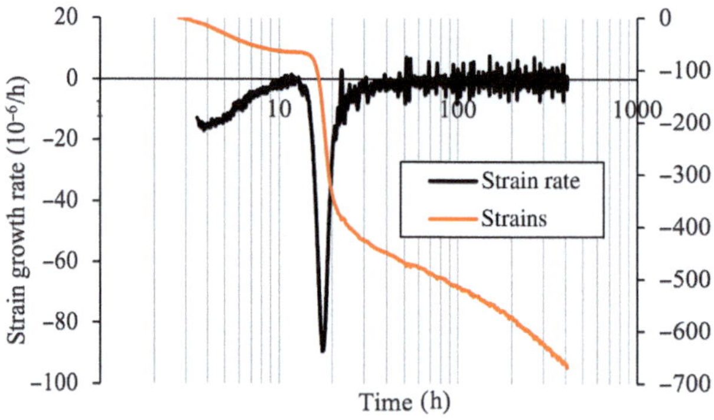

Figure 5.36 Shrinkage curve and rate of shrinkage for UHPC.

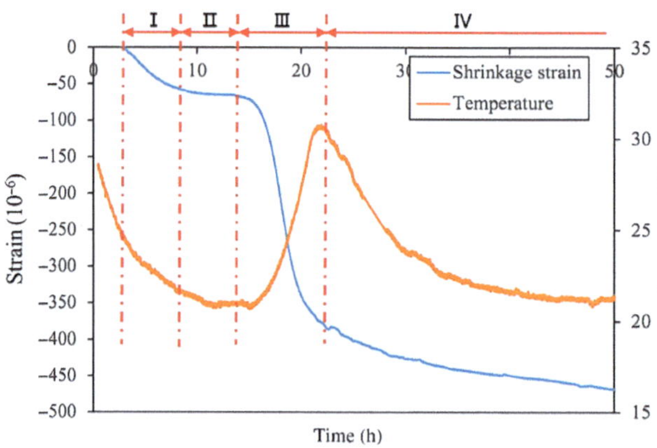

Figure 5.37 Temperature and shrinkage development curves for UHPC.

the temperature of the specimen increased rapidly from 20.7 to 30.6 °C, with an increase of nearly 10 °C, and then the temperature gradually decreased in Stage 4 of shrinkage to a stage of relative stability with the room temperature. Further analysis of the synergistic development curve of contraction and heat of hydration shows that (Figure 5.38): when the contraction is in the first stage and accompanied by a small contraction, the hydration reaction is in the latent period; when the contraction is relatively stable in the second stage, the hydration reaction is still in the latent period; when the contraction is accelerated in the third stage, the hydration reaction appears to have an exothermic peak, which is in the stage of intense hydration; when the contraction is smooth in the fourth stage, the hydration reaction tends to be in the latent period; and the contraction is stable in the fourth stage. When the contraction is in Stage 4, the hydration reaction tends to be steady.

Through the comprehensive analysis of shrinkage and temperature and the development of hydration reaction exothermic correlation can be obtained [28]: the first stage of hydration reaction is in the latent period, the cement hydration exothermic is

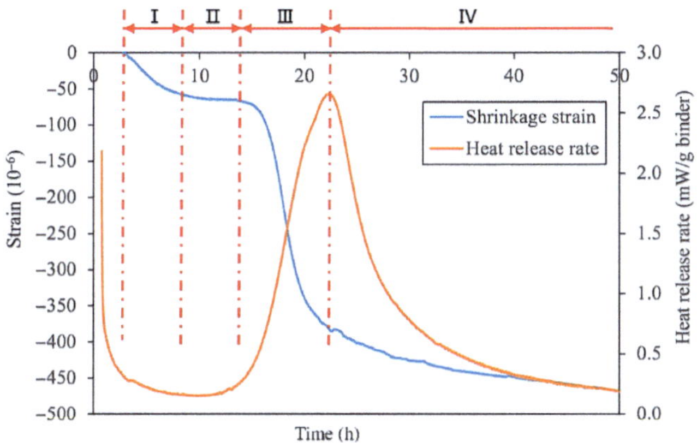

Figure 5.38 Heat of hydration vs. shrinkage development curves for UHPC.

basically negligible, the new slurry in the mixing process due to the input of mechanical energy so that the temperature is higher than the ambient temperature, molding through the environment through the exchange of heat to make the test block temperature drop, due to the effect of thermal expansion and contraction of the phenomenon of shrinkage. Therefore, the shrinkage of the first stage belongs to the thermal deformation caused by cold shrinkage; the second stage of the hydration reaction is still in the latent period, the test piece and the temperature exchange with the external environment ends and tend to stabilize, there is no temperature deformation, so the second stage of the latent period for the control of the hydration reaction; the third stage is in the accelerated period of the hydration reaction. On the one hand, the hydration reaction rapidly consumes the water in the capillary pores, resulting in self-drying and negative capillary pressure leading to the contraction of the apparent volume of the test piece, i.e., autogenous shrinkage; on the other hand, the exothermic hydration reaction causes a change in the temperature of the test piece, which brings about a thermal deformation through the thermal expansion and contraction effect. So this stage can be further distinguished by taking the apex of the hydration reaction peak as the boundary. Before the peak, the hydration reaction causes the temperature of the test piece to increase, and the linear shrinkage obtained in this stage is composed of two effects, namely, self-contraction and thermal expansion due to temperature increase, so the measured shrinkage value is smaller than the strict self-contraction; after the peak, the hydration reaction enters into the deceleration period, and the temperature of the test piece decreases, and the stage consists of self-contraction and cold shrinkage, and the measured shrinkage value is larger than the strict self-contraction. The effect of temperature on shrinkage can be confirmed by the relationship between temperature evolution and shrinkage rate. As shown in Figure 5.39, in the first stage of contraction, the rate of contraction is synchronized with the rate of temperature change, in the third stage of contraction rate and temperature change are not synchronized, the reason is that the third stage of the main controlling factor is the hydration reaction caused by the autogenous shrinkage, so the third stage of the contraction consists of autogenous

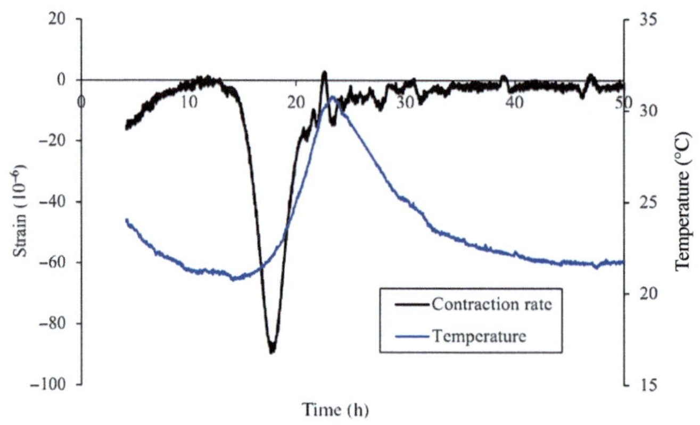

Figure 5.39 Temperature and shrinkage rate development curves for UHPC.

shrinkage and thermal deformation, but the main contraction is autogenous shrinkage; contraction of the fourth stage of the reaction, the rate of the hydration reaction is slower and relatively stable, the temperature has less influence on the In the fourth stage of contraction reaction, the rate of hydration reaction is slower and relatively stable, which has less influence on the temperature, and the continuous and smooth consumption of water in the capillary pores leads to the smooth development of autogenous shrinkage.

It can be seen that the shrinkage of UHPC can be divided into four stages, the first stage is the temperature-controlled thermal deformation stage; the second stage is the shrinkage latent period controlled by the hydration reaction; the third stage is the accelerated period of shrinkage controlled by the hydration reaction; and the fourth stage is the shrinkage smooth period controlled by the hydration reaction.

The decrease of relative humidity within the pore structure caused by self-drying is the essential cause of autogenous shrinkage, so the change of relative humidity by monitoring the test block is an effective means to study the shrinkage mechanism. Figure 5.40 shows the relationship between relative humidity and the shrinkage development curve. At the beginning of the relative humidity development, there will be a period of invalid data because the temperature and humidity probe need some time to reach equilibrium with the test environment. Counting from the highest point, the decreasing section of the relative humidity (RH) curve has a partially coincident section with the period of accelerated shrinkage, which is known to be mainly controlled by the hydration reaction, then the shrinkage and RH should be roughly consistent. However, in the measured data, there is no obvious turning point in the relative humidity curve when the shrinkage curve turns from the accelerated period to the stabilized period. This is because the apparent shrinkage of the specimen is related to the ability of the specimen to resist the shrinkage stress (i.e., the stiffness of the substrate) in addition to the negative capillary pressure, as shown in Eq. (5.9):

$$\varepsilon_{\mathrm{LIN}} = \frac{Sp_{\mathrm{c}}}{3}\left(\frac{1}{K} - \frac{1}{K_{\mathrm{s}}}\right) \tag{5.9}$$

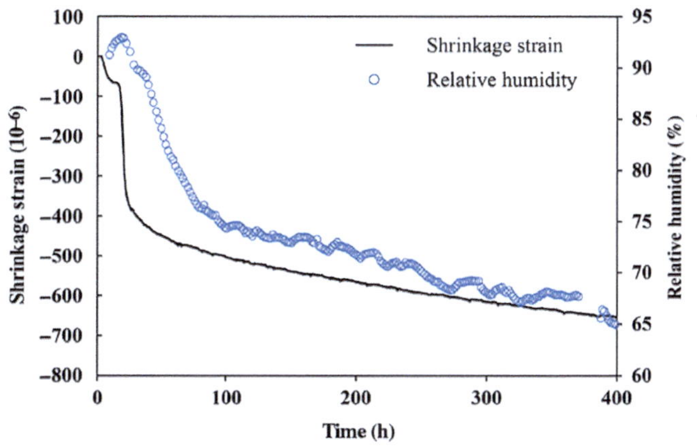

Figure 5.40 Curve of internal humidity and shrinkage development in UHPC.

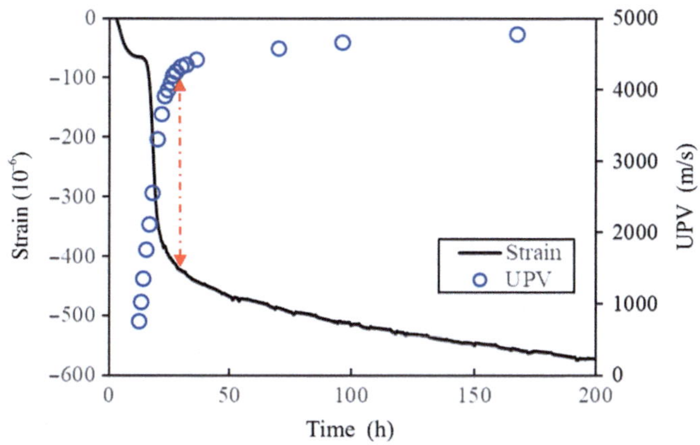

Figure 5.41 UPV vs. shrinkage development curve.

where $\varepsilon_{\mathrm{LIN}}$ is the linear contraction; S is the pore saturation, %; p_{c} is the negative capillary pressure, Pa; K is the matrix apparent stiffness, GPa; and K_{s} is the true stiffness of the matrix, GPa.

Figure 5.41 shows the relationship between ultrasonic pulse velocity (UPV) and shrinkage curves for UHPC. The transition point of shrinkage development from accelerated to stabilized phase is at the same time the turning point of UPV development. This is exactly the reason why the turning points of the shrinkage curve and the moisture curve do not coincide exactly.

5.4.3 Effect of Steel Fibers on Autogenous Shrinkage

Figure 5.42 shows the shrinkage curves of UHPC with different steel fiber admixtures. The introduction of steel fibers does not affect the stages of development of

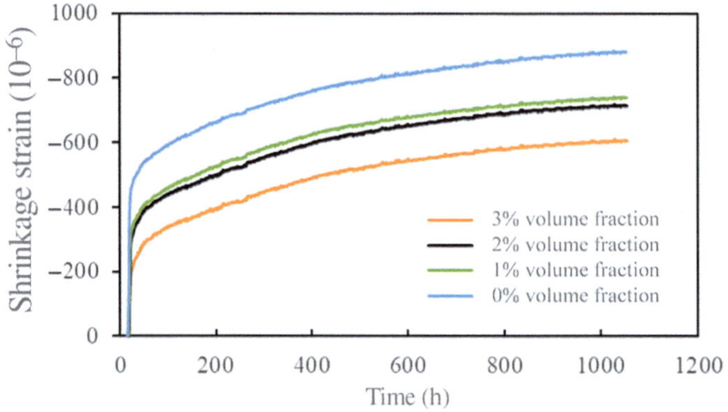

Figure 5.42 Shrinkage curves of UHPC with different volume additions of steel fibers.

shrinkage of UHPC and the duration of each stage. However, the total shrinkage values at different stages are significantly different. In the early stage of shrinkage, the effect of different dosages of steel fibers on shrinkage is not significantly correlated mainly because the early paste is still in the plastic stage and the hydration reaction is in the latent period; the paste and the surface of the steel fibers basically did not form an adhesive force. In the period of accelerated shrinkage, the incorporation of steel fibers significantly attenuates the shrinkage of the matrix. At the end of the accelerated shrinkage period, the shrinkage value of the matrix was as high as 648×10^{-6}, and when the fiber doping was 1%, 2%, and 3%, the shrinkage value was reduced to 530×10^{-6}, 440×10^{-6}, and 360×10^{-6}, respectively. The inhibitory effect of the steel fibers on the shrinkage in the fourth stage was relatively weak. Because in the fourth stage of the hydration reaction after a period of time, the matrix already has a certain strength, and the inhibition effect of the steel fiber on the matrix is not obvious.

The inhibitory effect of steel fibers on the shrinkage of the UHPC matrix is achieved through the interfacial bond strength between the two. Mangat has studied the effect of the interaction between steel fibers and the matrix [29]. Steel fibers are mainly bond to the matrix to develop interfacial bond strength. When the matrix produces deformation, this creates forces between the fibers and matrix parallel to the direction of the fibers (Figure 5.43). At this time, the pressure generated by the capillary pores of the matrix and the interfacial bonding strength form a force and reaction, and the interfacial bonding strength counteracts part of the capillary pressure, thus achieving the purpose of inhibiting the deformation of the matrix.

Mangat proposed an equation for the action of shear forces between steel fibers and the concrete matrix as in Eq. (5.10). The applicability of this equation to UHPC is discussed in this section, as there is a large difference in the proportioning and properties of UHPC compared to normal concrete:

$$\varepsilon_{\mathrm{f}}\left(\frac{\eta l}{2}+\frac{s}{2}\right)=\varepsilon_{\mathrm{m}}\left(\frac{\eta l}{2}+\frac{s}{2}\right)-\tau\pi d\times\frac{\eta l}{2}\times\frac{\eta l}{2}\times\frac{1}{AE} \tag{5.10}$$

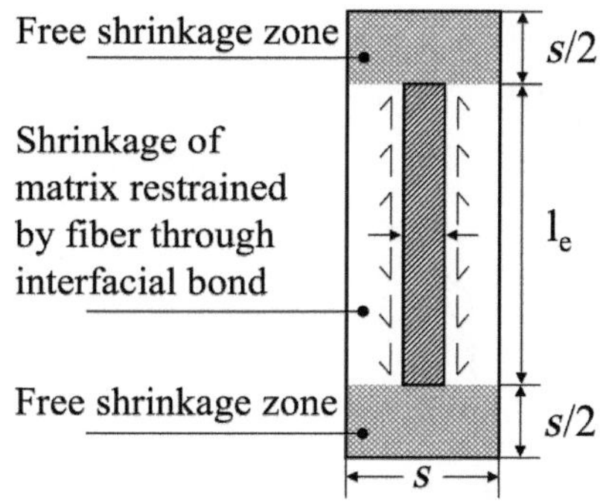

Figure 5.43 Effect of steel fibers on matrix shrinkage inhibition.

where ε_f is the shrinkage strain of matrix after steel fiber doping; ε_m is the shrinkage strain of the matrix without steel fibers; l is the fiber length, m; d is the steel fiber cross-section diameter, m; η is the orientation coefficient in the direction of inhibited contraction; s is the range of action of each fiber is the diameter centered on the fiber, m; τ is the average bond strength at the interface between the steel fiber and the specimen, MPa; A is the cross-sectional area of the specimen under the range of action of a single steel fiber, m^2; and E is the modulus of elasticity of the specimen, MPa.

Simplifying and distorting the above equation gives the following Eq. (5.11):

$$\varepsilon_f = \varepsilon_m - \frac{2\tau d\eta^2 l^2}{Es^2(\eta\, l + s)} \tag{5.11}$$

There is an exponential relationship between the development trend and its 28 d modulus of elasticity; the 28 d modulus of elasticity of UHPC can be obtained through the stress–strain curve (Figure 5.44), based on which the following

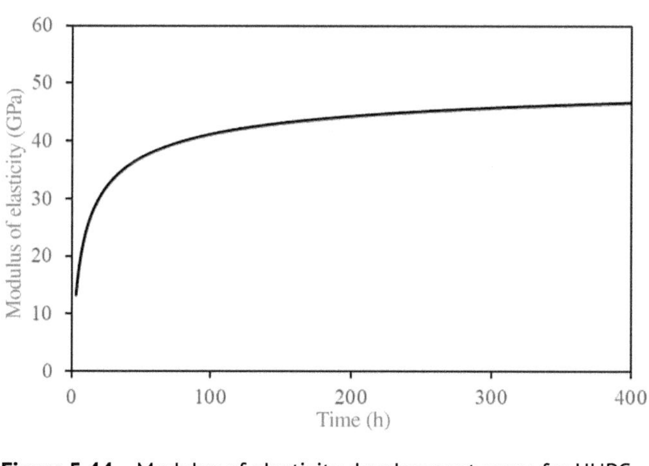

Figure 5.44 Modulus of elasticity development curve for UHPC.

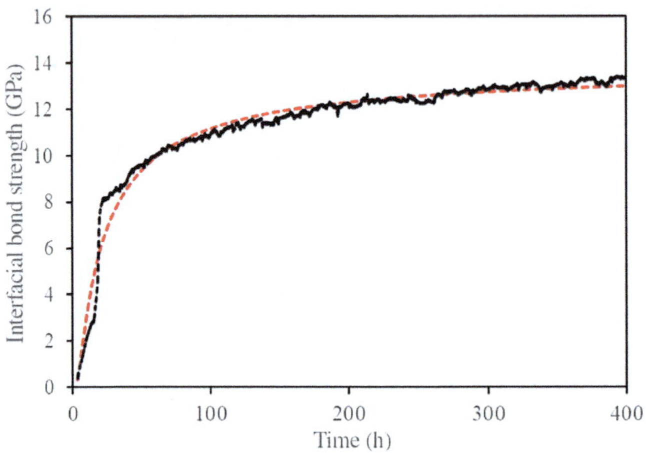

Figure 5.45 Interfacial bond strength development curve for UHPC.

development curve is obtained by simulating the evolution of its modulus of elasticity. In this paper, the interface bond strength τ is solved by the inverse solution method. Wille experimentally measured that when the steel fiber dosage is 1.5, its orientation coefficient is 0.783 [30], and experimentally measured the value of matrix autogenous shrinkage when the steel fiber dosage is 1.5%; based on this, the interface bond strength is solved by the inverse solution, and the interfacial bond strength is fitted, and the following development trend is obtained (Figure 5.45). Steel fiber orientation coefficient η is solved by the induction method. As analyzed above, in Eq. (5.11), the orientation coefficient is the only unknown. By measuring the matrix shrinkage at different steel fiber dosages and solving for the orientation coefficient in reverse, a linear relationship between them is found (Figure 5.46):

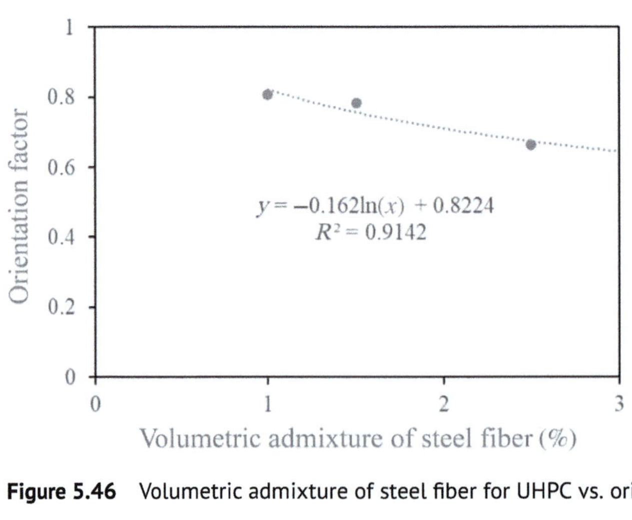

Figure 5.46 Volumetric admixture of steel fiber for UHPC vs. orientation factor.

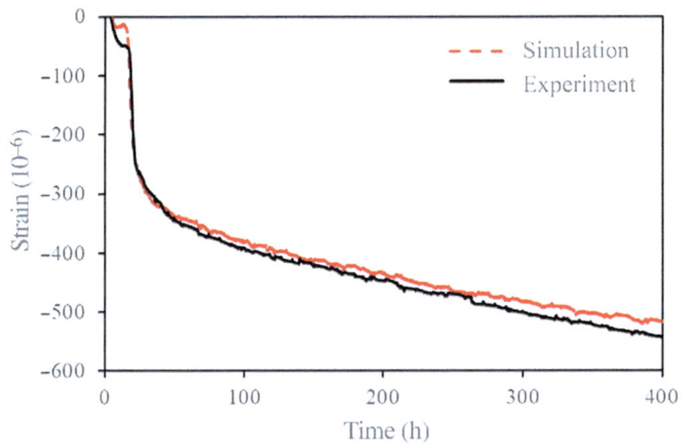

Figure 5.47 Simulation curve of the effect of steel fiber action.

Characterization of the unknown quantities through the above analysis yields the UHPC steel fiber inhibition model in Eq. (5.12):

$$\varepsilon_\text{f} = \varepsilon_\text{m} - \frac{2\tau d\eta^2 l^2}{Es^2\ \eta l + s)} \tag{5.12}$$

Among them

$$E = \left[\text{e}^{0.193\left(1-\left(\frac{28}{t}\right)^{0.5}\right)} \right]^{0.5} E_{28} \tag{5.13}$$

$$\tau = 14.0\text{e}^{-\left(\frac{16.43}{t}\right)^{0.82}} \tag{5.14}$$

$$\eta = -0.162\ln(x) + 0.8224 \tag{5.15}$$

The steel fiber doping of 3% matrix was selected to validate the model, and Figure 5.47 shows the comparison between simulation and measurement. As can be seen from the figure, the heat of hydration control stage has a high degree of match with the model. This is because the mechanism by which steel fibers act during this stage is consistent with the theoretical assumptions. However, as discussed in the previous section, the inhibition effect of steel fiber is not played out in the early stage of contraction, so the model simulation results do not match the actual effect. In the later stage of contraction, the structural strength of the matrix itself also has a certain degree of restriction on contraction, and the actual contraction will be smaller than the simulation results. However, the faster development of shrinkage occurs in the accelerated period of contraction. Therefore, the model has a certain degree of significance in guiding the shrinkage development of UHPC mixed with steel fibers.

5.4.4 Effect of Internal Curing on Shrinkage

Internal curing uses the water-absorbing effect of internal curing materials to store internal curing water. During hydration process, the internal water is constantly

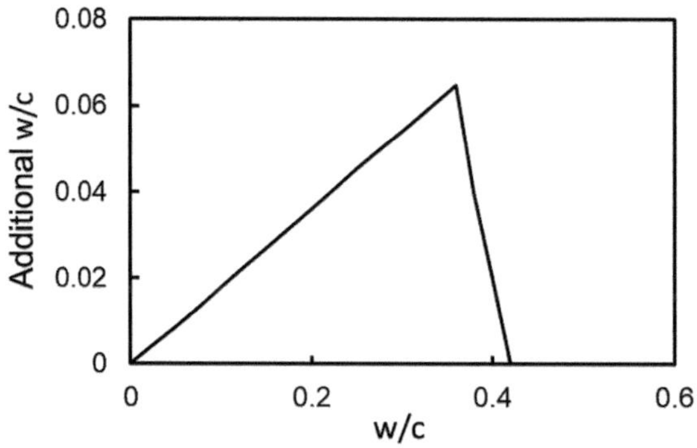

Figure 5.48 Calculation of water consumption for internal curing of cementitious materials with different water–cement ratios.

released to play the role of a "miniature reservoir." This effectively improves the internal humidity of the concrete. [18, 31, 32]. To reduce the autogenous shrinkage effect of concrete, it is necessary to choose the appropriate amount of internal curing water. Based on this, Jensen et al. proposed the theoretical minimum amount of introduced water to avoid self-drying [33], which provides a basis for the calculation of the amount of internal curing water required for low-shrinkage UHPC, and its calculation formula is shown in Eqs. (5.16) and (5.17):

$$\left(\frac{w}{c}\right)_{\mathrm{e}} = 0.18\left(\frac{w}{c}\right) \text{ when } w/c \leq 0.36 \tag{5.16}$$

$$\left(\frac{w}{c}\right)_{\mathrm{e}} = 0.42 - \left(\frac{w}{c}\right) \text{ when } 0.36 \leq w/c \leq 0.42 \tag{5.17}$$

where $\left(\frac{w}{c}\right)_{\mathrm{e}}$ is the additional introduced water–cement ratio for a sample with a particular water–cement ratio and $\frac{w}{c}$ is the effective water–cement ratio of the sample.

The amount of water introduced into UHPC with different water–cement ratios can be better calculated using the extra water–cement ratio formula, and the relationship between the extra water–cement ratio and the water–cement ratio of the sample can be shown using Figure 5.48. Since the water–cement ratio of the UHPC prepared in this section is 0.2, the additional introduced water–cement ratio required under ideal conditions is calculated to be 0.036. However, due to the composition of UHPC being quite different from that of high-performance concrete and ordinary concrete, the large admixture of ultra-fine powders (e.g., silica fume, etc.) and the absence of coarse aggregate limitation result in the need for a larger additional introduced water–cement ratio.

In addition to the effect of different water introduction and slurry components on the effect of internal curing, the determination of the type and amount of internal curing material needs special consideration, given their porous nature. At present, the commonly used internal curing materials include super absorbent polymer (SAP), pre-wetted plant fiber, rice husk ash, and pre-wetted ceramic sand, among which pre-wetted ceramic sand is the most widely researched and applied. Therefore, ceramic sand is selected as the internal curing material in this study. To ensure the fluidity performance of UHPC, ceramic sand is mixed using the method of equal particle size substitution; through the water absorption test of ceramic sand with different particle sizes, the water absorption rate of

Table 5.9 Introduced water–cement ratios corresponding to different dosages of internal maintenance materials.

Volume substitution	20%	40%	60%
Introducing the water–cement ratio	0.013	0.026	0.039

ceramic sand with the adopted quartz sand particle size distribution is calculated to be 10%. Ceramic sand is mixed using the volume substitution method, and the pre-wetted ceramic sand is mixed with 20%, 40%, and 60% of the volume of the quartz sand, respectively, and the actual introduction of water–cement ratios shown in Table 5.9, which takes into account the effect of silica fume on the theoretical amount of water, the maximum introduction of water–cement ratio of 0.039.

Due to the introduction of ceramic sand, the preparation method of UHPC was adjusted, mainly as follows: (1) First, the ceramic sand and the required mixing water were weighed, mixed and left to stand for 24 h. (2) During the mixing process, the pre-wetted ceramic sand was introduced together with the cementitious material, after which the water-reducing agent was added and fully stirred.

The evolution of autogenous shrinkage of UHPC with different amounts of ceramic sand is shown in Figure 5.49. Compared with the reference specimens, the introduction of different amounts of water, both in high and low dosages, compensates for the autogenous shrinkage of UHPC to different degrees. As the amount of introduced water increased, the autogenous shrinkage of UHPC is compensated more obviously. As can be seen from the figure, the 7 d autogenous shrinkage value of the specimen with 20% ceramic sand addition was 717×10^{-6}, which is 47.5% lower than that of the reference sample. As the dosage of internal curing material increased further to 40% and 60%, the 7 d shrinkage value decreased to 351×10^{-6} and 58×10^{-6} respectively. This trend indicates that the water in the internal curing material gradually enters into the capillary pores during the hydration process, forming an internal reservoir that saturates the otherwise unsaturated pores. This process decreases the unsaturation degree of pores, and consequently decreases the negative capillary pore pressure within the specimen. The 7 d shrinkage of the sample was only 58×10^{-6} when the dosage of ceramic sand was 60%, and the autogenous shrinkage was almost completely compensated, indicating that this introduced water–cement ratio is similar to the theoretical introduced water–cement ratio to prevent autogenous shrinkage.

After 7 d, the autogenous shrinkage development decelerated all the way. At the same time, the introduction of internal curing materials increased the internal humidity at this stage, resulting in a slower increase in shrinkage. After 14 d, the change in autogenous shrinkage of all samples became very slow, indicating that the hydration reaction was very slow at this time, resulting in a smaller change in autogenous shrinkage values at the later stage. Overall, the introduction of vitrified

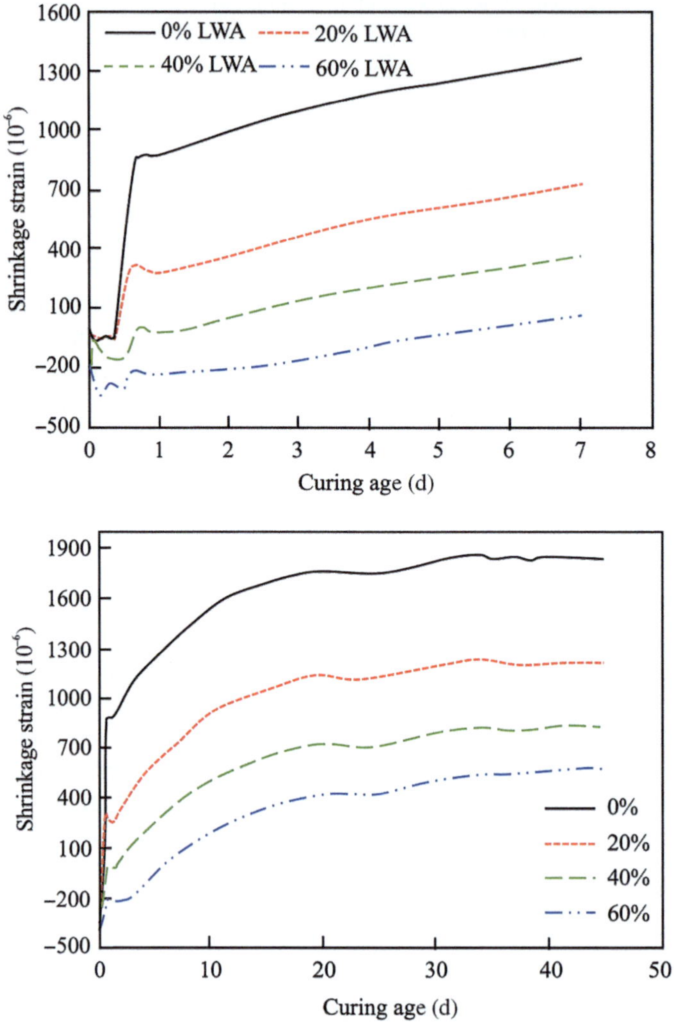

Figure 5.49 Autogenous shrinkage properties of UHPC with different amounts of ceramic sand introduction. (a) Early contraction and (b) Long-term contraction.

sand, with its internal curing effect, compensated for the autogenous shrinkage and significantly reduced the early autogenous shrinkage values. With the growth of hydration age, the autogenous shrinkage of UHPC continued to increase between hydration ages 1 and 14 d, but the rate gradually slowed down. After the hydration age of 14 d, the slowing down of the hydration reaction and the high strength of the matrix resulted in a slow change in the late autogenous shrinkage.

In addition, it should be noted that when a larger amount of water is introduced (40% ceramic sand and 60% ceramic sand), a weak expansion state is exhibited at the early stage of hydration due to the release of internal conservation water, which makes UHPC produce a certain amount of self-stress at the early stage, which is of great significance in preventing early shrinkage cracking.

5.5 Durability Properties

5.5.1 Freeze–thaw Durability

Although the water–cement ratio of UHPC is extremely low and the pore structure is highly dense, the moisture inside the matrix will still freeze at temperatures as low as a certain range to generate hydrostatic pressure. Therefore, the application of UHPC in cold regions also faces the challenge of freeze–thaw damage. The freeze–thaw durability of cementitious materials consists of two damage patterns: surface salt-freezing spalling and internal freezing damage, and the introduction of air voids is the most effective means to improve the freeze–thaw durability. In this section, the introduction of air voids and two kinds of freeze–thaw durability of UHPC will be discussed.

5.5.1.1 Characteristics of Entrained Air Void

Although air-entraining agents are not added during the preparation of UHPC, the need to use large quantities of surfactant-based water-reducing agents to improve its workability results in the introduction of a certain amount of air bubbles during the mixing process. When the paste develops strength, the air voids become embedded within the hardened cement paste. Figure 5.50 shows the polished section of UHPC, where some circular pores with a diameter of 500 μm can be clearly seen. The regular morphology of the pores further indicates that they are not due to trapped pores formed during the mixing process. Of course, there are also a very small number of trapped pores with irregular morphology. Microscopic porosimetry by the linear traverse method (LTM) of several UHPC-to-concrete ratios revealed that more than 90% of the introduced pores in UHPC concrete were within 1 mm in diameter (Figure 5.51). It is noteworthy that the concrete proportion in

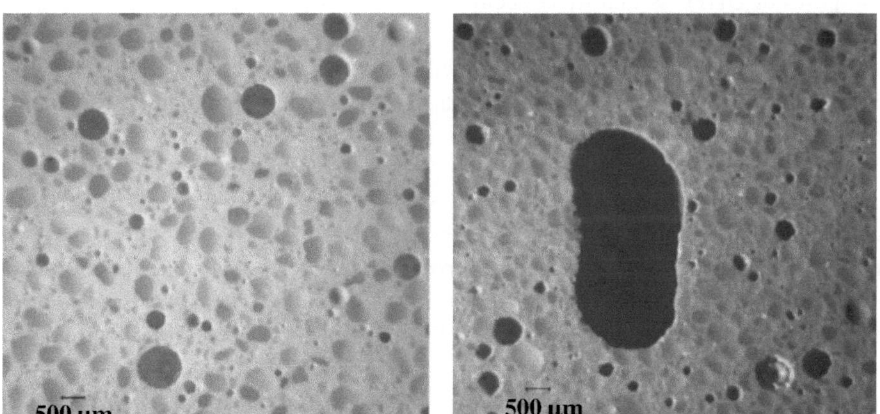

Figure 5.50 Distribution of entrained air voids after hardening of ultra-high-performance concrete (total air content 4.24%; specific surface area 13.16 mm^{-1}; powers pore spacing factor 519 μm).

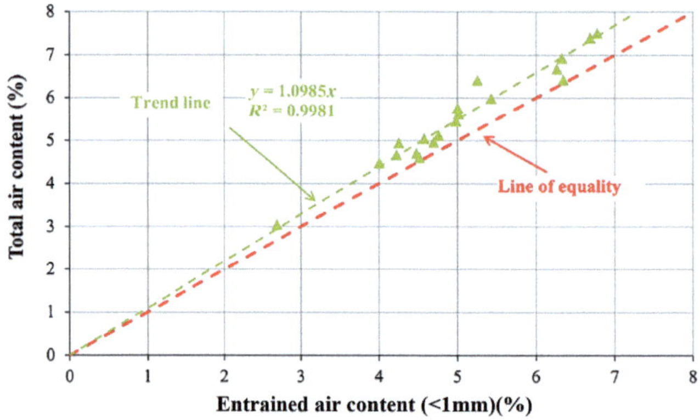

Figure 5.51 Relationship between introduced porosity and total porosity content of ultra-high-performance concrete.

Figure 5.50 has a total air content of 4.24%, but a T.C. powers spacing coefficient of 519 μm. This is because UHPCs, generally have a cement content of around 50%, which is considerably higher than that of normal concrete (around 30%). The total air content measured is relative to the total concrete volume, not the paste volume. At the same content, the spacing coefficient increases with increasing paste content. Therefore, when there is a large difference in the cementitious material content of two concrete ratios, it is necessary to use the normalization method to process the air content of the two before comparing them. Figure 5.52 shows the cumulative distribution and distribution frequency of air voids of UHPC and ordinary concrete before normalization, and the cumulative total air content of the two is relatively close. When normalized relative to the cement stone, the air content of ordinary concrete is significantly higher than that of UHPC, and the size of the pores of UHPC is mainly concentrated near 500 μm, while the distribution of pores of ordinary concrete is characterized by bimodal peaks at both 80 and 500 μm. In summary, the higher hardened cement paste content and a larger average size of stomata in UHPC are the main reasons for its high stomatal spacing coefficient.

5.5.1.2 Freeze–thaw Durability

The freeze–thaw durability of four groups of UHPC was evaluated using the unilateral salt-freezing method. After 72 freeze–thaw cycles, the relative dynamic modulus of the four proportions was maintained at about 100%, indicating no internal freeze–thaw damage (Figure 5.53). Furthermore, the cumulative spalling of the surfaces was within 150 g/m^2, which was much lower than the damage critical value of 1500 g/m^2 (Figure 5.54). By observing the morphological changes of the test surface of the test block before and after the test, it can be seen that the surface of the test block after 72 freeze–thaw cycles has a very slight degree of spalling (Figure 5.55).

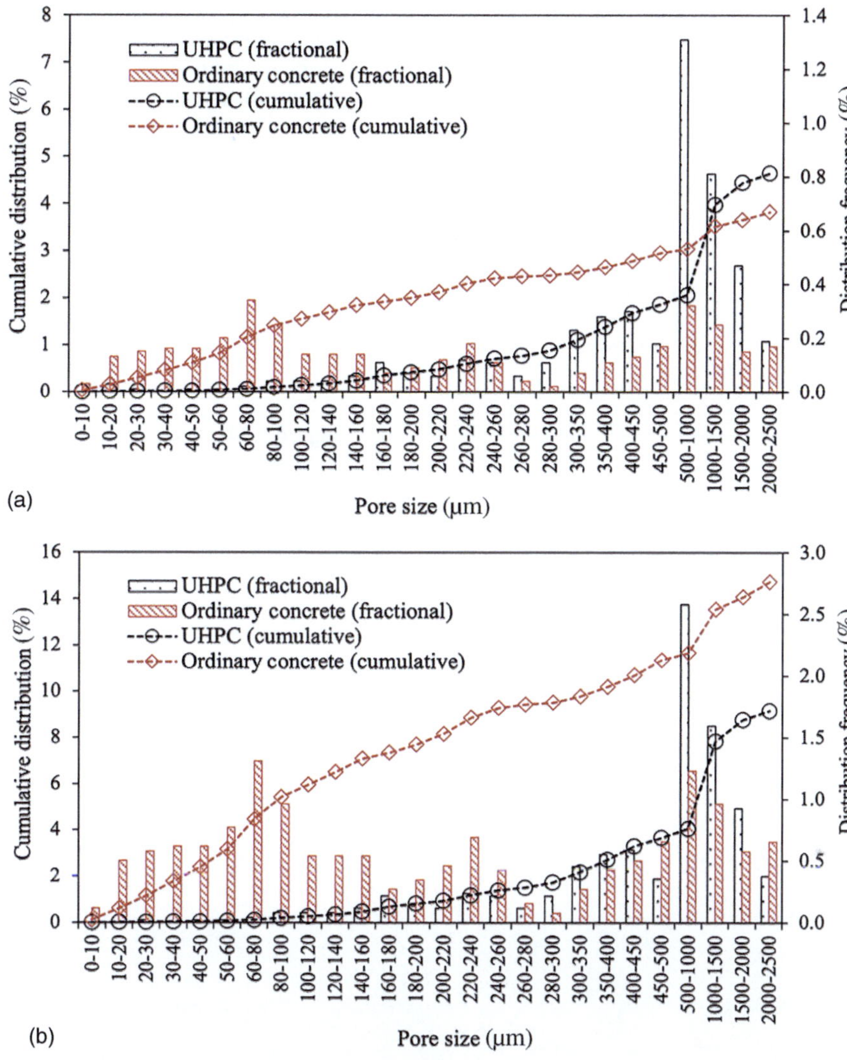

Figure 5.52 Pore size distribution of introduced air holes in ultra-high-performance concrete and normal concrete. (a) Pore size distribution relative to total concrete volume and (b) pore size distribution relative to volume of hardened cement paste.

The key factors in the freeze–thaw damage of concrete are the freezable water content of the internal pores and the rate of transport of external water to the interior. The low porosity of UHPC makes its matrix structure exceptionally dense, and the resistance of external water to enter the interior through capillary water absorption is very high. As can be seen in Figure 5.56, the capillary water absorption of UHPC is much lower than that of other concretes and is even close to that of ordinary concrete with a waterproofing treatment on the surface, which explains the extremely low surface mass loss rate of UHPC (Figure 5.57). The limited amount

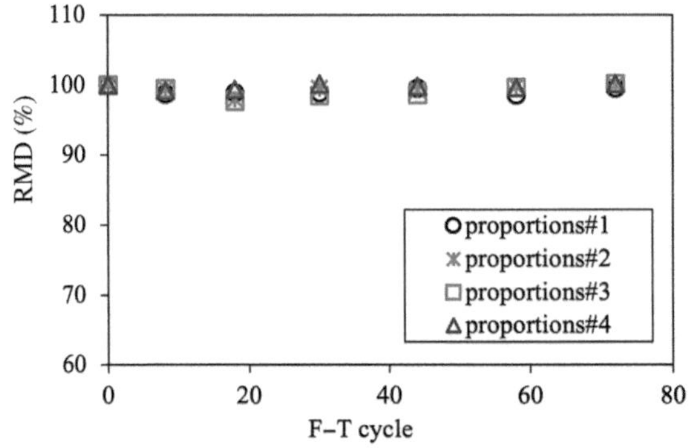

Figure 5.53 Relationship between relative dynamic modulus and number of freeze–thaw cycles for ultra-high-performance concrete.

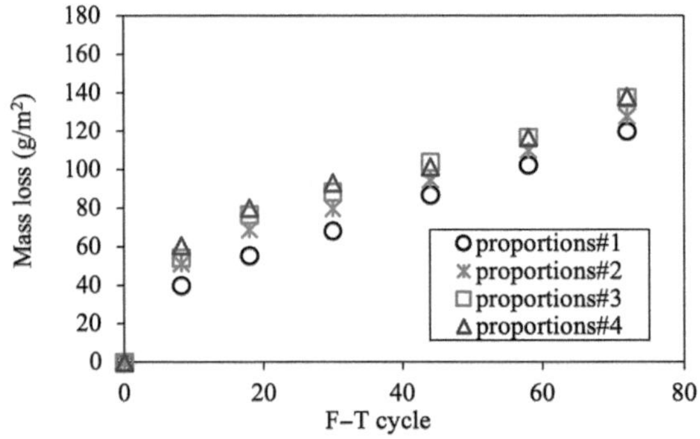

Figure 5.54 Relationship between the rate of mass loss and the number of freeze–thaw cycles in ultra-high-performance concrete.

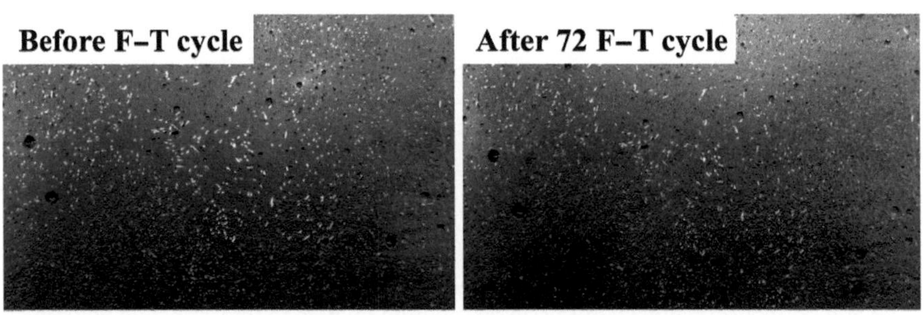

Figure 5.55 Changes in surface morphology of ultra-high-performance concrete before and after freeze–thaw tests.

of internal freezable water is also the main reason why UHPC did not experience internal frost damage during freeze–thaw cycles.

5.5.2 Resistance to Sulfate Attack

Under sulfate attack conditions, normal concrete cementitious paste generates hydration products such as calcite and gypsum, leading to degradation of concrete properties, and SO_4^{2-} on the molecular scale changes the polymerization degree and

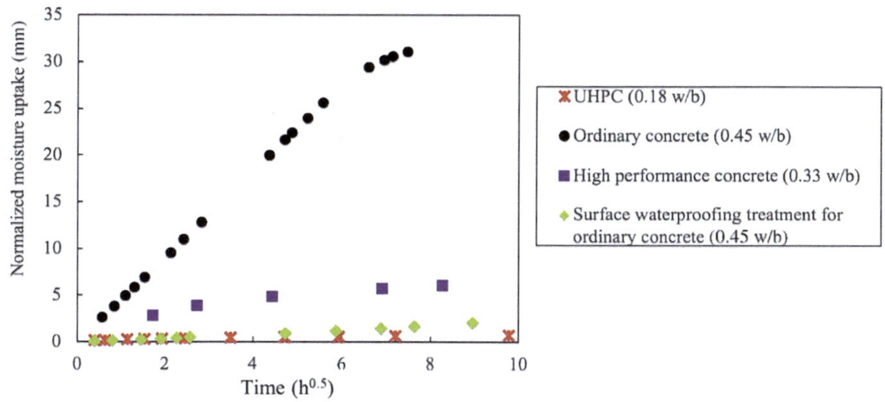

Figure 5.56 Comparison of capillary water absorption properties between ultra-high-performance concrete and other concretes.

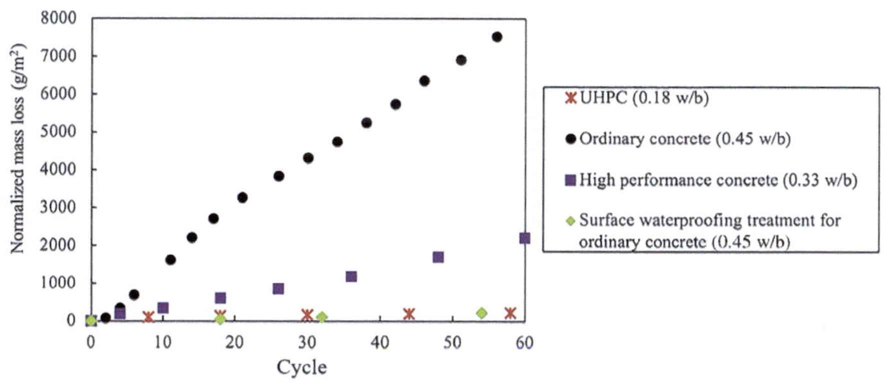

Figure 5.57 Comparison of mass loss rate of ultra-high-performance concrete with other concretes.

aluminum distribution of C-S-H gels through decalcification and de-alumination, affecting the microstructure of C-S-H [34, 35]. In this section, the physical phase composition, micro-morphology, microstructure formation, and evolution of C-S-H of the hydration products of UHPC cementitious pastes under sulfate attack are investigated.

Figure 5.58 shows the XRD pattern of UHPC cementitious slurry under the condition of sulfate erosion. As can be seen from the figure, the XRD spectrum of UHPC cementitious slurry did not change significantly after 28 and 180 d of 5% sulfate erosion, and the intensity of AFt diffraction peaks was basically unchanged, indicating that there was no significant effect of sulfate erosion on the types of hydration products of UHPC cementitious slurry within 180 d.

Under sulfate attack conditions, interconversion of aluminum-containing phase hydration products, such as AFt and AFm, mainly occurs in normal concrete cementitious pastes, while the molecular scale alters the distribution of aluminum in the

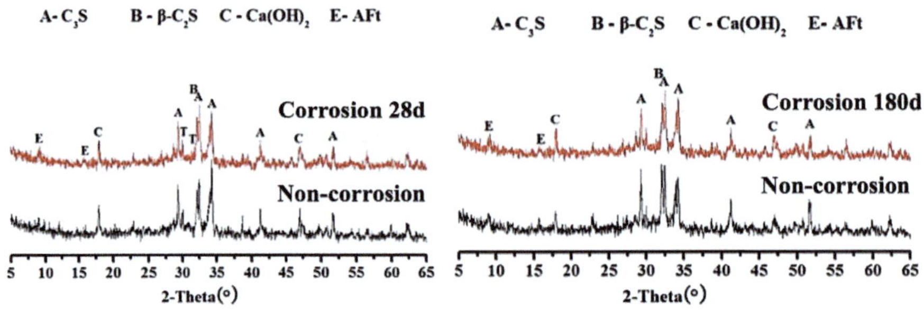

Figure 5.58 XRD patterns of cement paste under sulfate attack.

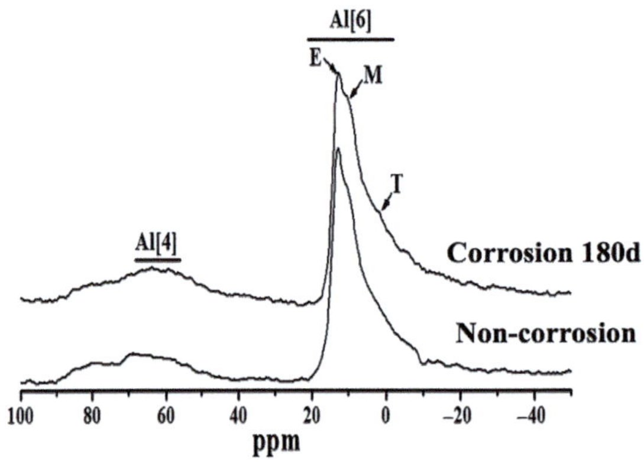

Figure 5.59 ^{27}Al NMR patterns of sulfate-eroded cement paste. *Source*: Adapted from Constantinides and Ulm [27].

Table 5.10 Deconvolution results of ^{27}Al NMR patterns.

	Al [4]	E	M	T
Non-eroded −180 d	12.65	38.68	21.36	27.31
Eroded −180 d	12.42	37.92	21.93	27.73

C-S-H gel through de-alumination, which affects the microstructure of the C-S-H, and thus may lead to degradation of concrete properties. Therefore, the effect of sulfate attack on Al^{3+} in UHPC gel slurry was first analyzed by ^{27}Al nuclear magnetic resonance (NMR) tests. Figure 5.59 shows the ^{27}Al NMR patterns of UHPC cementitious paste under sulfate erosion. ^{27}Al NMR pattern deconvolution calculation results are shown in Table 5.10. The ^{27}Al NMR spectrum of UHPC cementitious slurry after 180 d of sulfate erosion did not change significantly, and the AFt content was basically unchanged, indicating that there was no significant effect of sulfate erosion on the hydration product species of UHPC cementitious slurry within 180 d, which was in line with the results of the above XRD spectra.

The UHPC cementitious paste was tested using ^{29}Si NMR, and the test results are shown in Figure 5.60. The deconvolution of the ^{29}Si NMR maps yielded the

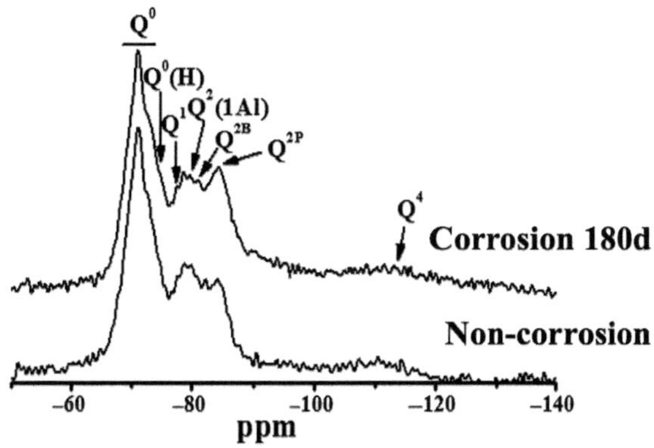

Figure 5.60 ^{27}Si NMR patterns of sulfate-eroded slurries. *Source:* Adapted from Constantinides and Ulm [27].

Table 5.11 ^{27}Si NMR deconvolution Q^n relative intensity values.

	Q^0	Q^1	Q^2 (1Al)	Q^{2B}	Q^{2P}	Q^4
Non-eroded –180 d	35.1	18.8	4.7	12.4	23.9	5.1
Eroded –180 d	34.1	19.2	4.3	12.8	24.6	5

Source: Adapted from [27].

Table 5.12 Effect of sulfate attack on the degree of hydration, average molecular chain length, and Al [4]/Si of ultra-high-performance concrete cementitious pastes.

	α_c	α_{SF}	MCL	Al [4]/Si
Non-eroded –180 d	50.35	82.59	6.61	0.039
Eroded –180 d	51.77	82.94	6.57	0.035

relative strength values of Q^n, as shown in Table 5.11, and the results of the degree of cement hydration, α_c; the degree of silica fume hydration, α_{SF}; the average molecular chain length (MCL); and the degree of substitution of Si^{4+} by Al^{3+}, Al [4]/Si, are shown in Table 5.12. As can be seen from Table 5.11, the UHPC pastes exhibited specific changes in its C-S-H structure compared with the uneroded reference specimens of the same age: the Q^2 (1Al) content slightly reduced, while the contents of Q^{2B} and Q^{2P} increased. This indicated that the sulfate erosion for 180 d had a weak de-alumination effect on the C-S-H structure. Additionally, the content of Q^0 was reduced, and the content of Q^1 slightly increased, which indicated that sulfate erosion promoted further cement hydration. As can be seen from Table 5.12, sulfate erosion for 180 d improved the degree of cement hydration in UHPC, with increase of 1.42%, 1.70%, and 1.98%, respectively. However, the reaction degree of silica fume did not change much. The average MCL of C-S-H of slurry with 180 d of

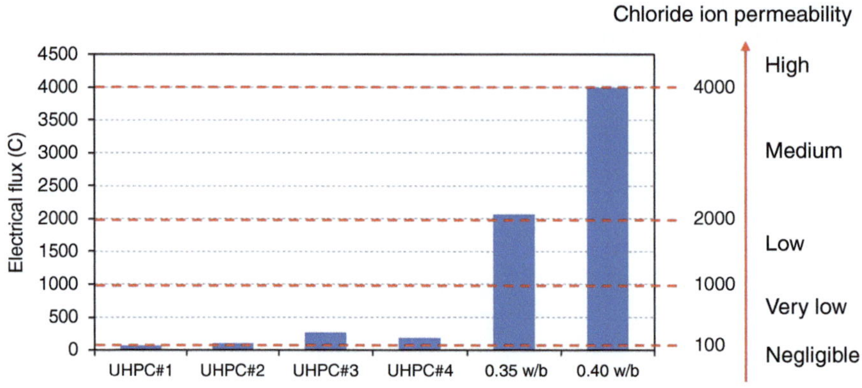

Figure 5.61 Chlorine ion permeability of ultra-high-performance concrete.

sulfate erosion was slightly reduced, which was because SO_4^{2-} improved the cement hydration activity, and had less effect on the silica fume hydration. As, cement hydration mainly produced Si-O dimer, it increased the content of Q^1 and lowered the average MCL of C-S-H. The weak de-alumina effect of sulfate erosion on the structure of C-S-H also lowered the average MCL of C-S-H chain length. The Al [4]/Si of the slurry with 180 d of sulfate attack also decreased slightly. According to the previous analysis, the sulfate attack with 180 d had a weak de-alumina effect on C-S-H, and the continued hydration of the cement in the later stage also inevitably led to an increase in Si in the slurry, which led to a decrease in Al [4]/Si.

5.5.3 Resistant to Chloride Ion Penetration

The chloride ion permeability of four different UHPC ratios was evaluated using the electrical flux method, as shown in Figure 5.61. The results showed that the highly dense matrix structure of UHPC hindered the intrusion of chloride ions, the total electrical conductivity was within 200C, and some ratios were even lower than 10C. According to the classification of chloride permeability resistance capability of concrete in accordance with ASTM C1202, UHPC was in the very low and negligible class, which was far superior to ordinary and high-performance concretes.

5.6 Novel Ultra-high-performance Concrete

With the gradual application of UHPC in the domestic engineering field, a batch of novel UHPC materials has emerged. Through optimization of physical and mechanical properties, these materials can not only meet complex working conditions but also demonstrate superior mechanical performance compared to traditional concrete materials. The author's team has conducted work in the application areas of green solid waste-based UHPC and lightweight low-shrinkage materials, achieving the following progress.

5.6.1 Green Ultra-high-performance Concrete

UHPC typically contains around 50% cementitious materials. However, due to its extremely low water–cement ratio, the overall hydration degree of cement is very low (generally around 30%). The majority of cement particles primarily serve a physical filling effect and do not fully utilize their hydration activity. Steel slag is an industrial waste residue produced during the steelmaking process. Due to its low activity and poor stability, the utilization rate of steel slag is low. Currently, the comprehensive utilization rate of steel slag in China is only 20%, much lower than in many developed countries. Large amounts of steel slag are casually discarded or landfilled, causing serious environmental and social hazards. This section investigates the use of steel slag as a partial replacement for cement in preparing UHPC, studies the impact of steel slag on the mechanical properties and volume stability of UHPC, and evaluates the influence of steel slag incorporation on the environmental load of UHPC, providing a new technical approach for the resource utilization of industrial waste [3, 5, 6].

Figure 5.62 illustrates the development trend of compressive strength of UHPC under different steel slag dosages. The addition of steel slag leads to a certain degree of strength reduction in UHPC, particularly affecting early strength. Without steel slag, the compressive strength of UHPC at 7 d is 134 MPa. When 30% steel slag is added, the compressive strength decreases to 118 MPa, although all mixtures achieve compressive strengths above 150 MPa at 28 d. However, with an increase in curing age, the gap between the experimental group and the control group gradually diminishes. The main reason for the decrease in strength of UHPC after adding steel slag is the low hydration activity of steel slag, which reduces the reactivity of the composite binder system (Figure 5.63). Analysis of the backscattered electron images of the hydration at 28 d in UHPC with steel slag addition reveals (Figure 5.64) that there is a certain degree of hydration of the steel slag cementitious components after 28 d of hydration reaction, with good bonding between the hydration products and the

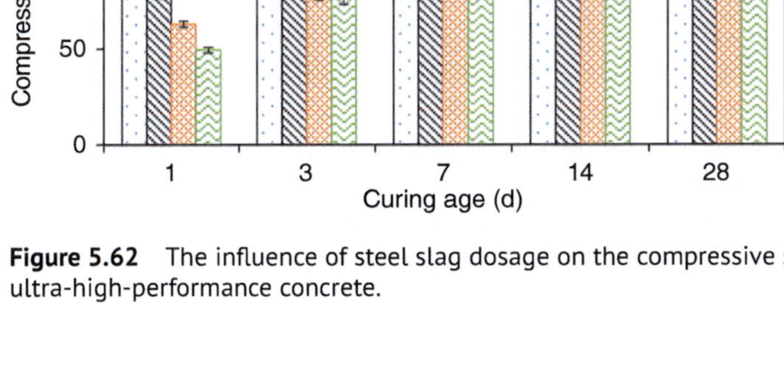

Figure 5.62 The influence of steel slag dosage on the compressive strength of ultra-high-performance concrete.

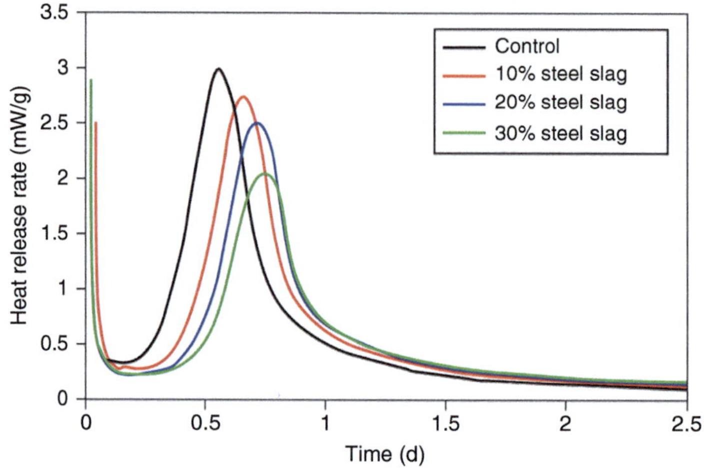

Figure 5.63 The impact of steel slag dosage on the hydration heat release rate of ultra-high-performance concrete.

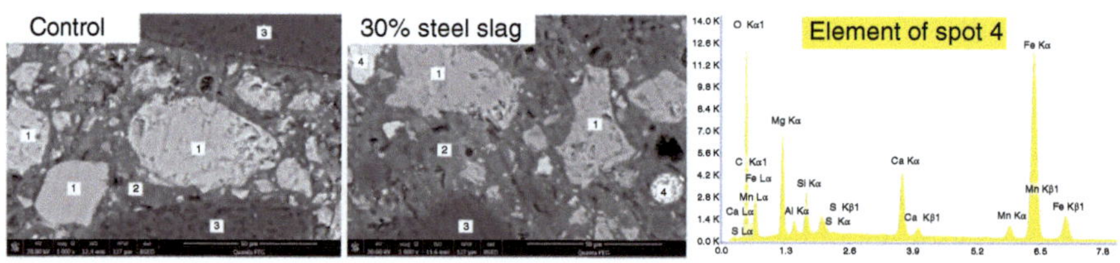

Figure 5.64 Backscattered analysis of UHPC with different steel slag dosages (1 represents cement, 2 represents fly ash, 3 represents quartz sand, and 4 represents steel slag, where the brighter components indicate the RO phase, while the rest are cementitious components within the steel slag). *Source:* With permission of [6] / Elsevier.

surrounding matrix. The RO phase has almost no reaction and integrates with the surrounding cement hydration to form a cohesive C-S-H gel with the matrix. Steel slag in UHPC serves not only a physical filling material but also participates in chemical reactions. The products contribute to the strength of the matrix, indicating that steel slag is not an inert filler. This is also the main reason for the relatively small loss in strength in the later stages after adding steel slag.

Figure 5.65 illustrates the pore size distribution of UHPC with different steel slag dosages at 28 d. As the dosage of steel slag increases, the porosity and pore size in the concrete increase. Steel slag has much lower hydration activity compared to cement, resulting in a lower degree of hydration and fewer hydration products, which cannot effectively fill the pores. Consequently, with an increase in steel slag dosage, the porosity and pore size roughen.

Figure 5.66 depicts the shrinkage development curve of UHPC with different steel slag dosages. The introduction of steel slag can suppress the development of shrinkage in UHPC but does not alter the trend of shrinkage development stages or

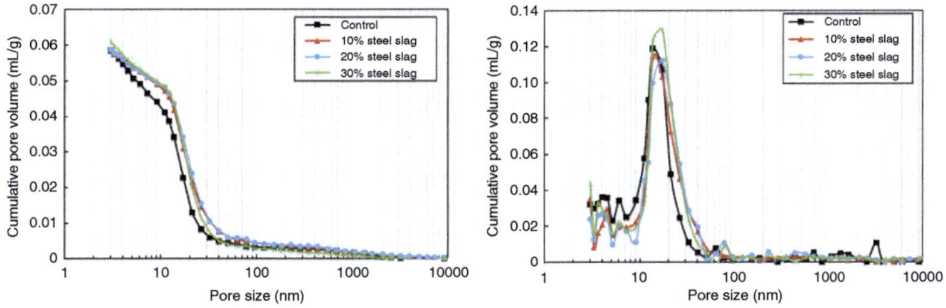

Figure 5.65 The influence of different steel slag dosages on the pore structure of UHPC.

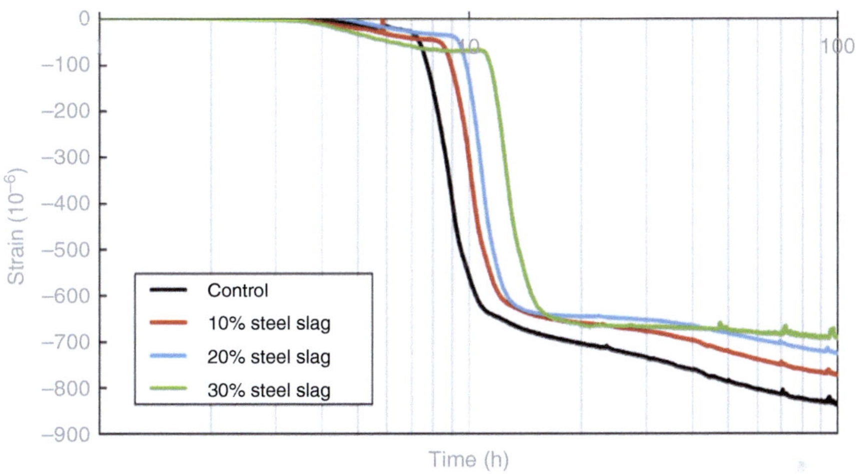

Figure 5.66 The influence of different steel slag dosages on the shrinkage performance of ultra-high-performance concrete.

their durations. Without steel slag, the shrinkage value of UHPC at 7 d is 820×10^{-6}, whereas, with 30% steel slag addition, the shrinkage decreases to 680×10^{-6}, representing a reduction of approximately 17%. Moreover, as the dosage of steel slag increases, the inhibitory effect on shrinkage becomes more pronounced.

Steel slag contains certain heavy metals that can leach out during storage and contaminate soil or groundwater, posing potential environmental risks. Therefore, it is necessary to investigate the leachability of heavy metal elements in UHPC incorporating steel slag. The leaching amounts of heavy metals in UHPC with different steel slag dosages were measured according to the Chinese standard GB 5086.2-1997 "Solid Waste Leaching Toxicity Leaching Method," as shown in Table 5.13. The addition of steel slag increases the leaching amount of arsenic element, which is due to the presence of arsenic element in the steel slag itself. However, since the steel slag used in this study does not contain heavy metal elements such as lead and zinc, the addition of steel slag reduces the leaching amount of these heavy metal elements through its dilution effect. Moreover, the leaching amounts of the four heavy metal elements are far below the specified limits.

Table 5.13 Dissolution of heavy metal elements in ultra-high-performance concrete with different steel slag admixtures.

	Concentration of heavy metal elements (mg/L)			
Steel slag dosage	**As**	**Cd**	**Pb**	**Zn**
0%	0.33	0.07	2.28	4.57
10%	0.43	0.08	2.03	4.62
20%	0.31	0.06	1.53	3.90
30%	0.37	0.07	1.77	4.08
Specified range value[a)]	5	5	5	100

a) GB5085.3-2007 Hazardous waste identification standard.

Table 5.14 Typical ecological indicators of ultra-high-performance concrete with different steel slag dosages.

Environmental impact indicator		**0%**	**10%**	**20%**	**30%**
Global warming potential (kg CO_2 eq./kg)		0.397	0.354	0.311	0.268
Primary non-renewable energy input (PEI(re)) ($\times 10^6$ J/kg)	Fossil	2.69	2.48	2.28	2.08
	Nuclear	0.27	0.254	0.237	0.22
	Biomass, $\times 10^5$	4.44	4.17	3.91	3.64
	Sum	2.96	2.73	2.52	2.30
Primary renewable energy input (PEI(re)) ($\times 10^6$ J/kg)	Biomass	0.0136	0.0129	0.0122	0.0115
	Wind, solar	0.0218	0.0214	0.021	0.0206
	Tidal power	0.11	0.11	0.10	0.09
	Sum	0.15	0.14	0.13	0.13
Land use (species. 10^{11} yr/kg)		1.90	1.86	1.81	1.76
Terrestrial acidification (species. 10^{11} yr/kg)		1.49	1.34	1.18	1.03
Terrestrial ecosystems (species. 10^{11} yr/kg)		9.56	8.49	7.43	6.36

The ecological and environmental performance of UHPC blended with steel slag was analyzed using Simapro life cycle assessment software. As shown in Table 5.14 and Figure 5.67, the addition of steel slag significantly reduces various ecological indicators of UHPC. Among these, the most important indicators, global warming potential (GWP) and primary energy consumption index, decreased from 0.397 kg CO_2 eq./kg and 2.96×10^6 J/kg in the control group to 0.268 kg CO_2 eq./kg and 2.30×10^6 J/kg, respectively. Additionally, the utilization of steel slag in UHPC reduces the land occupation index and soil acidification level. This indicates that the application of steel slag in UHPC plays a positive role in the safe

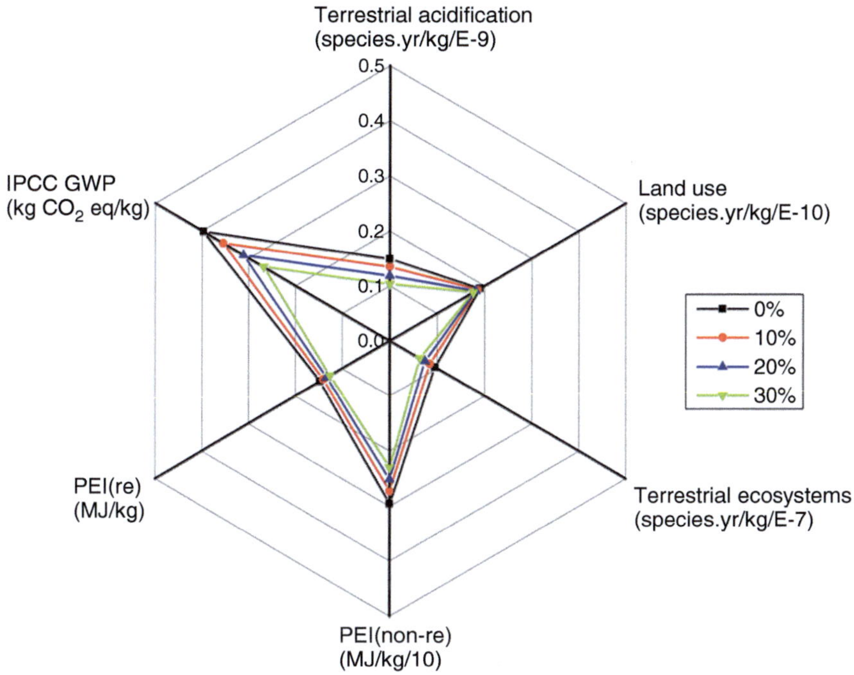

Figure 5.67 Impact of steel slag on the ecological and environmental performance of ultra-high-performance concrete.

and economically efficient disposal of steel slag, as well as the sustainable green development of UHPC.

5.6.2 Lightweight Ultra-high-performance Concrete

UHPC typically requires high cementitious material content and incorporates large quantities of river sand, quartz sand, and silica fume, leading to issues such as high self-weight, high cost, significant shrinkage, and poor volume stability. These limitations have severely restricted its widespread application in areas such as long-span bridges and high-rise buildings. Therefore, reducing the apparent density and shrinkage of UHPC is an important direction for its development. China possesses abundant natural lightweight aggregate resources, and the production of artificial lightweight aggregates has also begun to scale up. However, the domestic concrete engineering community still lacks sufficient emphasis on research concerning lightweight aggregate concrete. Studies often focuses on ordinary lightweight aggregate concrete for non-load-bearing structures with an LC50 below, while systematic research on lightweight UHPC (LUHPC) remains limited. Yichang City in Hubei Province is a production base for high-strength lightweight aggregates in China, where various densities, particle sizes, and water absorption rates of high-strength lightweight aggregates have been industrially produced. Leveraging this advantage, our team has used high-strength lightweight aggregates to replace quartz sand and industrial waste to replace silica fume, producing lightweight, low-shrinkage, steam-cured-free LUHPC [10, 11, 36]. This approach enables the material to achieve ultra-high mechanical performance and durability while reducing density, minimizing shrinkage, and lowering production costs.

Concrete is a multiphase composite material composed of a continuous cementitious matrix and a dispersed aggregate phase. Generally, based on the relative magnitudes of the elastic moduli between the matrix phase and the dispersed phase, it can be classified into composite hard matrix materials and composite soft matrix materials. The elastic modulus (E_a) of river sand and quartz sand aggregates in ordinary UHPC is higher than the elastic modulus (E_m) of the cementitious matrix, making it considered as a composite soft matrix material. LUHPC consists of a dispersed phase of lightweight aggregate and a cementitious matrix phase. The elastic modulus of fine lightweight aggregate is lower than that of the cementitious matrix; hence, LUHPC can be regarded as a composite hard matrix material. The distribution of fine lightweight aggregate in LUHPC significantly affects its performance. Considering the key influencing factors of LUHPC, its initial mix design is primarily based on the aggregate composition design method using elastic modulus. For UHPC where lightweight aggregate partially replaces ordinary aggregate, the elastic modulus satisfies Eq. (5.18).

$$E = E_{NA}E_{LC}/[E_{NA}E_{LC}E_{NA} - V_{NA}(E_{NA} - \mathrm{E}_{LC})] \tag{5.18}$$

where V_{NA} is the absolute volume content of ordinary aggregates in UHPC (%); E is the elastic modulus of UHPC (MPa); E_{NA} is the elastic modulus of ordinary aggregates (MPa); and E_{LC} is the elastic modulus of pure lightweight aggregate concrete (MPa).

As the volume of lightweight aggregate replacing ordinary aggregate increases, the elastic modulus of concrete gradually decreases, with density ranging between that of high-strength concrete materials with the same proportion and lightweight concrete materials. Therefore, in this section, the elastic modulus of LUHPC is used as the basis for the initial design of the absolute volume of lightweight aggregate. Fine lightweight aggregate with a continuous grading of 0.075–4.75 mm is used, with an apparent density of 1320 kg/m^3 and well-shaped particles. The elastic modulus of the aggregate can be designed based on the four-phase spherical model, which assumes that the concrete is under constant temperature conditions, the aggregate is spherical, and there are no significant interface transition zones within the concrete. The composition and structure of LUHPC conform to this model. The effective elastic modulus of LUHPC is calculated according to Eq. (5.19). For fine lightweight aggregate from crushing 800-grade high-strength ceramic particles, when using the above equation for calculation, it can be considered as dense aggregate:

$$E = \int_{a_{\min}}^{a_{\max}} E_0(a)\mathrm{d}S(a)\; S(a) = \left(\frac{a - a_{\min}}{a_{\max} - a_{\min}}\right)^m \tag{5.19}$$

where $E_0(a)$ is the elastic modulus of lightweight aggregate (MPa); $S(a)$ is the gradation of lightweight aggregate; a_{min} is the minimum particle size of aggregates (μm); and a_{max} is the maximum particle size of aggregates (μm).

Table 5.15 The mix proportion of lightweight ultra-high-performance concrete.

Cement	Fly ash microspheres	Silica fume	Ceramsite	Water–cement ratio	Steel fiber	Admixture
		(kg/m^3)			vol%	%
804	204	192	706	0.18	2.0	2.5

Table 5.16 Main performance comparison between LUHPC and UHPC.

	Compressive strength (MPa)			Flexural strength (MPa)			Apparent density (kg/m^3)	Specific strength
	3 d	7 d	28 d	3 d	7 d	28 d		
LUHPC	83.7	94.6	110.5	12.5	14.3	15.8	2065	0.0535
UHPC	95.4	111.0	133.4	18.8	21.4	24.9	2590	00515

In the preliminary mix design of LUHPC, the amount of cementitious materials and the dosage of fine lightweight aggregates were preliminarily determined by combining the elastic modulus-based design with the principles of dense packing of concrete. A dense packing method was applied to both the cementitious materials and lightweight aggregates. Furthermore, through optimization of preparation processes such as mixing sequence and time, the optimal mix ratio parameters of LUHPC were determined, as shown in Table 5.15. Table 5.16 presents a comparison of various properties between LUHPC and UHPC.

From Table 5.16, it can be observed that under the same mix proportion, the mechanical performance of UHPC is superior to LUHPC. This is because ceramic sand itself has lower strength and larger particle size compared to sand, leading to more interface defects compared to UHPC, thus resulting in a decrease in strength. When comparing and analyzing the performance differences between the two, the concept of specific strength is introduced, which refers to the ratio of compressive strength to the apparent density of the material. The greater the specific strength, the better it reflects the characteristics of lightweight and high-strength materials. From Table 5.16, it can be seen that the specific strength of LUHPC is 0.0535, while that of UHPC is 0.0515. LUHPC exhibits excellent characteristics of lightweight and high strength.

From the autogenous shrinkage curves of LUHPC and UHPC (Figure 5.68), it can be observed that both exhibit rapid early-age shrinkage, with the 7 d shrinkage already reaching approximately 73% of the 56-day shrinkage and gradually tend to plateau in the later stages. This is because early in the hydration process, there is sufficient moisture in the cement paste, resulting in significant shrinkage due to the rapid reduction of internal humidity as a result of the extensive participation of

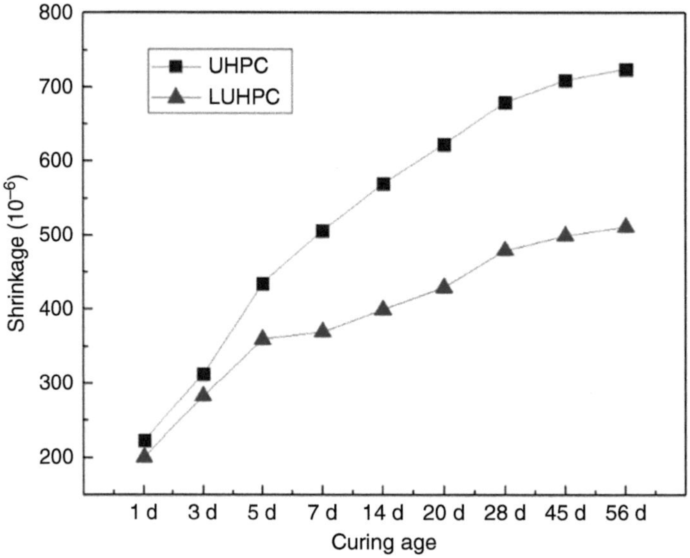

Figure 5.68 The development trend of autogenous shrinkage in LUHPC and UHPC.

cementitious materials in hydration reactions. At 56 d, the autogenous shrinkage of LUHPC is 512×10^{-6}, while that of UHPC is 725×10^{-6}, indicating smaller autogenous shrinkage in LUHPC. This is attributed to the high content of cementitious materials and low water–cement ratio in both LUHPC and UHPC, as well as the absence of coarse aggregates, leading to greater shrinkage during the hardening process compared to ordinary concrete and poorer volume stability. However, in LUHPC, fine lightweight aggregates release moisture under humidity gradients and migrate to the cement paste's capillary pores, providing internal moist curing to the cementitious matrix, delaying the decrease in internal relative humidity of concrete and reducing the self-drying effect of the cement paste. Compared to UHPC without internal curing, LUHPC exhibits smaller shrinkage and better volume stability.

Figure 5.69 presents SEM images illustrating the microstructural characteristics of the interface between cement paste and lightweight aggregates in LUHPC. When cement, silica fume, and fly ash microspheres are used as cementitious materials under low water–cement ratio, partially hydrated cementitious particles act as a skeleton, with hydration products embedded within them. This refines the structure of the matrix and enhancing the compatibility between the matrix and aggregates, resulting in a dense cementitious structure (as depicted in Figure 5.69(a)). In UHPC prepared with quartz sand, there is a distinct interface transition zone (as seen in Figure 5.69(b)), which becomes the weakest region in the UHPC microstructure. In contrast, when lightweight fine aggregates are used, their irregular surface characterized by numerous voids and low "terrain" areas, allow cementitious microparticles and fresh paste to fill these voids before concrete hardening. Meanwhile, the internal curing effect of fine lightweight aggregates significantly

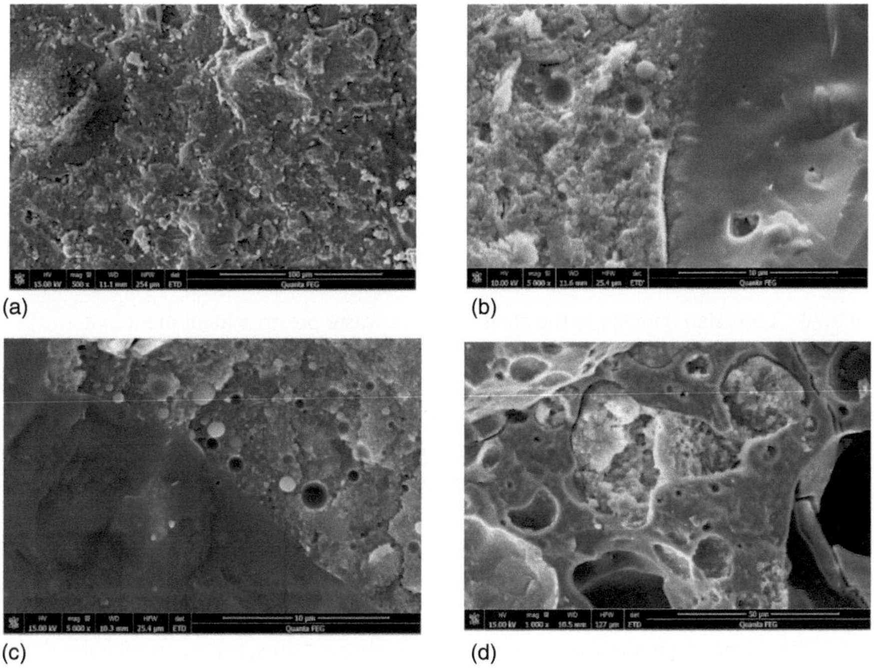

Figure 5.69 LUHPC microstructure SEM image. (a) LUHPC hardened cement paste, (b) UHPC ITZ, (c) LUHPC ITZ, and (d) UHPC interface coupling structure.

enhances the hydration degree of the surrounding cementitious materials. No obvious interface transition zone is observed between the fine lightweight aggregates and cement paste (as shown in Figure 5.69(c)). Additionally, the surface voids of the fine lightweight aggregates will absorb some cement paste during mixing. This paste subsequently co-cures and bonds with the external cement paste, forming a strong interface coupling effect (as depicted in Figure 5.69(d)). Furthermore, pre-saturation of the fine lightweight aggregates promotes the hydration degree of the surrounding cementitious materials in the later stages of hydration, forming a dense and high-strength arch structure.

5.7 Application of Ultra-high-performance Concrete

5.7.1 Ultra-high-performance Concrete for Municipal Solid Waste Pre-treatment Plants

With the acceleration of industrialization and urbanization in China, the safe and harmless disposal of a large amount of municipal solid waste has become a pressing issue for local governments. Wuhan University of Technology, in collaboration with Huaxin Cement Co., Ltd., has developed a complete set of technologies for the co-processing of solid waste in cement kilns. By constructing municipal solid

Figure 5.70 Corrosion damage to the steel structure waste pre-treatment plant in a strongly corrosive environment.

waste pre-treatment plants in cities such as Wuhan, Yangxin, and Wuxue, they have safely and ecologically disposed 3.5 millions tons of domestic waste, creating significant social, economic, and environmental benefits. At the same time, the leachate generated during the stacking of domestic waste (early fermentation leachate with $pH < 6.5$) and the high concentration of acidic gases (H_2S, acetic acid, and butyric acid, etc.) are highly corrosive. This imposes stringent requirements for the corrosion resistance of building materials used in waste pre-treatment plants. Initially, these facilities were typically constructed with steel structures, but the steel corroded extensively in a strongly corrosive environment. For example, the main plant at the Chenjia Chong facility in Xinzhou district, Wuhan, required an overall replacement and maintenance after only 1 year of service (Figure 5.70), consuming a large amount of material, labor, and financial resources.

In response to this situation, Huaxin Cement Co., Ltd., in collaboration with Wuhan University of Technology, leveraged a major special research and development plan of Hubei Province (Key Technology and Engineering Demonstration of Ultra-High-Performance Concrete Preparation and Application). This partnership led to the first domestically developed un-reinforced or lightly reinforced UHPC in China. This UHPC possesses advantages such as high impermeability, wear resistance, strong acid-base resistance, and fire resistance. It exhibits minimal degradation under various extreme harsh environments and long-term operation and has been successfully applied to the Huaxin Changshankou Environmental Protection Plant. In addition to its superior service performance in harsh environments compared to steel, UHPC is also a more cost-effective material. Taking a factory building with an area of 3000 m^2 as an example, the cost of steel beams for the large-span beams is about 250,000 yuan (including paint and labor), while the cost of the same amount of UHPC is about 120,000 yuan. Table 5.17 shows the mix proportions of UHPC used at the Huaxin Changshankou waste pre-treatment plant.

To ensure more accurate and efficient measurement of raw materials, a pre-mixed batch of 1.2 tons (i.e., one ton bag) is used for each mixing in the on-site actual mixing process. The corresponding other components are accurately weighed and then added to a forced concrete mixer for mixing. Before pouring and casting, the flowability is tested. Only after meeting the flowability requirements can the mixture be

Table 5.17 Mix proportion of ultra-high-performance concrete.

Premix	Admixtures	Water	Steel fiber
2200	35	130	156

Table 5.18 Test data of ultra-high-performance concrete.

Items	Results
Compressive strength after 1 d (MPa)	80
Compressive strength after 28 d (MPa)	155
Flexural strength after 28 d (MPa)	31
Acid gas corrosion resistance	pH > 2
Rating of sulfate resistance	>KS150
Carbonation depth (mm)	0
Anti-freeze–thaw cycle	Qualified after F500
Cracking resistance in the early stage	No cracks
Shrinkage rate after 56 d (10^{-6})	289
Creep coefficient 360 d	0.305

poured and cast. The on-site tested flowability range is 220–250 mm, which meets the casting requirements. Under these conditions, the UHPC is self-compacting and does not require vibration. The steel fibers are evenly distributed without segregation. The various test indicators of the UHPC used in the Huaxin Changshankou Environmental Protection Plant are shown in Table 5.18. It can be seen that the UHPC used has ultra-high mechanical and durability performance, which can meet the service requirements of the environmental protection plant.

The main structure of the waste pre-treatment plant includes "rib" 168 prestressed prefabricated beams with a span of 24.54 m (see Figure 5.71), all of which are cast with UHPC. The prepared prefabricated beams have a series of excellent properties such as light self-weight, corrosion resistance, fire resistance, cost-effectiveness, and convenient construction, with a designed service life of 150 years. Since its commissioning in 2017, no issues have been reported, and no maintenance is required.

This application case uses UHPC material to replace traditional steel beams, becoming the largest application of UHPC prefabricated beams in China and even the world, filling the gap in the large-scale application of UHPC as a special building material in strongly corrosive environments. To date, this material has also been successfully applied in several Huaxin projects, including the prefabricated beams of Shiyan Environmental Protection Plant, Yunyang Environmental Protection Plant, the floor repair of Wuxue Waste Reception Hall, and the bridge deck repair of Qu County Cement Plant's natural gas overhead bridge.

Figure 5.71 Ultra-high-performance concrete roof girder of Huaxin Wuhan Changshankou Environmental Protection Factory.

5.7.2 Pumpable Lightweight Ultra-high-performance Concrete for Steel Bridge Deck Pavement

UHPC exhibits characteristics such as high viscosity and low controllable workability in its fresh state, which limits its application in pumpable construction. The author's team, led by Ding Qingjun, used micro-expansive high-strength grade sulfoaluminate cement to replace 30–50% of the portland cement and selected 5–10 mm basalt crushed stones to replace 20–30% (by volume) of quartz sand to reduce the shrinkage of ultra-high-strength concrete and improve its early strength [10, 11, 36]. A super-dispersive, shrinkage-reducing, and viscosity-lowering water reducer was synthesized by molecular reconstruction. And fly ash microspheres with "rolling ball, dense filling, and water-reducing effects" were used to replace quartz powder. These modifications effectively reduced the viscosity and shrinkage of UHPC and improving the rheological performance of the concrete mix, making it easier for pumpable construction. Before pumping, 0.6–0.8‰ lithium carbonate is evenly mixed into concrete, using lithium carbonate to accelerate the early hydration rate of cement, increase its early hydration degree and early strength, and develop pumpable anti-cracking LUHPC. The mix ratio is shown in Table 5.19, and the key performance is shown in Table 5.20.

The developed anti-cracking LUHPC has been successfully applied to the bridge deck pavement of a steel box girder bridge on Xiongchu Avenue in Wuhan (span of 60 m, width of 26 m) (see Figure 5.72). The bridge deck pavement project utilized a

Table 5.19 Mix proportions for pumpable lightweight ultra-high-performance concrete for steel bridge deck pavement.

Cement	Fly ash microsphere	Silica fume	Expansive agent	Lightweight aggregate	Water	Steel fiber	Admixture
(kg/m³)							(%)
756	196	180	72	667	216	200	2.5

Table 5.20 Key performance indicators of pumpable lightweight ultra-high-performance concrete.

Slump (mm)	Spread (mm)	T_{500}(s)	Compressive strength after 28 d (MPa)	Flexural strength after 28 d (MPa)	Elasticity modulus 28 d (GPa)	Shrinkage after 56 d (10^{-6})
260	620	8	115.7	16.8	41.2	320

Figure 5.72 Construction effect image of pumpable lightweight ultra-high-performance concrete.

combination of lightweight UHPC and SMA-13 asphalt concrete. The construction plan included welding shear studs on the steel top plate (with a spacing of 30 cm), tying a steel mesh (with a steel bar diameter of ϕ10 and a grid spacing of 10 × 10 cm), and casting 5 cm of lightweight UHPC. Above this, a high-adhesion, high-elasticity asphalt waterproof bonding layer was applied (the amount of asphalt is 1.6–2 kg/m^2), followed by a surface paving of 4-cm-thick SMA-13 high-adhesion, high-elasticity asphalt concrete. The use of lightweight UHPC significantly reduces the self-weight of the bridge, enhancing its load-bearing capacity and durability. The cost calculation shows that the construction cost per square meter of the bridge deck is 600 yuan, offering significant economic benefits.

References

1. De Larrard, F. and Sedran, T. (1994). Optimization of ultra-high performance concrete by the use of a packing model. *Cement and Concrete Research* 24 (6): 997–1009.
2. Chen, B., Ji, T., Huang, Q. et al. (2014). Review of research on ultra-high performance concrete. *Journal of Architecture and Civil Engineering* 31 (3): 1–24. (in Chinese).
3. Peng, Y. (2009). *Study on Composition, Structure and Properties of Reactive Powder Concrete Containing Steel Slag Powder*. Wuhan: Wuhan University of Technology (in Chinese).
4. Shen, P. (2018). *Investigation on Design of Expansive UHPC Filled Steel Tube and Its Mechanical Behavior of ShortColumn*. Wuhan: Wuhan University of Technology (in Chinese).

5 Peng, Y., Chen, K., and Hu, S. (2011). Influence of the characteristics of steel slag powder particle on compressive strength of reactive powder concrete. *Journal of Building Materials* 14 (4): 541–545. (in Chinese).

6 Zhang, X., Zhao, S., Liu, Z. et al. (2019). Utilization of steel slag in ultra-high performance concrete with enhanced eco-friendliness. *Construction and Building Materials* 214: 28–36.

7 Peng, Y., Chen, K., and Hu, S. (2011). Study on interfacial properties of ultra-high performance concrete containing steel slag powder and fly ash. *Advanced Materials Research* 194–196: 956–960.

8 Hu, S. (2009). *Advanced Cement-Based Composite Materials.* Beijing: Science Press (in Chinese).

9 Wu, J., Wang, F., Hu, S. et al. (2013). Effect of aggregate-cement paste interface on damage and fracture of concrete. *Journal of Beijing University of Technology* 39 (6): 892–896. (in Chinese).

10 Ding, Q., Yan, P., Hu, S. et al. (2018, April 20). A lightweight low shrinkage ultra-high performance concrete and its preparation method. CN 201711233051. (in Chinese).

11 Ding, Q., Hu, J., Liu, Y. et al. (2019). Research on preparation and performance of lightweight ultra high performance concrete. *Concrete* 9: 1–5. (in Chinese).

12 Shen, P., Lu, L., He, Y. et al. (2018). Experimental investigation on the autogenous shrinkage of steam cured ultra-high performance concrete. *Construction and Building Materials* 162: 512–522.

13 Shen, P., Lu, L., He, Y. et al. (2020). Investigation on expansion effect of the expansive agents in ultra-high performance concrete. *Cement and Concrete Composites* 105: 103425.

14 Shen, P., Lu, L., Wang, F. et al. (2020). Water desorption characteristics of saturated lightweight fine aggregate in ultra-high performance concrete. *Cement and Concrete Composites* 106: 103456.

15 Wang, F., Yang, J., Hu, S. et al. (2016). Influence of superabsorbent polymers on the surrounding cement paste. *Cement and Concrete Research* 81: 112–121.

16 Wang, F., Zhou, Y., Peng, B. et al. (2009). Autogenous shrinkage of concrete with super-absorbent polymer. *ACI Materials Journal* 16 (2): 123–127.

17 Hu, S., Zhou, Y., Wang, F. et al. (2008). Effect of superabsorbent resin particles on self-shrinking and strength of concrete. *Journal of Huazhong University of Science and Technology (Urban Science Edition)* (1): 5. (in Chinese).

18 Zhou, Y. (2008). *Research on Internal Curing Mechanism and Control Technology of High Strength Concrete.* Wuhan: Wuhan University of Technology (in Chinese).

19 Box, P. and Wilson, B. (1951). On the experimental attainment of optimum conditions. *Journal of the Royal Statistical Society: Series B: Methodological* 13 (1): 1–45.

20 Khuri, A. and Mukhopadhyay, S. (2010). Response surface methodology. *Wiley Interdisciplinary Reviews Computational Statistics* 2 (2): 128–149.

21 Mao, R. (2017). *Study on Microstructure and Mechanical Properties of Reactive Powder Concrete.* Wuhan: Wuhan University of Technology (in Chinese).

22 Ding, Q., Liu, X., Liu, Y. et al. (2016). Effect of Curing Regimes on Microstructure of Ultra High Performance Concrete Cement Pastes. 1st International Conference on UHPC Materials and Structures. 371–379.

23 Shen, P., Lu, L., He, Y. et al. (2019). The effect of curing regimes on the mechanical properties, nano-mechanical properties and microstructure of ultra-high performance concrete. *Cement and Concrete Research* 118: 1–13.

24 Baby, F. and Graybeal, B.A. (2013). Development of direct tension test method for ultra-high- performance fiber-reinforced concrete. *ACI Materials Journal* 110 (2): 177–186.

25 Hu, C. and Li, Z. (2015). A review on the mechanical properties of cement-based materials measured by nanoindentation. *Construction and Building Materials* 90: 80–90.

26 Seddon, I. (1965). The relation between load and penetration in the axisymmetric boussinesq problem for a punch of arbitrary profile. *International Journal of Engineering Science* 3 (1): 47–57.

27 Constantinides, G. and Ulm, F.J. (2007). The nanogranular nature of C–S–H. *Journal of the Mechanics and Physics of Solids* 55 (1): 64–90.

28 Zhang, X., Liu, Z., and Wang, F. (2019). Autogenous shrinkage behavior of ultra-high performance concrete. *Construction and Building Materials* 226: 459–468.

29 Mangat, S. and Motamedi, M. (1984). A theory for the free shrinkage of steel fibre reinforced cement matrices. *Journal of Materials Science* 19 (7): 2183–2194.

30 Wille, K., Tue, N.V., and Parra-Montesinos, G.J. (2014). Fiber distribution and orientation in UHP-FRC beams and their effect on backward analysis. *Materials and Structures* 47 (11): 1825–1838.

31 Hu, S., He, Y., and Lü, L. (2006). Water-releasing factor technology to modify the internal relative humidity of concrete and its application. *Journal of Railway Science and Engineering* 3 (2): 11–14.

32 Hu, S. and Wang, F. (2006). *Light Aggregate Concrete.* Beijing: Chemical Industry Press (in Chinese).

33 Jensen, O.M. and Hansen, P.F. (2001). Water-entrained cement-based materials I. Principles and theoretical background. *Cement and Concrete Research* 31 (4): 647–654.

34 Liu, X. (2015). *Effect of Curing Regimes and Sulphate Attack on the Properties and Topics of Microstructure of Ultra High Performance Concrete.* Wuhan: Wuhan University of Technology (in Chinese).

35 Ding, Q., Liu, Y., Liu, X. et al. (2018). Effects of sulfate attack on microstructure of UHPC hydration products under different curing regimes. *Bulletin of the Chinese Ceramic Society* 37 (3): 772–780. (in Chinese).

36 Hu, J. (2019). *Research on Lightweight Ultra-High Performance Concrete.* Wuhan: Wuhan University of Technology (in Chinese).

6

High-performance Composite Structural Concrete

6.1 Concrete-filled Steel Tubes Composite Material

6.1.1 Overview

6.1.1.1 New Concrete-filled Steel Tubes Composite Design Principle

Concrete-filled steel tubes are a composite materials made by pouring concrete into steel tubes. The author's team innovatively proposed the use of high-strength concrete-filled steel tubes as the component material, which better combines the three roles of high strength, expansiveness, and steel tube constraints to form a new type of concrete-filled steel tube composite material. The steel tubes provide three-dimensional confinement to the core concrete, enabling the hooping effect to occur before loading. This approach not only addresses the deficiency of delayed hooping force in conventional concrete-filled steel tubes but also prevents structural damage caused by subsequent volume expansion of the core concrete. Due to the hooping effect of the steel tubes, the exchange effect between the core concrete and external moisture and harmful substances is cut off, which fundamentally improves the durability of high-strength concrete.

Compared with traditional reinforced concrete structures, concrete-filled steel tubes and concrete composite structures are characterized by high load-bearing capacity, good toughness, good fire resistance, improved durability, ease of fabrication and construction, and good economic efficiency. In the case of the same bearing capacity conditions, the use of concrete-filled steel tubes saves 50% of concrete than ordinary reinforced concrete structures and reduces the structural weight by 50%; compared with the all-steel structure, it can save more than 50% of steel, and the expansive high-strength concrete-filled tubes have a better technical economy. China has a serious shortage of iron ore resources, while limestone and other resources are relatively abundant in the country, making the use of concrete-filled steel tubes instead of all-steel structures highly significant for future development and application prospects.

6.1.1.2 Major Problems with the Established Technologies

The key technology of the expansive high-strength concrete-filled steel tube composite material is the precise design of the properties of filled concrete and its

High-Tech Concrete Materials: Design, Preparation, and Applications, First Edition.
Fazhou Wang and Shuguang Hu.

performance matching with the structural composite. According to the principle of composite steel-filled steel tubes and concrete materials, the core of the technology is that concrete should be closely combined with the wall of the steel tubes to achieve the hooping effect, so that this combination of materials can achieve the composite effect and ensure long-term structural safety. There is a degree of inherent volumetric shrinkage in ordinary concrete after hardening, so in order to ensure a tight bond between the core concrete and steel tube wall, the core concrete must be designed to counteract the volumetric shrinkage of filled concrete and preferably produce a micro-expansion property.

Major problems with established technologies include the following: The first problem is the lack of design method of concrete expansion properties, where the expansion induced by the expansion agent is not coordinated with the development of shrinkage of the core concrete and its strength growth to produce effective expansion stresses. Second, the concrete-filled steel tubes lacked moisture supply inside the closed steel tubes, and the expansion agent could not react continuously at a later stage, making it difficult to achieve precise control of expansion properties. Third, in large-scale structural engineering concrete preparation and construction, the large volume of concrete filling leads to several issues during high-pressure pumping: coarse aggregate and mortar separation occurs, aggregate accumulation and blockage phenomena develop in the pumping pipes, in addition to serious segregation. This causes severe separation in the concrete-filled steel tubes. As the quality of concrete expansion is not guaranteed, the composite performance design of concrete-filled steel tubes cannot be realized, which is easy to cause the concrete and steel tube wall debonding to form a cavity, and the modulus of elasticity and load-bearing capacity of concrete-filled steel tubes is greatly reduced, resulting in potential safety hazards to the project.

In the past, the research on concrete-filled steel tubes was mainly focused on structural design, while there was little research on the core concrete of steel tubes, which lacked a systematic and in-depth understanding of the expansion properties of concrete and its strength development law, not to mention the precise quantitative design and controllable level. Due to the lack of systematic theoretical guidance in the performance design and construction of concrete-filled steel tubes, trial-and-error method was adopted in actual project, which is not only a waste of materials but also cannot guarantee the quality of concrete and engineering, and the above-discussed problems in the existing technology have seriously restricted the application of concrete-filled steel tubes in the structural engineering of concrete-filled concrete composite materials.

6.1.1.3 Major Research and Technology Development

To solve the problems discussed above, several key technologies must be broken through: synergy of expansion and shrinkage and strength development, continuous and stable expansion of concrete under confinement, design and controllability of concrete-filled steel tube wall expansion stresses, and preparation of expansion concrete and its construction techniques and technology. The author's team carried out the following research and development work and achieved innovative results:

(1) The shrinkage mechanism of concrete under confined conditions has been investigated, and an expansion model for high-strength concrete filled steel tubes has been established, providing theoretical guidance for concrete expansion performance design [1–5]. Through the systematic analysis of the expansion source, the principle of material composite design, optimal matching of the expansion composition, research and development of expansion materials, and their preparation methods suitable for the expansion characteristics of concrete were developed [6–9].
(2) A internal curing mechanism and technology for concrete material systems have been developed to address the key issues of sustained and stable development of concrete hydration and expansion in later stages [10, 11]. Comprehensive use of new composite expansion materials and a system of self-supply of water technology, optimization of the matching of cementitious materials, and expansion of the composition, to promote the coordinated development of concrete expansion and strength, research and development of concrete-filled steel tubes, and concrete expansion performance of the continued stability of the precise regulation of the technology [12–16].
(3) For the construction technology characteristics of expansive high-strength concrete-filled steel tubes, the research and development of high-efficiency composite additives for expansive concrete-filled tubes adapted to the construction of large-span structures (bridges) [17, 18]. Key construction process parameters and key technologies for the preparation of high-quality, high-strength expansive concrete were developed, which promoted the large-scale popularization and application of the resulting technologies [19–23].

6.1.2 Design and Control of Concrete-filled Steel Tube Expansion Properties

The technical core of concrete-filled steel tubes is the composite effect of concrete and steel tubes, and the concrete in the steel tube should maintain a certain expansion stress to achieve the hooping effect. Concrete shrinks in volume after hardening, so concrete expansion techniques must be used to compensate for the volumetric shrinkage of concrete and to generate a certain amount of expansive self-induced stresses. However, at the same time, the expansion stress needs to be controlled from the structural safety point of view. Therefore, it is critical to study and address the design and control of concrete expansion properties. The design of concrete-filled steel tube expansion properties and the expansion characteristics and laws of concrete should be explored to study the shrinkage characteristics of concrete, compensating shrinkage mechanism, and expansion model.

6.1.2.1 Concrete-filled Steel Tube Expansion Modeling

Compensatory shrinkage modeling of core concrete under steel tube constraints is discussed below. In the whole process of volume deformation of expansion concrete, there are two kinds of deformation at the same time: expansion and shrinkage [24, 25]. Expansion deformation consists mainly of expansion ε_1, which is an increase in volume due to the formation of crystals of calcium alumina, calcium hydroxide, or magnesium hydroxide in the hardened cement paste. The total shrinkage S_{2m} consists of unconfined shrinkage $S_t + S_a + S_p$ and confined shrinkage $S_e + S_c$. During the whole process of deformation, the final value of deformation ε_{2m} of expansive concrete is

$$\varepsilon_{2m} = \varepsilon_1 - S_{2m} = \varepsilon_1 - [(S_t + S_a + S_p) + (S_e + S_c)] \tag{6.1}$$

where ε_1 is the expansion of expansive concrete, S_{2m} is the total expansion concrete shrinkage, $S_t + S_a + S_p$ is the unconfined shrinkage of expansive concrete, and $S_e + S_c$ is the constraint-induced shrinkage of expansive concrete.

For core concrete in rigidly constrained conditions, the final deformation value $\varepsilon_{2m} \leq 0$ ($\varepsilon_{2m} < 0$ indicates contraction). The expression for the final deformation value ε_{2m} during the whole process of deformation of concrete-filled steel tube constraints is the same as in the case of rigid constraints.

Figures 6.1 and 6.2 show the deformation curves of volumetric deformation vs. time for concrete-filled tubes with different expansion energies under the concrete-filled steel tube constraints – expansive concrete compensatory shrinkage model.

The curves in the Figures 6.1 and 6.2 are as follows: (1) expansion of expansive concrete ε_1, (2) total contraction of expansive concrete S_{2m}, (3) unconstrained contraction of expansive concrete $S_t + S_a + S_p$, (4) constraint-induced contraction of expansive concrete $S_e + S_c$, and (5) total deformation of expansive concrete ε_{2m}.

Under steel tube confinement conditions, the expansion curve ε_1 variation patterns of expansive concrete with larger expansion energy and smaller expansion energy are similar. Both generate significant expansion deformation from the beginning of hardening, with values gradually increasing over time and eventually leveling off. However, the magnitudes differ, with the former being greater than the latter, as shown in curve 1.

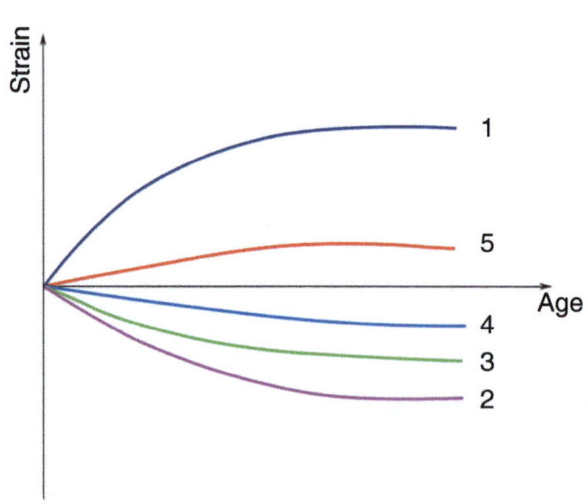

Figure 6.1 Compensatory shrinkage modeling of concrete-filled steel tubes with large expansion energy under steel tube constraints.

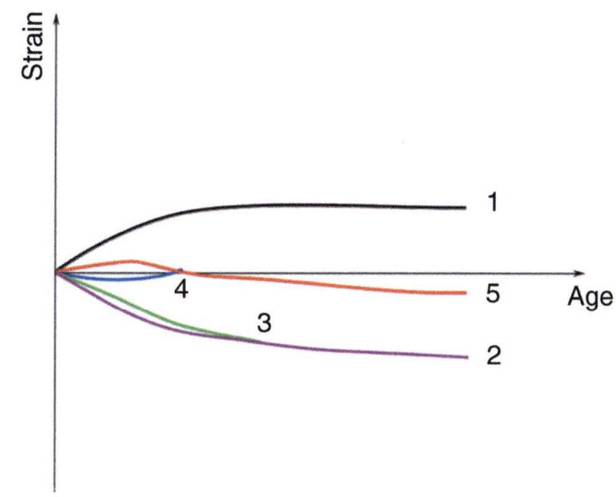

Figure 6.2 Compensatory shrinkage modeling of concrete-filled steel tubes with small expansion energy under steel tube constraints.

The unconfined shrinkage $S_t + S_a + S_p$ of expansive concrete, which consists of the superposition of concrete thermal shrinkage S_t, autogenous shrinkage S_a, and plastic shrinkage S_p, produces a large shrinkage deformation from the beginning of hardening for both expansive concrete with large and small expansion energies, and thereafter increases gradually with the prolongation of time and eventually tends to stabilize, as shown in curve 3.

Constrained shrinkage curve $S_e + S_c$ is caused by elastic shrinkage S_e and creep shrinkage S_c. For expansive concrete with large expansion energy, the resulting expansion not only compensates for the unconfined shrinkage of the concrete but also produces a certain number of self-induced stresses, and self-induced stresses with the extension of time gradually increase and tend to stabilize, so the resulting shrinkage $S_e + S_c$ with the extension of time also gradually increases and tends to stabilize. As for the expansion concrete with smaller expansion energy, the expansion generated in the early hydration compensates the unconfined shrinkage of the concrete and generates a certain number of self-induced stresses, so there exists the resulting shrinkage $S_e + S_c$. However, due to the smaller expansion energy and relatively early expansion stabilization time, the unconfined shrinkage of the concrete such as thermal shrinkage and self-induced stresses are still increasing. Thus, the expansion generated in the middle and late stages of hydration is not sufficient to compensate for the unconfined shrinkage, the core concrete is de-voided from the steel tubes, and the confinement-induced shrinkage $S_e + S_c$ ceases to exist with a value of 0. This is shown by curve 4.

For the total shrinkage S_{2m} of expansive concrete, expansive concrete with larger expansion energy exhibits significant shrinkage deformation from the beginning of hardening due to both unrestrained shrinkage and restraint-induced shrinkage. This deformation gradually increases over time and eventually stabilizes, so the total shrinkage S_{2m} of expansive concrete essentially follows the same pattern. For expansion concrete with small expansion energy, the shrinkage due to confinement, $S_e + S_c$, is present in early hydration but not in mid- and late hydration, so its total shrinkage is the same as that of unconfined shrinkage in mid- and late hydration, while it is slightly larger than that of unconfined shrinkage in early hydration, as shown in curve 2.

For the total deformation ε_{2m} of expansive concrete, expansive concrete with larger expansion energy generates expansion that not only compensates for the total shrinkage of core concrete but also produces certain expansion deformation. With extended age, the total deformation gradually increases and eventually stabilizes. For expansion concrete with small expansion energy, the expansion produced in the early hydration compensates for the shrinkage and still has some expansion deformation, but in the middle and late hydration due to the small expansion energy and relatively early expansion stabilization time, the expansion produced is not enough to compensate for the shrinkage, so it is reflected as shrinkage of expansion concrete, as shown in curve 5.

This model tells us that a new type of expansive material with larger expansion energy and a longer expansion stability period must be developed to replace the traditional expansive material to establish a suitable late expansion in the core concrete, to effectively solve the problem of de-bonding between the core concrete and the wall of the steel tubes, and to establish a higher expansive self-induced stress in the core concrete and improve the load-carrying capacity. At the same time, it makes the core concrete produce a tightening force before loading, which can make up for the defect that the tightening force of ordinary concrete-filled steel tubes appears too late, thus improving the working performance of the composite materials.

6.1.2.2 Design of Expansive Self-induced Stress Values for Concrete-filled Steel Tubes

The relationship between the self-stress σ_c of concrete-filled steel tubes with core and concrete expansion is given by the following equation:

$$\sigma_c = E_c(\varepsilon_0 - \varepsilon_R - \varepsilon_A) \tag{6.2}$$

where E_c is the modulus of elasticity of concrete, ε_0 is the free expansion rate of concrete, ε_R is the expansion of concrete-filled steel tubes under constrained conditions, and ε_A is the corrected value for concrete expansion and $\varepsilon_A = 1 - \sqrt{1-q}$. q is related to the air content of the concrete and the concrete pumping pressure. When the air content of concrete is 1.5% and the pumping pressure is 15 MPa, $\varepsilon_A = 1 \times 10^{-4}$.

The relationship between the free expansion of concrete-filled steel tubes core ε_0 and the restrained expansion of concrete under steel tube constraints ε_R is given by the following equation:

$$\varepsilon_R = \frac{KE_c}{KE_c + E_S}(\varepsilon_0 - \varepsilon_A) \tag{6.3}$$

where K is R/t, R is the radius of the steel pipe, t is the wall thickness of the tube, and E_s is the modulus of elasticity of the steel tube.

It is obtained by extrapolation based on the above formula:

$$\sigma_c = E_c\left(1 - \frac{KE_c}{KE_c + E_s}\right)(\varepsilon_0 - \varepsilon_A) \tag{6.4}$$

Given the relationship between the design value of self-stress for concrete-filled steel tubes and the free expansion of the concrete, the modulus of elasticity and air content of the concrete are determined, and after selecting the construction pumping pressure of the concrete, a value of ε_A can be obtained and the ε_0 of the concrete can be determined.

6.1.2.3 Concrete Expansion Performance Design and Expansion Fitting Techniques

The volume deformation characteristics and ideal expansion model of concrete-filled steel tubes are systematically studied, the composite design idea of concrete expansion performance is put forward, the high-energy composite expansion material and expansion performance fitting design method are invented, and the fine control technology of concrete expansion is created.

(1) Design of concrete-filled steel tubes with expansion properties
Studies have shown that the main factors in the shrinkage of concrete-filled steel tubes are autogenous shrinkage, thermal shrinkage, plastic shrinkage, elastic shrinkage, and creep shrinkage. The design introduces expansion components to compensate for concrete shrinkage and produce expansion. Under the concrete-filled steel tube constraint condition, both shrinkage and expansion deformation of concrete exist at the same time, and the concrete deformation model is obtained analytically by substituting Eq. (6.1).
According to the deformation characteristics of concrete-filled steel tubes and performance requirements of the combined materials, Figure 6.1 is taken as the ideal design expansion model, and the design principle of high-energy delayed expansion composition is proposed. The main problems of the existing expansion technology are as follows: the U-expansion class has a low early expansion rate and small expansion amount; the sulfoaluminate class has a fast hydration rate, which has an influence on the early structure and strength of hardened cement paste; and the oxide class has an unstable expansion performance. The idea of expansion composite design is proposed here: the components with early, middle, and late expansion characteristics are selected to be calcined RO minerals (CaO and/or MgO), sulfoaluminate (representing the composition $3(CaOAl_2O_3){\cdot}CaSO_4$, i.e., $C_4A_3\overline{S}$), and RO solid fusion phases (containing CaO, MgO, and FeO overcooked solid fusion phases, respectively), as the main ones [9].
(2) Expansion performance fitting techniques
The expansion components corresponding to the above were taken and tested to measure the free expansion rate $\varepsilon_{01} = f_1(t)$, $\varepsilon_{02} = f_2(t)$, and $\varepsilon_{03} = f_3(t)$ of the concrete prepared under the construction set conditions in the closed state. The composite expansion characteristic $\varepsilon = \varepsilon_1 + \varepsilon_2 + \varepsilon_3 = k_1 f_1(t) + k_2 f_2(t) + k_3 f_3(t)$ is obtained by fitting, as shown in Figure 6.3, and the composite expansion component (HBEA) is formulated accordingly. The fitted ratios for the experiments in Table 6.1 are $k_1 : k_2 : k_3 : \overline{S} = 20\% : 37\% : 15\% : 28\%$. Their comparative results show that the fitted expansion concrete has good design strength and stable late expansion.

6.1.2.4 Precise Control Technology for Concrete Expansion

(1) Calculation of expansion properties

Research has been conducted to understand the mechanism of coordinated development between concrete strength and expansion performance, achieving precise expansion control through hydration control. The specific steps are as follows: (a) From the concrete strength (f_c) and expansion stress (σ_c) designed by the

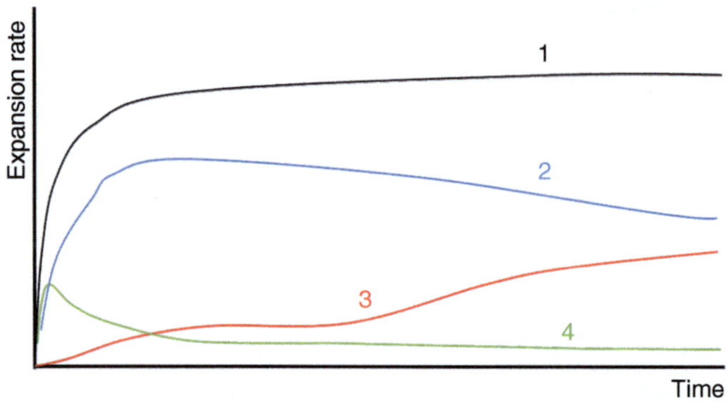

Figure 6.3 Fitting design for expansion composite properties.

Table 6.1 HBEA concrete performance test results.

Cement (kg/m^3)	Dosage (kg/m^3)	Water-binder ratio	Sand rate (%)	28 d strength (MPa)	Limiting the rate of expansion/10^{-4}	
					14 d	90 d
460	0	0.39	40	52.1	−3.4	−5.2
414	46	0.39	40	52.7	3.6	2.8
377	83	0.39	40	52.9	4.5	3.5
520	0	0.33	39	59.2	−3.8	−5.0
468	52	0.33	39	61.1	4.0	3.0
426	94	0.33	39	58.4	4.3	3.4

structural requirements, the expansion rate (ε_0) is calculated and the expansion characteristic (ε) is set. The relationship between the design value of self-stress for concrete-filled steel tubes (σ_c) and free expansion rate of concrete (ε_0) is shown in Eq. (6.4). (b) Fitting techniques were used to determine the expansion composition (HBEA) and its raw material ratio ($k_1 : k_2 : k_3$) based on the expansion characteristics (ε). (c) Control is achieved by a combination of the following factors: cement properties, expansion composition, raw material pretreatment, particle grading, and water release factor.

(2) Precise control of expansion properties

(1) Concrete expansive self-induced stress setting
Set the range of expansive self-induced stresses of concrete according to the design requirements, generally taking the value of 1.0–3.0 MPa.

(2) Concrete raw materials selection
According to the design requirements and engineering construction conditions, select a variety of raw materials for the preparation of concrete – a. cement: selection of cement types, mineral composition of clinker, and fineness of the product; b. admixtures: types, varieties, mixing ratios, active ingredients, and fineness of the product; c. coarse and fine aggregates: modulus of fine aggregates and strength and gradation of coarse aggregates; d. chemical admixtures: use of special WUT-G admixtures; and e. water-release factors: use of developed water-release factors.

(3) Concrete performance test
a. Strength: according to the design requirements of the concrete strength test mix, so that it reaches the strength index. b. Workability: according to the project construction conditions, with additives to adjust the workability of the concrete. c. Air content: measure the air content of the concrete; the expansion rate of the concrete needs to be calculated for this parameter. d. Modulus of elasticity: measure the elasticity of the concrete; concrete self-stress needs to be calculated for this parameter.

(4) Calculate the expansion of concrete
Eq. (6.4) gives the relationship between the design value of concrete self-induced stresses in concrete-filled steel tubes and free expansion rate of the concrete and determines the modulus of elasticity of the concrete and air content, and after selecting the construction pumping pressure of the concrete, the value of ε_A can be obtained and the ε_0 of the concrete can be determined. According to the expansion rate of the concrete, we get the characteristic curve of the expansion with change in time: $\varepsilon_0 = f(t)$.

(5) Composite matching of expansion components
a. Set the expansion components with early, middle, and late expansion characteristics selected in this technology as BEA_1, BEA_2, and BEA_3, respectively, and measure them in a closed state; according to the construction of the set condition parameters for the preparation of expansion of concrete free expansion rate, $\varepsilon_{01} = f_1(t)$, $\varepsilon_{02} = f_2(t)$, and $\varepsilon_{03} = f_3(t)$. Set the release factor of the baseline mixing to 15% of the amount of sand. b. Mathematical fitting of $\varepsilon_{01} = f_1(t)$, $\varepsilon_{02} = f_2(t)$, and $\varepsilon_{03} = f_3(t)$ is performed to obtain composite expansion characteristic curves:

$$\varepsilon_0 = \varepsilon_{01} + \varepsilon_{02} + \varepsilon_{03} = f_1(t) + f_2(t) + f_3(t) \tag{6.5}$$

c. Adjustment of physical property parameters and proportions of expansion components: Through adjustment of the physical property parameters of expansion components, calcination temperature and time of calcined RO minerals, $C_4A_3\overline{S}$ clinker composition and content, and RO solid-melt phase composition and fineness. In addition, the proportion of matching between the three, physical parameters of water-release factor, and dosage are adjusted to meet the fitting equation to determine the BEA_1, BEA_2, and BEA_3 ratio, to obtain the composite high-energy composite expansion material HBEA composition.

Table 6.2 C50 concrete-filled steel tube mixing ratio.

Cement	Water	Fly ash	Aggregates	Sand	Water-reducing agent	HBEA	Water release factor	Initial slump	3 h slump
(kg/m^3)								(mm)	
460	185	70	1050	525	8.5	66	77	220	200

(6) Verification test

a. According to the above high-energy composite expansion material HBEA composition, conduct a concrete-filled steel tube expansion performance test. b. Adjust and modify the physical property parameters, composition, and proportions of related materials c. Determine the HBEA ratio of the high-energy composite expansion material, the concrete ratio as shown in Table 6.2, and relevant construction process parameters. Test raw materials: cement: 42.5 ordinary portland cement; expansion components: BEA_1 (RO mineral calcination temperature 950–1050 °C), BEA_2 $m(C_4A_3\overline{S}):m(C_4AF):m(C_2S) = 65\%:10\%:20\%$, and BEA_3 (RO solid-melt phase $_{Km}(MgO/(FeO{+}MnO)){>}1$); and specific surface area: 450–500 m^2/kg. Water release factor: particle size in the range of 2–7.5 mm, 24 h water absorption rate of 15%–20%. Admixture: class fly ash. Aggregate: coarse aggregate is 5–20 mm continuous gradation interval particle size of crushed stone, crushing value <10%, and needle and flake content of <10%. Fine aggregate: sand with a fineness modulus of 2.3–3.1. Water-reducing agent: polycarboxylic acid high-efficiency water-reducing agent with a reduction rate of 28–30%.

Table 6.3 shows the experimental results. It can be seen that the admixture of HBEA can basically meet the design requirements in the range of 5–16%, and the optimal range is within 8%–12%; at this time, the strength of the concrete is higher, and the value of the self-stress of the concrete is approximately between 1 and 2.5 MPa. When the dosage of the expansion component is too high, the early expansion reaction of the expansion component is too much, too fast on the production of hydration products of cement, resulting in poor development of the microstructure of hardened cement paste, resulting in a decline in strength.

Table 6.3 Effect of HBEA admixture on mechanical and expansion properties of concrete.

Samples	A	B	C	D	E	F	G	H	I
HBEA dosage (%)	0	5	8	10	12	14	16	18	20
Free expansion rate/10^{-6}	−120	95	189	302	400	486	577	712	816
Expansion stress (MPa)	—	0	0.76	1.58	2.34	3.09	3.83	4.65	5.40
Concrete strength (MPa)	58.2	60.8	62.6	61.7	60.3	59.3	57.4	54.1	47.4

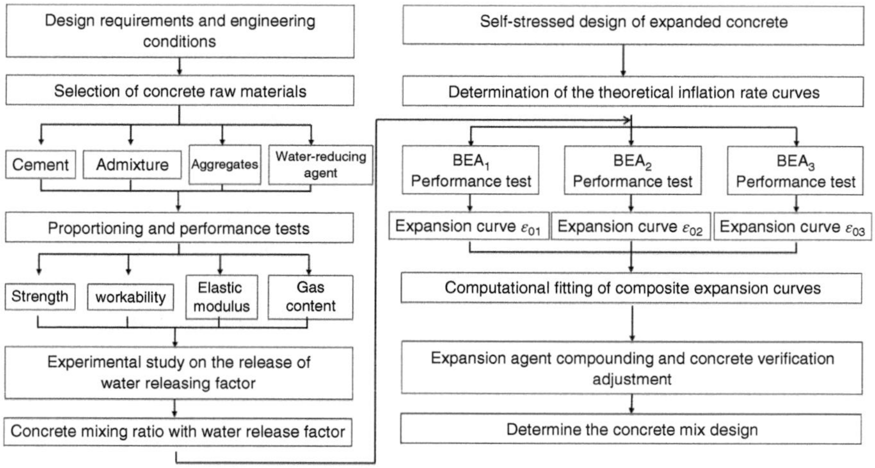

Figure 6.4 Flowchart for precise control of concrete expansion.

(3) Expansion performance control process

Using the principles of materials science and technology, adopting the research and technical achievements already made, and according to the characteristics of materials preparation and construction, the following concrete expansion precise control technology program is designed, as shown in Figure 6.4.

6.1.3 Design of Cementitious Material Composition Matching

6.1.3.1 Composition of Cementitious Material and Shrinkage Deformation

The type and mineral composition of cement directly affects the shrinkage characteristics of concrete-filled steel tubes, which include autogenous shrinkage and chemical shrinkage. Chemical shrinkage represents the volume of the products of cement hydration less than the sum of the volumes of cement and water involved in the reaction, and autogenous shrinkage represents the macroscopic volume shrinkage of the hardened cement paste at constant temperature and in the absence of humidity migration. Chemical shrinkage can be calculated on the basis of the relevant parameters of the reactants and products, and the process in which chemical shrinkage occurs does not necessarily manifest itself as macroscopic volume contraction, but may also manifest itself as expansion. For example, in the hydration system of expanded cement or the hydration system of cement mixed with an expansion agent, the process of generating calcium alumina is a process accompanied by chemical contraction, but on the macro level, the hardened cement paste or concrete shows expansion deformation. Autogenous shrinkage and chemical shrinkage are not linearly proportional; there is no simple relationship between them, and most of the chemical shrinkage translates into micropores in the hardened hardened cement paste.

Cement clinker minerals have different shrinkage values. Calcium aluminate cement and early-strength cement early autogenous shrinkage value is relatively large, and the autogenous shrinkage value of slag cement after hydration for a period of time is relatively large; in cement minerals, when the C_3A and C_4AF content is higher, concrete shows greater autogenous shrinkage value, while the autogenous shrinkage value of the C_3S and C_2S mineral is smaller. The reason for this is closely related to the hydration rate, hydration degree, and water content of the hydration products of various mineral phases in cement.

Among the four main mineral phases of cement, C_3A has the fastest hydration rate, which is due to the fact that both Ca^{2+} and Al^{3+} are in an irregular coordination state in the C_3A crystals and, at the same time, there are large voids in the structure, which make it easy for OH^- ions to enter the interior of the crystal lattice, and thus, the hydration reaction activity is higher. In fact, the hydration reaction of C_3A in the cement hydration system can be represented by the following equations:

$$C_3A + 3C\bar{S}H_2 + 26H \rightarrow C_6A\bar{S}_3H_{32} \tag{6.6}$$

$$C_6A\bar{S}_3H_{32} + 2C_3A + 4H \rightarrow 3C_4A\bar{S}H_{12} \tag{6.7}$$

$$C_3A + CH + 12H \rightarrow C_4AH_{13} \tag{6.8}$$

In the ordinary portland cement system, when gypsum and $Ca(OH)_2$ are present at the same time, C_3A hydrates to form C_4AH_{13}, which then reacts with gypsum to form ettringite (AFt), and subsequently, after gypsum in the slurry is consumed, C_3A reacts with AFt to form monosulfate-type hydrated calcium aluminate hydrates (AFm). Tables 6.4–6.6 show data related to the volume change of the system for the reactions (6.6)–(6.8), respectively. From Tables 6.4–6.6, it is easy to find that the hydration reaction of C_3A in the abovementioned cement system has a different degree of absolute volume reduction.

Hydration of the other main mineral phases C_3S, C_2S, and C_4AF in cement clinker is also a volume reduction process, and their final chemical shrinkage values are shown in Table 6.7:

In addition to the variety and mineral composition of cement, the fineness of cement, particle grading, dosage, admixture, water-reducing agent, and water–binder ratio will have a certain effect on the shrinkage and deformation of concrete-filled steel tubes. For example, the increase in cement fineness, increase

Table 6.4 Volume changes before and after reaction in Eq. (6.6).

Mineralogical phase	Molar mass (g/mol)	Densities (g/cm^3)	Molar volume (cm^3/mol)
C_3A	270.190	3.001	90.033
$3C\bar{S}H_2$	516.509	2.320	222.633
26H	468.385	0.998	469.229
Total reactants	1255.084		781.896
$C_6A\bar{S}_3H_{32}$	1255.084	1.780	705.103
Total generators	1255.084		705.103

Source: Adapted from [3].

Table 6.5 Volume changes before and after reaction in Eq. (6.7) [3].

Mineralogical phase	Molar mass (g/mol)	Densities (g/cm^3)	Molar volume (cm^3/mol)
$C_6AS_3H_{32}$	1255.084	1.780	705.103
$2C_3A$	540.380	3.001	180.067
4H	72.059	0.998	72.189
Total reactants	1867.523		957.359
$3C_4ASH_{12}$	1867.523	1.990	938.454
Total generators	1867.523		938.454

Source: Adapted from [3].

Table 6.6 Volume change before and after reaction in Eq. (6.8).

Mineralogical phase	Molar mass (g/mol)	Densities (g/cm^3)	Molar volume (cm^3/mol)
C_3A	270.190	3.001	90.033
$Ca(OH)_2$	74.100	2.242	33.050
12H	216.177	0.998	216.567
Total reactants	560.467		339.650
C_4AH_{13}	560.467	2.046	273.970
Total generators	560.467		273.970

Source: Adapted from [3].

Table 6.7 Final chemical shrinkage values for C_3S, C_2S and C_4AF monazite hydration.

Mineralogical phase	Chemical shrinkage value (cm^3/g)
C_3S	0.0532
C_2S	0.0400
C_4AF	0.1113

Source: Adapted from [3].

in cement content, and decrease in the water–binder ratio will increase the early shrinkage value of concrete.

6.1.3.2 Cementitious Material Composition and Expansion Deformation

For the commonly used portland cement, the influence of cementitious material composition on the expansion and deformation of concrete-filled steel tubes is mainly reflected in its influence on the process, quantity, morphology, and structure of ettringite formation. Ettringite is one of the main products of portland cement hydration, accounting for about 7% of all hydration products, while in concrete-filled steel tubes, the content of ettringite in the hydration products can usually reach more than 25%. The formation of ettringite is determined by the content of Ca^{2+}, SO_4^{2-}, OH^-, and AlO_2^- in the liquid phase in the pore space of hardened cement paste, and its equilibrium constant is expressed by the following equation:

$$K_p = \frac{[3CaO \cdot Al_2O_3 \cdot 3CaSO_4 \cdot 32H_2O]}{[Ca^{2+}]^6 \left[SO_4^{2-}\right]^3 [OH^-]^4 \left[AlO_2^-\right]^2 [H_2O]^{29}} \tag{6.9}$$

From the above equation, it can be seen that the formation of ettringite is determined by the concentration product of these four ions. Tests have shown [4, 5] that Ca^{2+} concentration is a major factor in determining the formation of ettringite, and cements with a high C_3S content can produce a large amount of ettringite in the early stages of hydration, whereas belite cements with a high C_2S content do not form ettringite in the early stages of hydration. High Ca^{2+} concentration not only increases the amount of AFt generated but also affects its morphology. After the Ca^{2+} concentration exceeds the supersaturation level (pH 13.9), AFt is formed in a solid-phase reaction, and its morphology shows agglomerated and outwardly radiating needle-like crystals. Crystals of this form have a large specific surface area and are cross-extruded into each other, generating high expansion stresses. CaO, in addition to providing Ca^{2+} ions, more importantly provides OH^- ions to the structural units of the ettringite, thus favoring the formation of the ettringite nuclei. Ettringite is formed by three processes in tandem: $[Al(OH)_6]^{3-}$ octahedra, aluminum-oxygen octahedra alternating with calcium polyhedral to form calcium–aluminum polyhedral columns, and SO_4^{2-} entry into inter-column grooves. Among them, the slowest $[Al(OH)_6]^{3-}$ forming process is the controlling step for the formation of ettringite, and among the ions in the liquid phase, AlO_2^- is the determining factor for the rate of formation of ettringite, and the rate of dissolution of AlO_2^- and level of its concentration are the controlling steps for the rate of formation of ettringite. For example, CA forms ettringite at a significantly faster rate than C_3A due to the faster dissolution of AlO_2^- and the higher equilibrium concentration of AlO_2^- in CA.

In concrete-filled steel tubes, $CaSO_4$ and aluminum-phase substances required for the generation of ettringite are mainly brought in by the expansion agent, while $Ca(OH)_2$ is generated by the hydration of C_3S and C_2S in the cement clinker minerals. Factors affecting the generation rate and content of $Ca(OH)_2$ are likely to have an impact on the rate of generation of ettringite and amount of generation. It is now generally accepted that a lower OH^- concentration results in the formation of less ettringite, which is well crystallized and has a coarse, short crystal form, producing less expansion stress, while a higher OH^- concentration results in the formation of a large amount of acicular ettringite, which produces greater expansion stress. The change in $Ca(OH)_2$ content in the cement hydration system by the addition of mineral admixtures is multifaceted.

It has been shown that reactive mineral admixtures accelerate the hydration of mineral phases such as C_3S due to the dilution effect, which accelerates the rate of $Ca(OH)_2$ production in the early stages of hydration. However, on the other hand, the mineral admixtures undergo pozzolanic reaction with $Ca(OH)_2$ in the middle and later stages of cement hydration, which consumes $Ca(OH)_2$ and reduces its content, and the decrease in the proportion of cement clinker due to the incorporation of mineral admixtures also causes a corresponding decrease in the amount of $Ca(OH)_2$ generated. Practice shows that increasing the dosage of active cementitious materials such as fly ash and ground granulated blast furnace slag, or concrete admixtures, reduces the expansion rate of concrete. 10% of the fly ash can be made to limit the expansion rate of concrete, and the free expansion rate is significantly reduced; when the content of fly ash is increased to 30%, the expansion inhibition becomes more aggravated. Of course, the reduction in the expansion rate is related to the chemical composition and physical properties of the fly ash and expansion agent used, and the magnitude of inhibition will vary with the use of different fly ash and expansion agents. The limiting expansion of concrete in water may not be reduced in case of low Al_2O_3 content and high SO_3 content in cement and high Al_2O_3 content in admixtures. The effect of finely ground slag powder on expansion properties is similar to that of fly ash. The reason for this is related to the decrease in $Ca(OH)_2$ production due to the decrease in the proportion of cement clinker after the addition of reactive mineral admixtures and the consumption of part of the $Ca(OH)_2$ by the pozzolanic reaction, which leads to a decrease in alkalinity of the pore solution in the hardened cement paste.

6.1.3.3 Matching of Cementitious Material Composition

The clinker composition of the cement is closely related to the chemical shrinkage and autogenous shrinkage of concrete-filled steel tubes, as well as to the rate and amount of production of ettringite in the hydration system, morphology of ettringite, and resulting expansion deformation. Among the four main mineral phases of cement clinker, C_3S, C_2S, C_3A, and C_4AF, C_3A has the highest chemical shrinkage value, followed by C_4AF. For the autogenous shrinkage of hardened cement paste, when the content of C_3A and C_4AF is high, the formulated concrete shows larger autogenous shrinkage values, while the content of C_3S and C_2S mineral phases has less effect on autogenous shrinkage. Meanwhile, in the hydration system of composite cement with an expansion agent, C_3S and C_2S have an important effect on the generation rate and amount of ettringite, ettringite morphology, and magnitude of the expansion rate.

When preparing high-strength micro-expansive concrete-filled steel tubes, ordinary Portland cement or Portland cement with high C_3S and C_2S contents and relatively low C_3A and C_4AF contents should be selected. Slag cement, composite cement, or pozzolanic cement should not be used. Meanwhile, when adding active mineral admixtures such as fly ash and ground granulated blast furnace slag to concrete, their dosage should not exceed 30%. The coordinated development of concrete expansion and strength is the key to ensuring that concrete-filled steel tube expansion concrete has excellent structural and service properties. All factors that affect the coordinated progress of cement hydration reaction speed and expansive agent hydration reaction speed, and influence the coordinated formation of main hydration products C-S-H gel and AFt (for ettringite-type expansive agents), such as concrete expansive agent dosage, types and dosages of admixtures, cementitious material content, water–binder ratio, chemical admixtures, and concrete mix proportions, are related to whether the expansion and strength performance of concrete-filled steel tubes can develop in coordination.

The expansion of concrete increases with increasing mineral admixture, but the strength of concrete decreases with increasing content of mineral admixture especially when the content is too high. The effect of admixture on the coordination of expansion and strength properties of concrete, on the other hand, is related to the strength class of concrete [6–8]. Adding mineral admixtures to medium-strength expansive concrete generally reduces both early strength and concrete expansion performance. For high-strength expansive concrete, adding a certain amount of mineral admixtures, although reducing early strength somewhat, improves concrete expansion performance. It was analyzed that the early strength should be moderate; if the early strength is too low, it will cause too much expansion in the plastic state, leading to ineffective expansion and a reduction in the effective expansion, and if the early strength is too high, it may limit concrete expansion. Generally, C60 below the strength level of concrete-filled steel tubes in the content of 10%–20% is appropriate, but the maximum should not exceed 25%; the content in the C60 above the strength level of concrete-filled steel tubes can be slightly increased, but the maximum should not exceed 40%.

The effect of water–binder ratio on concrete strength is obvious and the law is very clear, i.e., an increase in the water–binder ratio decreases the strength of concrete, but its effect on the expansion properties of concrete and coordinated development of concrete expansion and strength is not. If the water–binder ratio is increased, on the one hand, the hydration reaction rate of the expansion agent may be accelerated due to sufficient water, generation of AFt may be increased, and expansion performance is improved by increasing the expansion rate. On the other hand, although the rate of hydration reaction of the cement clinker may be accelerated, and the rate of C-S-H gel production may also be accelerated, the increase in porosity of the concrete, which has to be filled with more AFt or consume expansion energy, results in a decrease in expansion properties. Therefore, when preparing concrete-filled steel tubes, it is not necessary to use too low a water–binder ratio in order to improve the expansion properties, and the water–binder ratio should still be designed for the concrete strength as the main consideration.

Particle size matching of expansion agent on the expansion properties is also very critical; it is not only related to the size of the concrete expansion energy but also related to controlling the expansion reaction rate of an important technical indicator. The study of adjustable expansion of high-performance expansion agent is an important technical route. If there are too many micronized powders in the expansion agent, the formation of expansion is accelerated, and it will form expansion products during the plastic state of concrete, which ineffectively consumes its expansion energy and reduces the expansion rate of the concrete in the early stage of hardening (7–14 d).

6.1.3.4 High-efficiency Composite Water-Reducing Agent for High-strength Concrete-filled Steel Tubes

The high-strength concrete-filled tubes have a low water–binder ratio, high cement fineness, high cementitious material dosage, and further deterioration of the concrete's workability with the addition of high-energy composite expansion materials. Contrarily, the construction of an concrete-filled steel tubes arch bridge requires long-distance pumping and jacking grouting for dense construction, with high resistance along the way, thus putting forward special requirements on the working performance and strength of concrete, and the general-purpose admixtures cannot meet such special requirements. First, concrete must have good pumpability and applicability to steel tube concrete construction processes. Initial setting time must satisfy the time required for concrete pumping and top-up, and slump loss of fresh concrete over extended periods should be minimal. Second, the concrete is required to have suitable compensatory shrinkage and certain self-stress properties. Third, the core concrete should have early-strength characteristics and suitable stiffness to ensure that the line of the arch rib is guaranteed during the sequential pouring of each concrete-filled steel tube according to the requirements of the bridge structure, especially the pouring of trussed arch ribs composed of multiple steel tubes.

Currently, naphthalene-based retarding water reducers are commonly used in concrete-filled steel tube (CFST) construction. However, these admixtures exhibit significant slump loss over time. Aminosulfonic acid-type water reducers tend to cause concrete bleeding. Unfavorably, both types of admixtures introduce considerable air entrainment, which can lead to the formation of gas films during CFST construction. These gas films consume the expansive energy of the concrete, thereby adversely affecting the bond between the concrete and the steel tube.

Aiming at the above-discussed main problems, the author's team developed a composite special water-reducing agent (WUT-G) with low air-entraining capacity and high water reduction rate dedicated to concrete-filled steel tubes with a high-energy delayed expansion component [17], whose technical scheme is to take advantage of the characteristics of polycarboxylic acid's high water reduction and low air-entraining capacity and to compound the delayed expansion component to retain plasticity. The average molecular weight of the polymer of polycarboxylic acid water-reducing agent is 11,000–15,000; meanwhile, in order to ensure the very small air-entraining quantity of WUT-G, the polyoxyethylene group (i.e., the EO addition number) in the molecular structure of the polycarboxylic acid used as one of the side chains is greater than 60, and the molecular ratio of carboxylate group to ester group is 1.9–2.1, in order to ensure the adaptability of WUT-G to cement. The polycarboxylate superplasticizer is compounded with other main components including sodium gluconate, tributyl phosphate, and zinc sulfate to form an aqueous solution with a solid content of 30–36%. The typical solid mass ratio is polycarboxylate superplasticizer : sodium gluconate : zinc sulfate = 87–94.5% : 5–10% : 0.5–3%. Within the dosage range, this admixture maintains concrete air content $\leq$ 1.5%.

Table 6.8 Properties of expansive concrete with WUT-G.

Water reduction rate (%)	Water secretion rate (%)	Air content (%)	Solidification time difference (min) Initial setting	Final setting	Compressive strength ratio (%) 7 d	28 d
≥30	≤95	≤1.5	≥600	≥720	≥128	≥116

Table 6.9 Results of C60 concrete-filled steel tubes implementation using WUT-G.

Cement: Fly ash: HBEA: Water: Sand: Gravel (kg/m^3)	WUT-G dosage (%)	Extensibility (cm)	Slump (cm) 0 h	1.5 h	3 h	5 h	28 d strength (MPa)	Air content (%)
450:60:60:175:660:1010	1.4	64	24.0	22.5	21.0	18.5	72.9	1.5
460:70:55:180:655:1050	1.4	62	23.0	22.0	20.5	18.0	76.0	1.4

The experimental results give 0.9%–1.3% of WUT-G admixture in C50 steel tube concrete and 1.2%–1.6% in C60 steel tube concrete, and their performance indexes are examined according to GB8076-1997. The prepared concrete has a high water reduction rate, good plasticity preservation, good adaptability to expansion, low air content, long and easy-control setting time, and good pumpability. The effects are shown in Tables 6.8 and 6.9.

The advantages of WUT-G are its low air-entraining capacity and high water-reducing rate, which can effectively improve the fluidity and workability of freshly mixed concrete-filled steel tubes, significantly improve the pumping performance of concrete-filled steel tubes, and reduce the air content of concrete. Experimental studies have shown that for steel-filled steel tube arch bridges with a diameter of 1 m and span of 200–300 m and construction pumping pressure of 10–20 MPa, when the air content of the core concrete is less than 1.5%, the amount of air adsorbed into the wall of the steel tubes accounts for 0.02%–0.05% of the volume of the concrete, and the air content of the core concrete is less than 1.5%. Calculated with a larger value of 0.05%, the annular gap formed on the wall of the steel tubes is 0.13 mm, and thus, the concrete-filled steel tubes with an expansion rate of 3×10^{-4} developed through the optimized design of the mix ratio can avoid the problem of debonding of the concrete-filled steel tubes with the wall of the steel tubes.

6.1.4 Engineering Application of Concrete-filled Steel Tubes in Large-span Arch Bridges

6.1.4.1 Application Overview

Since the mid-1990s, Wuhan University of Technology has carried out systematic and in-depth research work on the concrete material properties of concrete-filled steel tubes, forming a more complete system of performance design and preparation technology of high-strength concrete-filled tubes, which greatly promotes the rapid development of the application of this structural form in large and extra-large bridge structures. In recent years, more than 20 large-scale concrete-filled steel tube arch bridges with spans of more than 200 m have been built, and in 2013, the world's first span concrete-filled steel tube arch bridge (single span of 530 m), the First Yangtze River Bridge in Hejiang, Chongqing, was opened to traffic. This project successfully addressed the technical challenges commonly encountered in concrete-filled steel tube construction, including segregation and bleeding, pump blockage, and inadequate compaction at the arch crown. It achieved tight bonding between the core concrete and steel tubes, resulting in significant technical, economic, and social benefits. This achievement not only demonstrates that China's application of concrete-filled steel tubes technology in long-span bridges continues to maintain a world-leading position, but also marks that this technology has reached an even higher level of advancement.

Concrete Preparation and Construction Process

(1) Dense preparation technology of concrete-filled steel tubes for span arch bridges

Achieving dense concrete within steel tubes is crucial for realizing effective steel-concrete composite behavior and ensuring structural safety. Beyond the design and precise control of concrete expansion properties, it is essential to address the unique construction characteristics of long-span structures by resolving several key technical challenges. These include managing ultra-long setting times and their control mechanisms, ensuring compatibility between expansion components and set-controlling agents as well as plasticizers, and developing a comprehensive suite of supporting technologies for complex construction processes.

The high-strength concrete-filled tubes have a low water–binder ratio, high cement fineness, high content of cementitious material, and further deterioration of the concrete's workability after the addition of the combined expansion components. Concrete-filled steel tubes in large-span arch bridges need to be pumped over long distances and jacked up for dense construction, with high resistance along the way, and general-purpose admixtures are unable to meet such special requirements. Research and development of special pumping agent WUT-G adapted to high-energy delayed expansion components, formulated concrete with a high water reduction rate, good plasticity, adaptation to expand, low air content, long setting time, ease in controlling, and pumpable performance.

(2) Construction process and key technology of concrete preparation

On-site construction organization, equipment selection and arrangement, construction program and quality control, and other systematic research and testing developed for large-span concrete-filled steel tube arch bridge concrete construction of pre-wet clearing tube, grading pumping, top push suction, overdense, structural balance of the filling and other series of supporting key technologies effectively solved the construction of the key technical challenges [23, 26]. The mixing ratios and properties of the concrete-filled steel tubes and expansive concrete-filled tubes

Table 6.10 C50 concrete-filled steel tubes.

Cement	Water	Fly ash	Aggregates	Sand	WUT-G	HBEA	Water release factor	Initial slump	5 h slump
			(kg/m^3)					(mm)	
430	170	70	1050	565	8.5	66	77	220	200

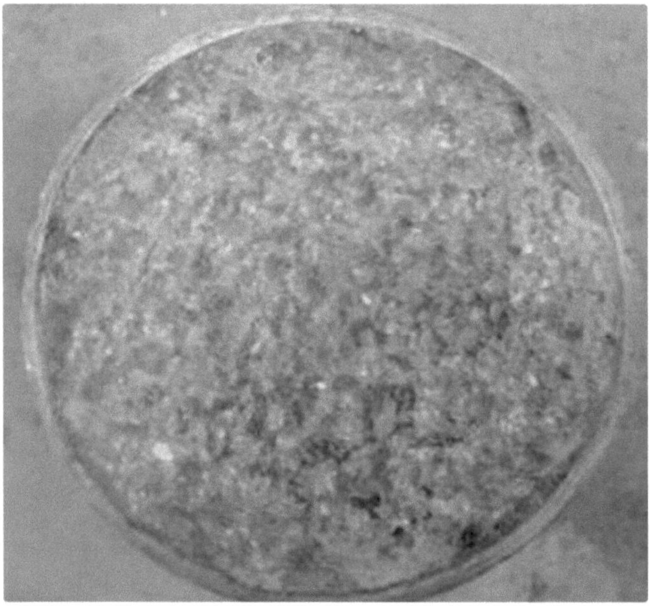

Figure 6.5 Tightly bound.

formulated using this technology are shown in Tables 6.10 and 6.3, and the quality testing results of the actual materials are shown in Figures 6.5 and 6.6.

Figure 6.5 shows the concrete-filled steel tubes prepared using the present technique, and the technical comparison shows that there is no gap between the concrete and wall of the steel tubes. Figure 6.6 shows the concrete-filled steel tubes with a general concrete expansion agent formulated a concrete combination of materials; a gap can be seen between the concrete and steel tubes, measured to reach 0.2–0.5 mm; the concrete structure is not dense; there are more voids and the need to increase the replenishment of grouting measures in order to make it dense.

6.1.4.2 Concrete-filled Steel Tube Expert System

The concrete-filled steel tube expert system developed by integrating the abovementioned research results of concrete-filled steel tube technology and combining it with the application of the technology in actual projects. The system includes subsystems such as concrete proportion design, expansion

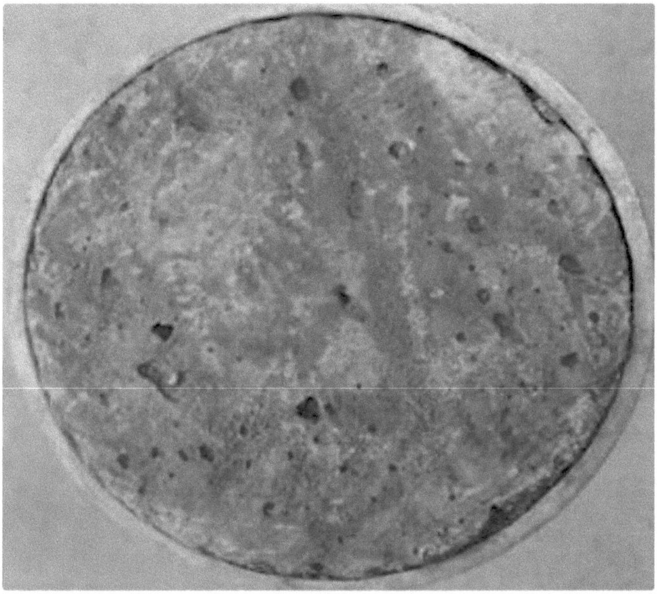

Figure 6.6 Delocalized state.

property calculation, bearing capacity calculation, construction technology, quality monitoring, and literature search and helps in its functional design. The system is characterized by complete functions, comprehensive knowledge, novel content, advanced technology, and comprehensive practicality. It has important innovations in engineering quality and structural safety testing and construction guidance. It plays an important role in promoting the application of the concrete-filled steel tube technology and improving the design, construction, building, and engineering safety of large-span structures. The function and technical principle of the expert system are shown in Figure 6.7.

6.1.4.3 Engineering Applications

(1) Mechanical properties of concrete-filled steel tubes and concrete composite members

Table 6.11 shows the mechanical property test and material composite effect of steel-filled steel tube expansion concrete; the ultimate destructive load value of concrete-filled steel tube composite members is increased by 13.8% and 12.6% than that of steel-filled concrete tube ordinary concrete; and the corresponding hoop coefficients are 1.14 and 1.13, respectively, which indicates a good hoop composite effect and improves the load carrying capacity of the composite members.

(2) On-site monitoring, quality testing, and safety assessment of engineering structures

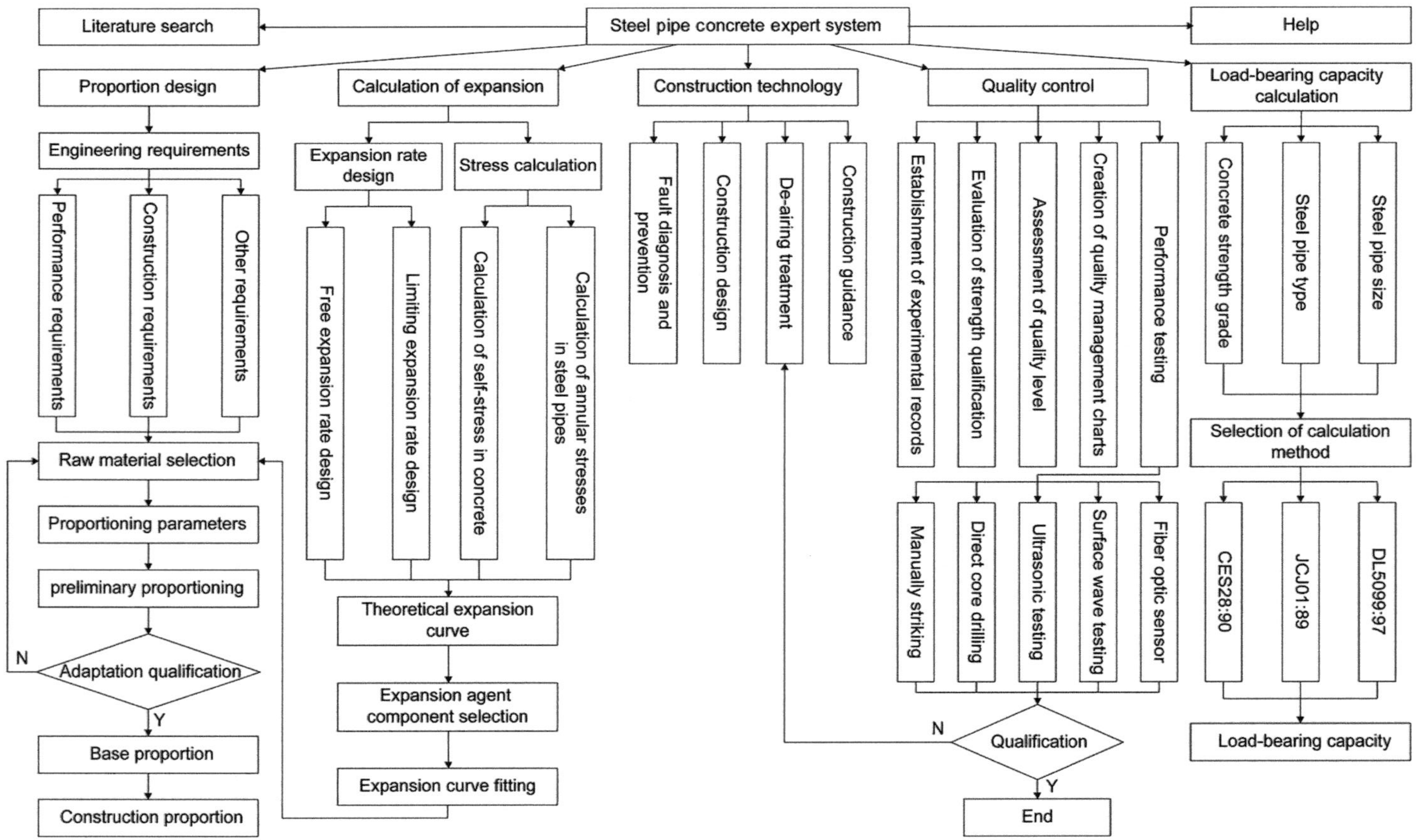

Figure 6.7 Functional and technical principles of expert systems.

Table 6.11 Influence of concrete expansion properties on bearing capacity of short concrete-filled steel tube columns.

Type of concrete	Concrete 28 d strength (MPa)	Concrete 28 d limiting expansion rate/10^{-4}	Ultimate loads for concrete-filled steel tubes and short columns (kN)	Increased values of ultimate loads for short columns (kN)	Percentage increase in ultimate load for short columns (%)	Combined member composite effect (hoop factor)
C50 General Concrete	62.2	−3.0	1740	0	0	0
C50 Expanded concrete	60.5	4.5	1980	240	13.8	1.14
C60 General Concrete	69.1	−3.4	2008	0	0	0
C60 Expanded concrete	70.8	4.3	2261	253	12.6	1.13

The matching state of the combination of concrete and steel tubes is not only affected by the deformation of concrete itself but also related to the coordination of the deformation of concrete-filled steel tubes and concrete. Due to the difference in the coefficient of thermal expansion and modulus of elasticity between steel and concrete, respectively, α_s is 12.0×10^{-6}/°C, E_s 210 GPa, α_c 10×10^{-6}/°C, and E_c 34.5GPa. In case of large temperature differences (cold winters and hot summers), the effect increases and the safety of the combined material and structure has to be evaluated [25, 27–29]. The ANSYS finite-element method analysis (Figure 6.8) was used to obtain the expansion rate and stress variation as shown in Tables 6.12 and 6.13 [30, 31].

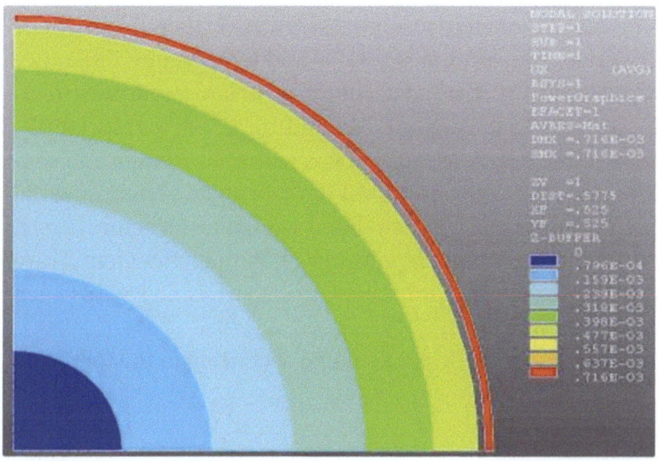

Figure 6.8 Finite-element analysis of temperature field.

Table 6.12 Concrete free expansion required for summer temperature change.

t_s/t_c (°C)[a)]	70/40	60/40	50/40
$\varepsilon_0/10^{-6}$	231.8	169.7	107.6

a) t_s is the temperature of the steel tubes, and t_c is the temperature of the concrete-filled steel tubes.

Table 6.13 Stress changes in steel tubes due to winter temperature variations.

t_s/t_c (°C)	−30/0	−20/0	−10/0
σ_c (MPa)	0.524	0.384	0.243
σ_t (MPa)	16.375	12.00	7.594

(1) **Summer:** The gap between the concrete and steel tubes resulting from temperature shrinkage is such that, for the steel tubes to be unattached to the concrete, the concrete-filled steel tubes should have a limiting expansion rate that satisfies $\varepsilon_R \geq \delta/R$. According to $\varepsilon_0 = [\varepsilon_R(KEc + E_S)/KE_c]$, the free expansion rate of concrete ε_0 is obtained.
(2) **Winter:** The shrinkage rate of steel tubes is greater than that of concrete-filled tubes, at which time no debonding occurs, but the radial compressive stress σ_c and circumferential tensile stress σ_t on the surface of the steel tubes are increased.

Analysis of the calculation results: When the summer temperature goes up to 70 °C, the temperature difference between the inside and outside of the steel tubes is 30 °C, the free expansion rate of concrete reaches 231.8×10^{-6}, then there will be no debonding. When the winter temperature is −30 °C and temperature difference between the inside and outside is 30 °C, the radial compressive stress and circumferential tensile stress of the steel pipe increase by 0.524 and 16.375 MPa, respectively. The results show that the required concrete expansion is within the design and control range, and the resulting stress changes are within the safe range of steel material properties, so the combined material and concrete-filled steel tubes structure is safe. Table 6.14 shows the results of long-term monitoring of Wuhan Jianghan Third Bridge using ultrasound [32]. The actual results show that there is no gap between the concrete-filled steel tubes and concrete.

The concrete-filled steel tube clearance X is calculated according to the following formula:

$$X = \frac{[(tV_g - 2d)V_h + (2d - D)V_g]V_k}{V_g(V_h - V_k)} \tag{6.10}$$

Table 6.14 Ultrasonic test results.

Detection time/maximum or minimum temperature	Average speed of sound (m/s)	Calculate clearance (mm)	Condition of concrete-filled steel tubes
June 2002	4251–4775	<0	Gapless
August 2007/42 °C	4190–4793	<0	Gapless
February 2008/−12 °C	4407–4856	<0	Gapless
October 2009/39 °C	4280–4641	<0	Gapless

where t is the time of an ultrasonic wave passing through the concrete-filled steel tubes; d is the thickness of the steel pipe wall; D is the distance between two probes; V_g, V_k, and V_h is the speed of sound in steel, air, and concrete, respectively.

(3) Engineering applications

The concrete-filled steel tubes and concrete composite material technology realizes the fine design and control of expansive high-strength concrete-filled tubes; improves the bearing capacity and stability of the structure; realizes the coordinated development of concrete strength, expansion, and workability; and realizes the construction of the whole concrete-filled tubes of the large-span structure at one time on the basis of quality assurance. It simplifies the complex process of segmented construction and late replacement of cement slurry; reduces the consumption of raw materials and power; shortens the construction cycle; saves project costs; and effectively solves the key problems of concrete-filled steel tubes and concrete-filled steel tubes such as concrete pump pipe blockage, lack of densification, uneven expansion, and easy to de-vacuum, which significantly improves the construction efficiency, quality of the project, and safety of the structure. Due to this technological breakthrough, the practical design strength of CFST has been elevated from C50 grade to C80 grade, steel tube diameters have increased from less than 1000 mm to 1800 mm, and arch bridge spans have extended from previously less than 200 m to 530 m. The application scope has expanded from being limited to southern regions to cold climate areas, including Tibet and Jilin. The development and application of this technology have made important contributions to national capital construction and promoted scientific and technological progress in the technical field of this discipline. Table 6.15 and Figure 6.9 show the projects in which the team of this achievement has directly participated in the promotion and application works.

The results have been successfully applied to the world's top 4 spans, including Sichuan Hejiang Yangtze River Highway Bridge, Wushan Yangtze River Bridge, Hunan Maocaojie Bridge, and Guangxi Yonghe Bridge, and the world's first concrete-filled steel tubes railway bridge – Yichang Yangtze River Railway Bridge (550 m, 2 × 275 double spans), which is the world's first concrete-filled steel tubes railway bridge. As well as more than 40 arch bridges such as Zhejiang Chun'an Nanpu Bridge, Chongqing Fengjie Meixi River Bridge, Hubei Wuhan Jianghan

Table 6.15 Selected concrete-filled steel tubes bridge projects in which the author's team was directly involved.

No.	Bridge name	Main span of a bridge
1	Sichuan Hejiang Yangtze River Highway Bridge	530 m, World's first span, center-bearing, concrete C60
2	Chongqing Wushan Yangtze River Bridge	460 m, World's second longest span, center-bearing, concrete C60
3	Hubei Hurong Expressway Zhijing River Bridge	430 m, Top-bearing, concrete C55
4	Hunan Xiangtan Xiangjiang Fourth Bridge	400 m, center-bearing, concrete C55
5	Hunan Maocaojie Bridge	368 m, center-bearing
6	Guangxi Nanning Yonghe Bridge	349 m, center-bearing
7	Zhejiang Chun'an Nanpu Bridge	308 m, center-bearing
8	Chongqing Fengjie Meixi River Bridge	288 m, top-bearing
9	Hubei Wuhan Qingchuan Bridge	280 m, bottom-bearing
10	Guangdong Dongguan Waterway Bridge	280 m, center-bearing
11	Yichang Yangtze River Railway Bridge	2 × 275 m, Railway world's first span, bottom-bearing
12	Zhejiang Xiangshan Sanmenmen Middle Bridge	270 m, center-bearing
13	Zhejiang Xiangshan Sanmenmen North Bridge	270 m, center-bearing
14	Sichuan Yibin Rongcheng Bridge	260 m, center-bearing
15	Zhejiang Plain Mountain Bridge	260 m, center-bearing
16	Hubei Zigui Qinggan River Bridge	256 m, center-bearing
17	Wuhan Jianghan Fifth Bridge	240 m, center-bearing
18	Zhejiang copper tile door bridge	238 m, center-bearing
19	Guizhou Shuibai Beipanjiang Railway Bridge	236 m, railway, world's second span, bottom-bearing
20	Hubei Enshi Nanlidu Bridge	220 m, top-bearing
21	Hubei Zigui Longtan River Bridge	208 m, center-bearing
22	Jiangxi Gao'an Gyunzhou Bridge	170 m, center-bearing
23	Zhejiang Huaguangtan Bridge	165 m, center-bearing
24	Sichuan Luding Cat Ziping Bridge	160 m, center-bearing
25	Changchun Yitong River Bridge	158 m, bottom-bearing
26	Changsha Heishipu Xiangjiang Bridge	Triple arch 144 m + 162 m + 144 m, center-bearing
27	Guangdong Chaozhou Hanjiang North Bridge	Five-link arch 5160 m, center-bearing
28	Lhasa River Bridge of Qinghai-Tibet Railway	Triple arch 3 × 108 m, bottom-bearing
29	Beijing-Zhuhai Expressway Zhengzhou Yellow River Second Bridge	Octagonal arch 8 × 100 m, bottom-bearing
30	Sichuan Yalu Ganhaizi Expressway Bridge	1811 m, The world's longest steel tube truss road bridge

Note: Bridges with unspecified concrete strength are C50 concrete.

The Yangtze River Bridge in Hejiang, Sichuan (with a span of 530m, the world's longest)

Wushan Yangtze River Bridge in Chongqing (with a span of 460m, the world's second longest)

Hubei Hurong Expressway Zhijing River Bridge (430m span)

Yiwan Railway Yangtze River Bridge

Guizhou Shibei Beipan River Railway Bridge (Span 236m, the world's second longest)

The Ganhaizi Super Large Bridge on Sichuan Ya'an-Lugu Expressway is 1811m long, the world's longest steel truss highway bridge.

Figure 6.9 Some influential and representative concrete-filled steel tube arch bridges.

Third Bridge, Guangdong Dongguan Waterway Bridge, Hubei Zigui Qinggan River Bridge, Hubei Wuhan Jianghan Fifth Bridge, Hubei Enshi Nanlidu Bridge, Hubei Zigui Longtan River Bridge, Zhejiang Ningbo Xiangshan Sanmenkou Cross-sea Bridge (Main Bridge), and Qinghai-Tibet Railway Lhasa River Special Bridge. These engineering projects have successfully resolved critical technical challenges that frequently arise in the construction of concrete-filled steel tube (CFST) arch bridges with long spans, including segregation and bleeding, pump blockage, inadequate concrete compaction, uneven expansion, and void formation within steel tubes, particularly at arch crowns. These achievements have generated substantial economic and social benefits while significantly advancing the application of CFST technology in large-span composite structural engineering. Using the research results of the project, it is of great significance to participate in guiding the design, construction, and application of concrete-filled steel tube arch bridges, to ensure the quality of concrete works, to reduce reconstruction and maintenance costs in the future, to prevent the emergence of major engineering accidents, to prolong the service life of the project, as well as to improve the concrete-filled steel tube technology, to push forward the development of the high-performance concrete technology of our country, and to enrich the theory and practice of materials science.

6.2 Steel-concrete/Asphalt Composite Bridge Deck Paving Structural Materials

6.2.1 Overview

6.2.1.1 Durability Challenges of Steel Bridge Deck Pavement Structures

Steel structure bridges have the characteristics of light self-weight, large-span capacity, good integrity, convenient erection, and short construction period. Steel bridge structures are commonly used for long-span bridges and high-gradient interchange bridges. Bridge pavement is a material structure laid on the top surface of the bridge structure, with evenly distributed loads, protecting the bridge structure, and meeting the requirements of driving comfort and safety. It is an indispensable and important component of bridge structures.

Under the comprehensive effects of traffic loads, wind loads, climate temperature changes, and local deformation of steel bridge decks, the stress and deformation of pavement materials are very complex. Due to factors such as vehicle overload rolling, impact, long-term fatigue, high temperature, heavy rainfall, low temperatures, dryness, and large day-night temperature difference, the difference in material properties such as elastic modulus and temperature deformation between asphalt concrete and steel plates is significant. When subjected to changes in traffic loads and climatic environments, their coordination and deformation capabilities are poor, leading to deformation of pavement materials; poor bonding performance between layers; easy delamination; and damage such as rutting, sliding, potholes, cracking, and grooving on the bridge deck, resulting in a significant decrease in durability [33], as shown in the common durability failures in Figure 6.10. Some bridge pavements often need to be redeployed and repaired after only a few years or even one to two years, causing inconvenience to traffic and economic losses. The high durability of steel bridge deck materials has become a global technological challenge for the construction of long-span bridges. With the widespread application of steel bridge structures in river, lake, and urban interchange projects, this problem needs to be urgently addressed.

6.2.1.2 Performance Characteristics of Steel Bridge Deck Paving Materials

In the past, solving the durability problem of steel bridge decks mainly focused on the technical development of asphalt concrete, with three main technical routes:

(1) Stone mastic asphalt (SMA), characterized by good temperature adaptability and skid resistance, but poor adhesion to steel plates, prone to displacement, rutting, and premature failure.
(2) Pouring asphalt concrete (GA), characterized by better conformity with steel plates, good fatigue resistance, but poor high-temperature stability and prone to rutting.

Figure 6.10 Common forms of durability damage to steel bridge deck paving materials.

(3) Epoxy asphalt concrete (EA), characterized by good bonding to steel plates and high-temperature stability, but prone to cracking at low temperatures, with shallow construction depth, poor skid resistance, and demanding construction conditions, complex processes, and high costs.

Under the combined effects of traffic loads, wind loads, temperature changes, and local deformation of steel bridge decks, their stress and deformation are far more complex than those of highway pavements and airport runways. Therefore, there are higher requirements for their strength, deformation characteristics, temperature stability, and fatigue durability. Additionally, due to the special position of the pavement, specific requirements such as lightweight, good bonding, and impermeability are also demanded. From the perspective of materials science, steel's modulus of elasticity, coefficient of thermal expansion, and thermal conductivity differ significantly from those of asphalt concrete. Under conditions of large temperature fluctuations and heavy loads, the different deformations of the two materials will result in significant shear forces between pavement layers, leading to delamination and material structure damage. Therefore, the fundamental problem should be addressed by improving the synergy of material performance, reducing structural deformation, and solving the intrinsic issues.

6.2.1.3 Major Research and Technology Development

Based on the requirements for the service environment and functions of pavement materials, they should meet the following criteria: high strength, good deformability, temperature stability, heat insulation, impermeability, and fatigue resistance, and for large-span structures, they should also have a relatively lightweight. In response to the aforementioned issues, the author innovatively proposes a technical route using a composite structure of concrete and asphalt for steel bridge deck

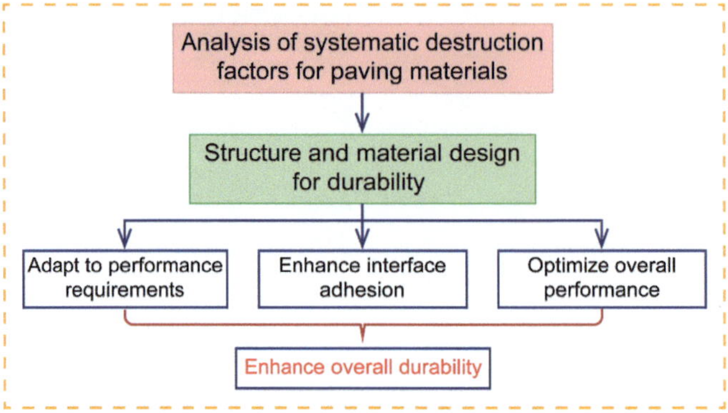

Figure 6.11 General idea of the research.

pavement. The research and development plan and overall approach are illustrated in Figure 6.11.

The author's team conducted the following research and development work and achieved innovative results:

(1) By simulating the mechanical and environmental response of steel bridge deck pavement materials, the causes and patterns of their failure were identified. Based on the performance characteristics of pavement materials and structural-functional requirements, a multilayered composite gradient structure of pavement materials was developed to improve the durability of steel bridge deck pavement layers.
(2) Based on the functions performed by each layer of the composite steel bridge deck pavement structure, corresponding technologies were developed for preparing high-toughness lightweight concrete, high-viscosity and high-elasticity modified asphalt, interface reinforcement, and stress-absorbing materials. This aimed to enhance the strength, fatigue resistance, resistance to rutting, cracking, and deformation coordination of the pavement structure, providing material and technological support for steel bridge deck pavement in different climatic regions and for large-span, high-gradient structures.
(3) According to the design and construction characteristics of the overall composite structure, research was conducted on the structural connection technology between pavement materials and bridge deck steel plates. Techniques were developed for mastering the composite pavement method, material construction processes, and construction and quality evaluation methods, advancing the application of technological achievements in steel bridge deck pavements of different structural bridge types.

6.2.2 Design of a New Type of Steel Box Girder Pavement Structure

By analyzing the common causes of poor durability in asphalt pavement layers on steel bridge decks, the concept of gradient structure design is proposed to address the significant differences in elastic modulus performance between materials and the poor interface shear resistance. A structural design method is created to achieve synergistic deformation between pavement layer materials and steel bridge decks, namely "shear elements and steel mesh reinforcement + high-toughness lightweight aggregate concrete + asphalt concrete" [34–36]. This design significantly extends the service life of steel bridge deck pavement layers while reducing the weight of the pavement layers and the bridge itself, thereby enhancing driving safety and comfort.

6.2.2.1 Material Mechanics Analysis

(1) Stress analysis of steel bridge deck pavement materials

Using SHELL63 elements to simulate orthotropic plates and SOLID65 elements to simulate pavement materials, the mechanical response of pavement materials to wheel loads and environmental temperature changes was studied by setting and adjusting parameters such as elastic modulus, Poisson's ratio, and coefficient of linear thermal expansion [37]. The failure patterns of steel bridge deck pavement materials were identified (Figures 6.12 and 6.13).

Research results indicate the following: (1) Pavement material failure depends on the maximum tensile stress on the surface and interlayer shear stress. Cracking depends on transverse tensile stress, and interlayer displacement depends on interlayer shear stress. (2) Differential material strains caused by temperature changes lead to deformation of pavement structures. (3) Delamination of pavement structures originates from non-coordinated interlayer deformation and weak adhesive strength. (4) Homogeneous materials and their structures cannot meet the complex functional requirements of pavement structures, which is the fundamental reason for poor durability.

(2) Stress analysis of steel bridge deck pavement layer structure

The failure of bridge deck pavement structures mainly depends on the tensile stress of the upper layer, interlayer shear stress, and tensile strength and interlayer

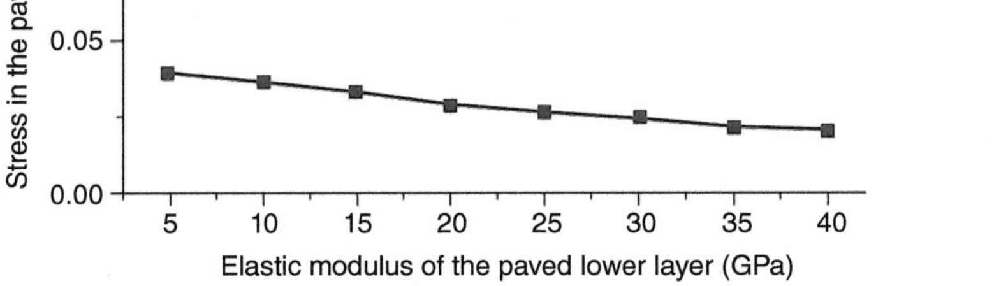

Figure 6.12 Stress analysis for change in modulus of elasticity of paving material.

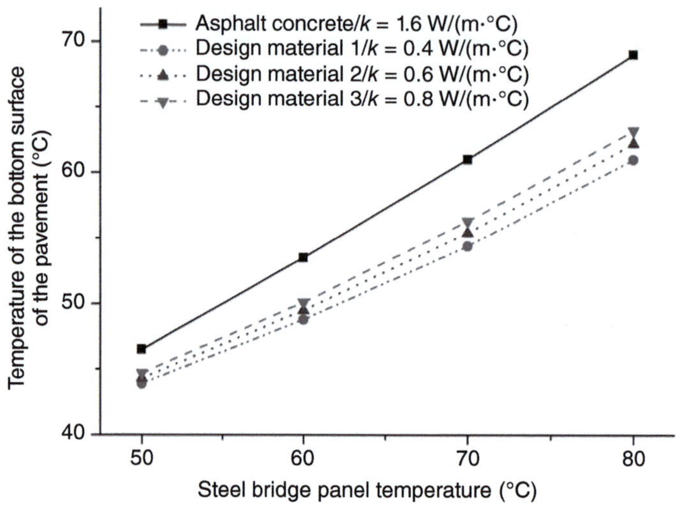

Figure 6.13 Analysis of the effect of changes in thermal conductivity of paving materials.

bonding strength of pavement layer materials. Take the Wuhan Airport Second Highway Channel main bridge steel box girder section project as an example, which had a span of 145 m, width of 15 m, top plate thickness of 20 mm, transverse diaphragm thickness of 10 mm, spacing of 3000 mm, longitudinal stiffener thickness of 8 mm, spacing of 640 mm, designed speed of 60 km/h, standard axle load coupling conditions, pavement total thickness of 100 mm, upper layer 50 mm, and lower layer 50 mm. The stress analysis results are shown in Figure 6.14 and Table 6.16: (1) The maximum transverse tensile stress is located on the surface

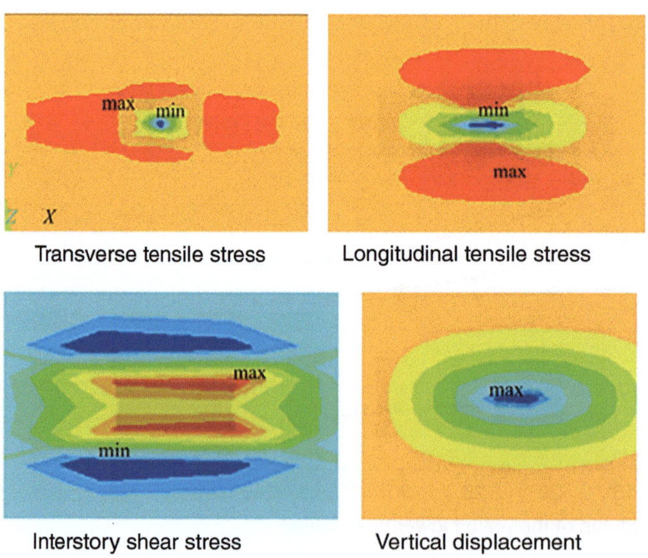

Figure 6.14 Mechanical response analysis of steel box girder segment durable paving structure.

Table 6.16 Mechanical analysis of SMA in the upper layer and different modulus paving materials in the lower layer.

Modulus of the paving material of the following layer	400 MPa	800 MPa	3000 MPa	30 GPa
Maximum transverse/longitudinal tensile stress of upper layer (MPa)	0.42/0.32	0.35/0.26	0.28/0.21	0.11/0.08
Maximum tensile stress in transverse/longitudinal direction of lower layer (MPa)	0.37/0.30	0.49/0.39	0.77/0.54	1.02/0.82
Maximum shear stress between upper and lower layers (MPa)	0.23	0.17	0.12	0.09
Maximum shear stress between pavement and steel plate (MPa)	0.44	0.48	0.61	0.81

of the pavement layer above the longitudinal stiffener plate along the bridge, and the maximum longitudinal tensile stress is located on the surface of the pavement layer above the transverse diaphragm. The maximum interlayer shear stress occurs in the area near the transverse diaphragm on both sides of the longitudinal stiffener. The maximum vertical displacement occurs at mid-span when the load is applied. (2) Pavement layer failure depends on the maximum tensile stress on the surface and interlayer shear stress. Cracking of the pavement layer depends on transverse tensile stress, and interlayer displacement failure depends on interlayer shear stress. (3) As the modulus of the lower layer increases, the transverse tensile stress and interlayer shear stress of the upper layer decrease, while the interlayer shear stress between the lower layer and steel plate increases.

(3) Gradient design of a multilayer composite pavement structure

With the increase of the elastic modulus ratio n ($n = E_{\text{Steel plate}}/E_{\text{Pavement subbase}}/E_{\text{Pavement wearing surface}}$) of the bridge deck pavement system, the maximum transverse tensile stress, maximum longitudinal tensile stress, maximum interlayer shear stress between pavement layers and steel plate layers, and steel bridge deck surface deflection all decrease, enhancing the structural safety and durability of the bridge deck pavement system. The elastic modulus of lightweight aggregate concrete (approximately 25–28 GPa) falls between the elastic modulus of steel (210 GPa) and SMA asphalt concrete (1.2–1.8 GPa), conforming to the design concept of gradient materials [38, 39]. The designed structure is illustrated in Figure 6.15.

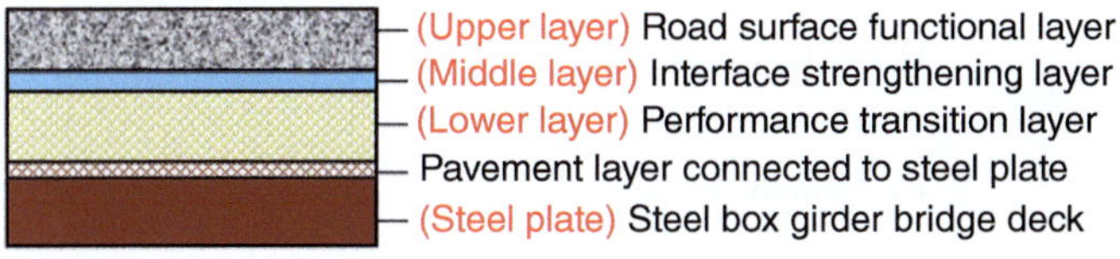

Figure 6.15 Gradient design model for multilayer combination pavement structures.

Figure 6.16 Double-layer asphalt pavement on steel bridge deck.

(4) Collaborative gradient combination design method of pavement structure

Based on the principle of material gradient design, the modulus of the lower layer material is increased to reduce the tensile stress of the upper layer, thereby enhancing the stability and durability of the pavement structure. High-adhesion materials and shear-resistant structural designs are used between layers to strengthen interlayer interface bonding, to improve the integrity of the pavement structure, and to achieve coordinated deformation between pavement layers, enhancing interlayer shear force. Shear studs are welded on the steel plate (elastic modulus 210 GPa); reinforcing steel mesh is tied; and lightweight, high-strength, and high-toughness cement-based composite material with good compliance with the steel plate is poured as the lower layer (elastic modulus approximately 25–28 GPa, I20 > 20, thickness 5–8 cm). The upper layer is paved with SMA13 asphalt concrete (elastic modulus 1.2–1.8 GPa, thickness 4–5 cm), forming a gradient composite structure of elastic modulus [40, 41], as shown in Figures 6.16 and 6.17.

In the lower layer pavement structure, the shear studs and steel mesh form a bridge anti-sliding skeleton, improving the anti-sliding ability of the lower layer lightweight aggregate concrete while evenly distributing the multidirectional stresses exerted by vehicle loads on the steel bridge deck, further enhancing the coordinated deformation capability and fatigue resistance characteristics between the pavement layer and steel bridge deck. A 2 mm thick high-adhesion and high-elastic modified asphalt waterproof bonding stress-absorbing layer is hot-sprayed between the upper and lower pavement layers, improving the interface bonding strength (≥0.8 MPa) and shear strength (≥1.4 MPa, greater than the maximum shear stress of 0.8 MPa produced between layers during emergency braking of the standard vehicle – 20 tons for the main vehicle and 55 t for the heavy vehicle). It also prevents water penetration causing corrosion of shear studs, steel mesh, and steel plates; dissipates the energy generated by stress concentration at concrete crack locations under repeated vehicle loadings; and prevents crack reflection to the SMA13 asphalt concrete layer. The use of high-adhesion SMA paving technology gives the surface-wearing course superior high-temperature rutting resistance, low-temperature crack resistance, and durability.

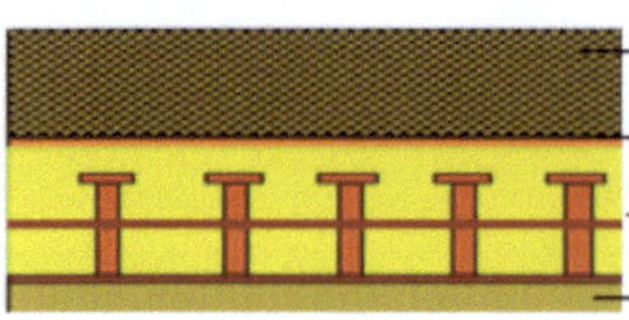

Figure 6.17 "Transition layer – asphalt combination" pavement on steel bridge deck.

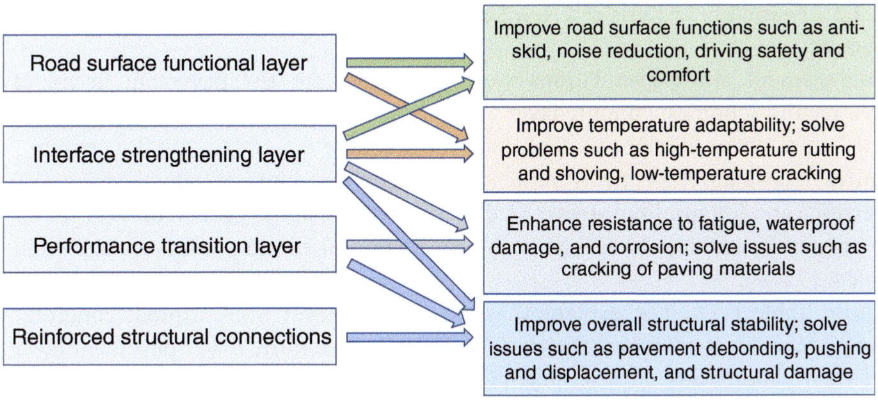

Figure 6.18 Multilayer composite materials with reinforced coupling effects.

(5) Enhanced interlayer connection technology

The use of a multilayer composite structure allows pavement materials to provide targeted performance for different areas. By strengthening the connection between pavement structure and steel plates, the original adhesive interface is transformed into a structural reinforcement connection, enhancing the integrity of pavement materials. This fundamentally solves the problem of pavement structures easily debonding from steel plates and improves interlayer shear strength and material durability [42]. The integration of pavement structures with steel plates also enhances the stability and load-bearing capacity of bridge steel plate structures. The roles of each material layer and structural reinforcement connection are depicted in Figure 6.18.

(6) Technical scheme of a multilayer composite pavement structure

To achieve high strength, high toughness, high durability, excellent fatigue performance, and long service life of lightweight concrete pavement layers, polymers, fibers, and high-efficiency mineral filler are introduced into lightweight concrete. Its 28 d compressive strength exceeds 58 MPa, and the toughness index η_{30} exceeds 25. A high-strength interface adhesive is applied between lightweight concrete and the SMA13 surface layer to improve waterproofing and resistance to sliding and cracking of the pavement surface. In the new steel box girder bridge deck pavement system, from the upper pavement layer to the steel plate, the elastic modulus gradually increases, reflecting the design concept of material elastic modulus gradient changes. This greatly enhances the durability performance of the bridge deck pavement system [43]. The schematic diagram of the new steel bridge deck pavement structure is shown in Figure 6.19.

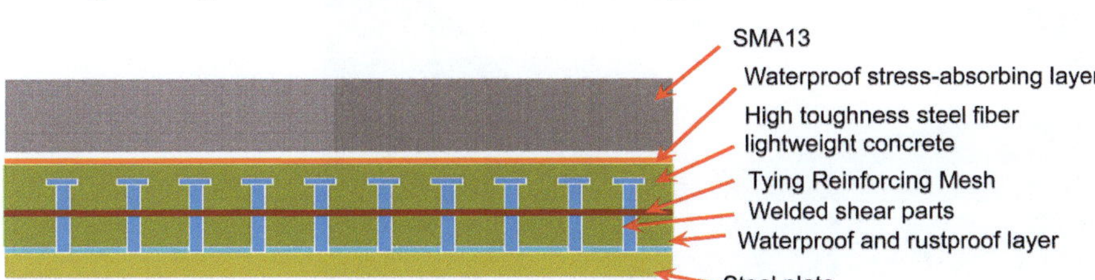

Figure 6.19 Technical scheme of a new steel bridge deck pavement structure.

6.2.2.2 Feasibility Analysis of a New Steel Box Girder Bridge Deck Pavement Structure

Finite-element stress calculations were conducted for the pavement layers of three paving schemes: "double-layer SMA asphalt concrete," "epoxy asphalt concrete," and "shear elements + high-toughness concrete + SMA asphalt concrete." Figures 6.20–6.23 respectively represent the contour plots of lateral stress (SX direction) and longitudinal stress (SZ direction) of the upper pavement layer under the most unfavorable vehicle load position, and the calculation results are shown in Table 6.17.

From Table 6.17, it is evident that the "double-layer SMA asphalt concrete" scheme faces an issue where the shear strength between the lower pavement layer and steel plate is less than its maximum shear stress, failing to meet the design requirements. Both the "double-layer epoxy asphalt concrete" scheme and "shear elements + high-toughness concrete + SMA asphalt concrete" paving scheme, based on modulus gradient design, demonstrate tensile and shear strengths of the pavement layers greater than the maximum tensile and shear stresses calculated by

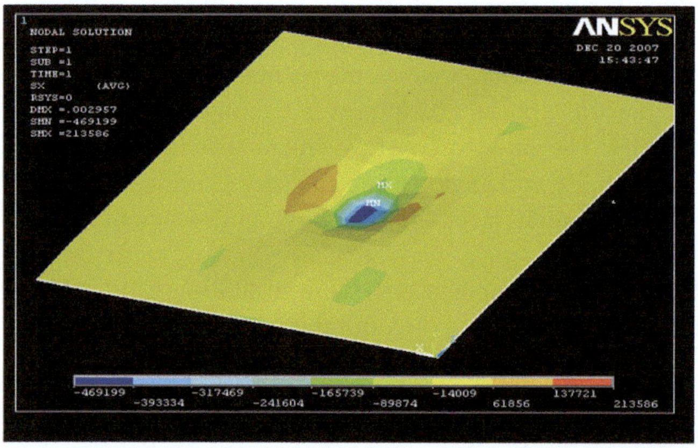

Figure 6.20 Cloud of transverse stresses (SX) in the upper layer of the pavement.

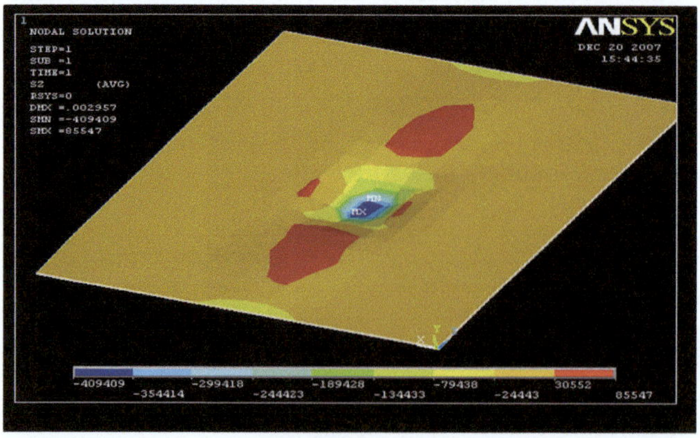

Figure 6.21 Cloud of longitudinal stress (SZ) in the upper layer of the pavement.

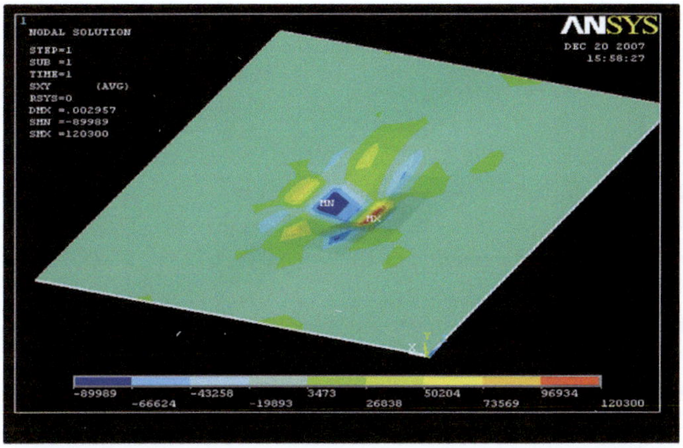

Figure 6.22 Cloud of transverse shear stresses in the upper layer of the pavement.

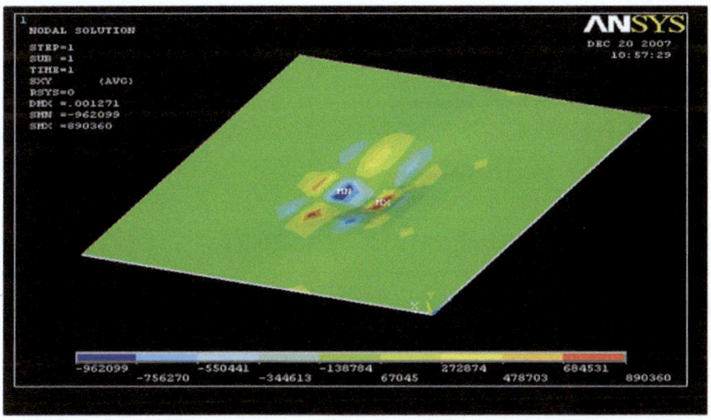

Figure 6.23 Cloud of transverse shear stresses at the interface between the lower layer of pavement and the steel plate.

finite-element analysis. Additionally, their toughness indices meet design requirements, making them suitable as materials for steel box girder bridge deck pavement layers. However, the "double-layer epoxy asphalt concrete" scheme is costlier and involves more complex construction processes compared to the "shear elements + high-toughness concrete + SMA asphalt concrete" paving scheme, rendering it less economically practical. Therefore, the feasibility of the "shear elements + high-toughness concrete + SMA asphalt concrete" paving scheme is higher.

6.2.2.3 Analysis of Influencing Factors on the Pavement Structure of a New Type of Steel Box Girder Bridge

Based on actual conditions, the author's team conducted a comparative analysis of key influencing factors including shear stud spacing, diameter, height, pavement layer thickness and modulus, vehicular load, and span [44]. Please refer to Table 6.18 for details.

Table 6.17 Metrics analysis of steel box girder bridge deck paving options.

Design index	Stress/flexure ratio	Theoretical calculation value	Mechanical properties
Double-layer SMA asphalt concrete	Surface maximum transverse tensile stress (MPa)	0.19	0.68
	Surface maximum longitudinal tensile stress (MPa)	0.09	0.68
	Sub-layer surface maximum transverse tensile stress (MPa)	0.5	0.74
	Maximum shear stress between sub-layer and steel plate layer (MPa)	0.8	0.35
	Reinforced rib local deflection-span ratio	1/178	1/200
Epoxy asphalt concrete	Surface maximum transverse tensile stress (MPa)	0.19	0.89
	Surface maximum longitudinal tensile stress (MPa)	0.09	0.89
	Sub-layer surface maximum transverse tensile stress (MPa)	0.5	1.12
	Maximum shear stress between sub-layer and steel plate layer (MPa)	0.8	0.94
	Reinforced rib local deflection-span ratio	1/178	1/200
Shear connector + high toughness concrete + SMA	Surface maximum transverse tensile stress (MPa)	0.1	0.68
	Surface maximum longitudinal tensile stress (MPa)	0.04	0.68
	Sub-layer surface maximum transverse tensile stress (MPa)	0.07	0.7
	Maximum shear stress between sub-layer and steel plate (MPa)	0.87	1.46
	Reinforced rib local deflection-span ratio	1/225	1/200
	Concrete toughness index	23	28

(1) Analysis of shear stud influencing factors

The influencing factors mainly include shear stud spacing, height, and diameter. (1) Shear Stud Spacing: When the spacing of shear connectors is between 300 and 500 mm, the maximum lateral tensile stress and maximum lateral shear stress on the pavement layer surface vary within an allowable range. However, when the shear stud spacing reaches 500 mm, the maximum lateral shear stress between the bonding layer and steel plate exceeds the allowable range. Therefore, for the

Table 6.18 Factors influencing pavement schemes and material properties.

Project	Calculation parameters	Project	Calculation parameters
Steel box girder top plate thickness	12 mm	Transverse diaphragm thickness	10 mm
Transverse diaphragm spacing	3000 mm	Transverse diaphragm height	1000 mm
Thickness of longitudinal stiffening ribs	8 mm	Height of longitudinal stiffening ribs	260 mm
Spacing between longitudinal stiffening ribs	640 mm	Thickness of SMA13	30, 40, 50 mm
Elastic modulus of SMA13	1200 MPa	Poisson's ratio of SMA13	0.3
Thickness of lightweight aggregate concrete sub-base	40, 50, 60 mm	Elastic modulus of lightweight aggregate concrete	25,000 MPa
Poisson's ratio of lightweight aggregate concrete	0.2	Height of shear connectors	40, 45, 50 mm
Diameter of shear connectors	8, 10, 12 mm	Diameter of steel reinforcement mesh	10 mm
Elastic modulus of steel	210,000 MPa	Poisson's ratio of steel	0.3
Grid size of steel reinforcement mesh	100 × 150 mm	Spacing of shear connectors	300, 400 mm, 500 mm

steel box girder bridge deck pavement scheme in this study, a spacing of 400 mm for shear connectors is deemed most reasonable. Although the maximum longitudinal tensile stress is less than the maximum lateral tensile stress, sometimes, under the combined influence of temperature and vehicle load, the pavement layer may experience fatigue cracking. The maximum longitudinal tensile stress (or maximum longitudinal tensile strain) occurs on the surface of the pavement layer at the top of the transverse diaphragm. Hence, the pavement layer surface at the transverse diaphragm is prone to lateral cracks, especially during low-temperature seasons, as shown in Table 6.19.

(2) Shear stud height

The height of shear studs in the pavement scheme also significantly affects the design parameters. Considering a spacing of 400 mm for shear connectors and concrete pavement layer thickness of 500 mm, the heights of shear connectors were set at 35, 40, and 45 mm for calculations. Load positions were considered at the most unfavorable locations, i.e., along the lateral load center of the bridge deck directly above the center of the reinforcing ribs, and longitudinally at the mid-span,

Table 6.19 Influence of shear stud spacing on design criteria.

Spacing of shear connectors (mm)	300	400	500
Maximum transverse tensile stress on pavement surface (MPa)	0.204	0.213	0.249
Maximum longitudinal tensile stress on pavement surface (MPa)	0.081	0.086	0.091
Maximum transverse shear stress on pavement surface (MPa)	0.097	0.12	0.137
Maximum transverse shear stress between sub-layer and steel plate (MPa)	0.81	0.89	0.92
Local deflection-span ratio of reinforced ribs	1/242	1/250	1/261

Table 6.20 Influence of shear stud height on design criteria.

Height of shear connectors (mm)	35	40	45
Maximum transverse tensile stress on pavement surface (MPa)	0.219	0.213	0.205
Maximum longitudinal tensile stress on pavement surface (MPa)	0.083	0.086	0.078
Maximum transverse shear stress on pavement surface (MPa)	0.126	0.12	0.118
Maximum transverse shear stress between bonding layer and steel plate (MPa)	0.92	0.89	0.83
Local deflection-span ratio of reinforced ribs	1/253	1/257	1/261

with a tire pressure of 0.91 MPa (considering a 30% impact coefficient). The results are shown in Table 6.20.

From the calculation results, it is evident that higher shear stud heights result in lower maximum tensile stress on the surface of the steel bridge deck pavement layer under wheel loads, as well as lower maximum shear stress between the pavement layer and steel bridge deck layer. Thus, selecting shear studs with a height of 45 mm in this pavement scheme is deemed feasible.

(3) Shear stud diameter

The diameter of shear studs in the pavement scheme is also a crucial factor to consider. Combining existing standards and engineering practices, preliminary diameters of shear studs were proposed as 8, 10, and 12 mm. The calculation results are detailed in Table 6.21, indicating that the diameter of shear studs significantly affects the maximum lateral tensile stress, maximum longitudinal tensile stress, and maximum lateral shear stress on the pavement layer surface. However, it has a relatively minor impact on the maximum lateral shear stress between the bonding layer and

Table 6.21 Influence of shear stud diameter on design criteria.

Diameter of shear connectors (mm)	8	10	12
Maximum transverse tensile stress on pavement surface (MPa)	0.244	0.213	0.195
Maximum longitudinal tensile stress on pavement surface (MPa)	0.136	0.086	0.089
Maximum transverse shear stress on pavement surface (MPa)	0.138	0.120	0.110
Maximum transverse shear stress between bonding layer and steel plate (MPa)	0.885	0.891	0.893
Local deflection-span ratio of reinforced ribs	1/260	1/257	1/256

steel plate and on the local flexural-to-span ratio of the reinforcing ribs. Considering existing standards, welding requirements for steel box girders, and economic considerations, a diameter of 10 mm is deemed appropriate.

(4) Pavement layer thickness

The thickness of lightweight aggregate concrete and SMA13 in the pavement scheme evidently affects its performance. SMA13 primarily serves as the upper layer to provide functions such as smooth driving, skid resistance, and noise reduction, typically set at 40 mm according to existing standards. On the other hand, the ordinary concrete pavement layer thickness is generally 50–80 mm, while the thickness of lightweight aggregate concrete in the scheme needs to be determined based on a comprehensive consideration of existing standards and economic factors. Initially, thicknesses of 5, 6, and 8 mm were proposed. Calculation results (Table 6.22) show that with the increase in the thickness of the lightweight aggregate concrete pavement layer, there is a slight decrease in the maximum lateral tensile stress and maximum longitudinal tensile stress on the pavement layer surface, while the maximum lateral shear stress between the bonding layer and steel plate increases and the

Table 6.22 Influence of lightweight aggregate concrete pavement layer thickness on design criteria.

Lightweight aggregate concrete thickness (mm)	50	60	80
Pavement surface maximum horizontal tensile stress (MPa)	0.213	0.208	0.203
Pavement surface maximum longitudinal tensile stress (MPa)	0.086	0.083	0.079
Interfacial shear stress between bonding layer and steel plate (MPa)	0.891	0.909	1.135
Local deflection span ratio of reinforced rib	1/257	1/298	1/369

Table 6.23 Determination of parameters for the bridge deck pavement scheme with lightweight aggregate concrete shear components.

Project	Calculation parameters	Project	Calculation parameters
Spacing of shear connectors	400 mm	Thickness of SMA13	40 mm
Thickness of lightweight aggregate concrete surface layer	50 mm	Height of shear connectors	45 mm
Diameter of shear connectors	10 mm	Diameter of steel reinforcement mesh	10 mm

local flexural-to-span ratio of the reinforcing ribs decreases. Therefore, considering existing standards and economic requirements, a thickness of 50 mm is deemed sufficient to meet the requirements.

(5) Determination of parameters for concrete pavement and shear connectors

For the high-toughness lightweight aggregate concrete shear stud bridge deck pavement scheme, the following design parameters can be chosen: spacing of shear connectors, height of shear connectors, diameter of shear connectors, diameter of reinforcing mesh, thickness of the lightweight aggregate concrete surface layer, and thickness of SMA13. Based on the aforementioned analyses, the parameters for the pavement scheme are determined as shown in Table 6.23.

6.2.2.4 Structural Connection Design Between Pavement Structure and Steel Plates

In order to thoroughly solve the problem of weak adhesion between pavement structure and bridge steel plates, a method for strengthening the structural connection between pavement structure and steel plates has been invented, which includes two forms: (1) adopting structural connection technology directly in bridge structure design (Figure 6.24(a)) and (2) when structural connection design is not adopted in

(a) (b)

Figure 6.24 Connection methods between pavement and bridge steel plate surface (a) Design on steel plate and (b) Anchoring on the steel plate surface.

bridge structure design scheme, structural connection treatment is carried out during construction (Figure 6.24(b)). The bridge deck anti-push displacement skeleton formed by the connection of pavement structures not only can improve the shear resistance of bridge deck composite structures but also increase the slip amount at the limit yield, enhance the overall stiffness of the bridge, evenly transfer the load to the steel bridge deck, and improve the synergistic deformation ability and fatigue resistance characteristics between the pavement layer and steel bridge deck. Under the same bearing capacity conditions, it can also reduce the weight of the bridge deck and the amount of steel used, thus reducing the cost of bridge construction [45].

6.2.3 Design and Performance of Pavement Materials and Structures

Based on the functions undertaken by the gradient design of the composite structure of steel bridge pavement, the performance of each layer of material and structural design is carried out. The preparation technology of high-toughness lightweight aggregate concrete as the "performance transition layer" material of the pavement structure has been developed. The preparation technology of rubber composite high-adhesion high-elastic asphalt as the "interface reinforcement layer" material of the pavement structure has been developed. The preparation technology of high-strength asphalt mastic as the "pavement functional layer" material of the pavement structure has been developed, providing technical support for the preparation of highly durable bridge pavement materials.

6.2.3.1 Performance Transition Layer Materials

The design of performance transition layer materials with requirements such as elastic modulus, thermal insulation, lightweight, fatigue resistance, and durability leads to the selection of lightweight aggregate concrete as the experimental material. To address issues like the low strength of lightweight aggregates and brittleness of concrete, a surface enhancement technology for lightweight aggregates using polymer emulsion combined with ultra-fine ground granulated blast furnace slag powder is invented [46–49]. The material selection criteria are established as follows:

$$d/D_P \geq 5, d/R \geq 2 \tag{6.11}$$

where d represents the average pore diameter on the surface of lightweight aggregates in μm, D_P represents the average particle diameter of the polymer emulsion in μm, and R represents the average particle diameter of the ground granulated blast furnace slag powder in μm.

The dosage for polymer emulsion is

$$W = 0.25\, C_f \cdot C_s \cdot \alpha_{\max} \tag{6.12}$$

where W represents the amount of polymer emulsion in L, C_f represents the amount of cement in kg, C_s represents the amount of water required for the hydration of cement in kg, and $\alpha_{\max}$ represents the maximum degree of hydration in %.

The main mechanisms of surface enhancement and toughening include optimization of the hardened cement paste structure by ground granulated blast furnace slag

(a) (b)

Figure 6.25 Surface repair effect of lightweight aggregates: (a) original surface defects and (b) repaired condition.

powder, filling and repairing the surface pore defects of aggregates (Figure 6.25); toughening of lightweight aggregates and their interface hardened cement paste by polymer solution; the demulsifying water promotes the curing effect within the hardened cement paste, improving the density and strength of concrete; and enhancing the ability of lightweight aggregates to impede cracks originating from within themselves or at the aggregate-hardened cement paste interface.

The composite toughening and material performance optimization design is based on Eq. (6.13):

$$f_{mb} = ke^{dfa} \tag{6.13}$$

where f_{mb} represents the upper limit of cement mortar strength in MPa, f_a represents the cylindrical compressive strength of lightweight aggregates in MPa, k represents the constant of lightweight aggregate particle shape, and d represents the water–binder ratio parameter.

By optimizing the structure of aggregates and hardened cement paste, the highest design value of concrete strength is achieved; high-modulus aggregates are used to deflect crack paths and dissipate fracture energy during the crack propagation process in concrete; high-modulus hybrid fibers and polymers are introduced to form fiber bundles and interpenetrating polymer networks for toughening effects, comprehensively improving the toughness of concrete (Figure 6.26); the strength and density relationship in Eq. (6.14) is established:

$$E = 0.043 \times \rho^{1.5} \times (f_{cu})^{0.5} \tag{6.14}$$

where f_{cu} represents the compressive strength of a $150 \times 150 \times 150\ mm^3$ cubic specimen in MPa and ρ represents the density of concrete in kg/m^3.

By adjusting the strength, density, elastic modulus, and thermal conductivity of concrete, the performance comparison of high-toughness lightweight concrete developed by the author's team with other similar materials is shown in Table 6.24.

6.2.3.2 Interfacial Reinforcement Layer Materials

Rubber compounded high-viscosity elasticity asphalt [50–53] (RHVE); the modification principle and mechanism of action are shown in Figure 6.27:

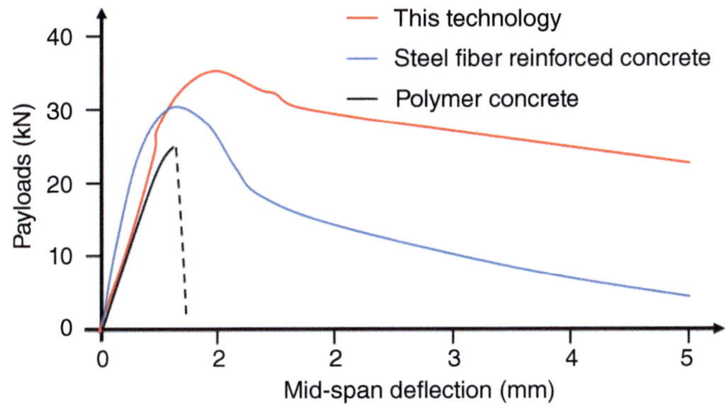

Figure 6.26 Toughening effect of high-toughness lightweight concrete.

Table 6.24 Performance comparison of high-toughness lightweight concrete with other toughened concretes (C50).

Technical content indicators	Steel fiber concrete	Polymer concrete	Organic fiber concrete	High-toughness lightweight concrete
Elastic modulus adjustment range (GPa)	35–40	30–38	35–40	20–35
Bending toughness index/I_{20}	15–20	5–10	7–15	25–28
Fatigue resistance to bending/10,000 times (stress ratio 0.65)	50–100	40–45	45–50	200
90-day drying shrinkage/10^{-4}	3.5–4.5	3.5–6.0	3.0–4.0	2.5–3.5
Thermal conductivity (W/(m·K))	1.4	0.8	1.2	0.4
Apparent density (kg/m^3)	2500 ± 100	2450 ± 100	2450 ± 100	<2000

(1) The introduction of stabilizing and elasticity-enhancing components (styrene-butadiene-styrene block copolymers, rubber particles) forms a cross-linked composite elastic network under vulcanization, enhancing the elasticity and weather resistance of asphalt. (2) The introduction of viscosity-enhancing components (terpene resin, rich in C=C bonds and carboxyl active groups) forms a hydrogen bond network structure with the asphalt molecular chains and the aforementioned composite elastic network, improving the interfacial bonding performance of asphalt and also enhancing the stability of asphalt during storage against separation. (3) The introduction of plasticizing and volumetric components (dibutyl phthalate) weakens the dipole orientation force between polymer molecules in asphalt at high temperatures, increases molecular chain mobility,

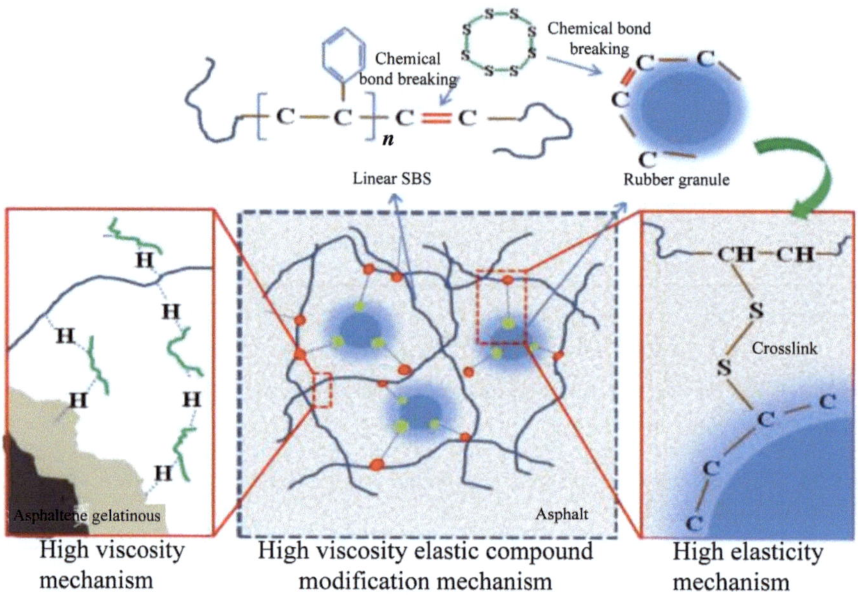

Figure 6.27 Principle of high-viscosity and elasticity asphalt modification.

Table 6.25 Performance comparison of RHVE with domestic and foreign asphalts.

Technical indicators	Rubber asphalt	HVA asphalt	Japanese TPS asphalt	RHVE asphalt
Softening point (°C)	60–70	80–90	80–85	≥90
Extension at 5 °C (cm)	10–20	40–50	40–50	≥70
Viscosity at 60 °C (Pa·s)	4000–7000	≥50,000	20,000–50,000	≥50,000
Elastic recovery (25 °C) (%)	80–90	95–100	90–100	≥95
Adhesive toughness (25 °C) (N·m)	10–15	≥25	≥25	≥30

achieves control over the modulus of asphalt, and promotes the uniform dispersion of the modifier. Table 6.25 compares the performance of RHVE with other asphalts.

An interface reinforcement layer composed of a single-size distribution of fine aggregate embedded in an asphalt mat is added on the surface of the concrete (Figure 6.28). By controlling parameters such as asphalt performance, asphalt thickness, fine aggregate particle shape, and embedding density, the following interface functions are achieved: (1) improve the adhesion and shear strength between the cement and asphalt concrete paving layer and coordinate interlayer deformation; (2) enhance the waterproof performance of the concrete panel to prevent water and deicing salt (Cl^-) infiltration that leads to corrosion of the steel mesh and steel plates; and (3) absorb and dissipate the energy from stress concentration at the expansion (crack) joints of the concrete layer, preventing crack reflection to the asphalt concrete layer. This reinforcement layer increases

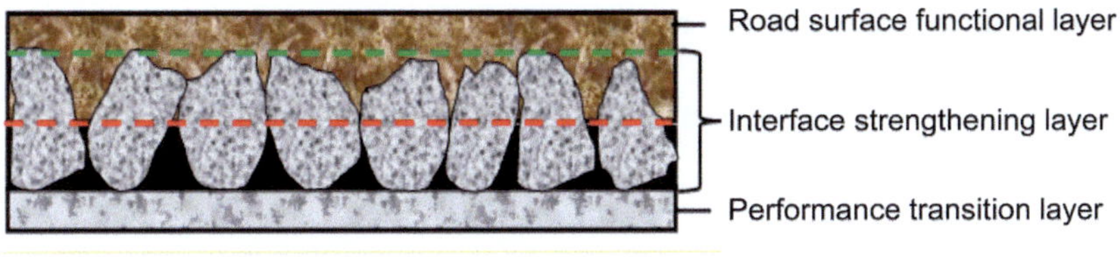

Figure 6.28 Schematic diagram of the interface reinforcement layer structure.

the interface bonding strength (≥0.8 MPa) and shear strength (≥1.4 MPa), and the fatigue cracking times at the initial crack reach 50 times that of the scheme without a stress absorption layer.

6.2.3.3 Pavement Functional Layer Materials

The high-viscosity and high-elasticity asphalt (RHVE) developed has been used to create a high-strength asphalt mastic (RHVE-SMA), and a pavement functional material with a discontinuous gradation and skeleton framework structure has been developed [54–58], the technical principle of which is shown in Figure 6.29. The high-strength mastic, primarily composed of high-viscosity and high-elasticity asphalt, mixed with slag and fibers, can increase the thickness of the asphalt film on the aggregate surface and interfacial bonding performance, enhancing the pavement's resistance to water damage, low-temperature cracking, and fatigue resistance. The use of large-diameter crushed stones with a small amount of stone chippings to form a discontinuous skeleton framework can reduce the anisotropic deformation of pavement materials at high temperatures, enhance pavement construction depth, thereby improving the pavement's anti-slip and noise reduction performance, solving the problem of asphalt concrete's difficulty in balancing high-temperature resistance, low-temperature performance, and service durability.

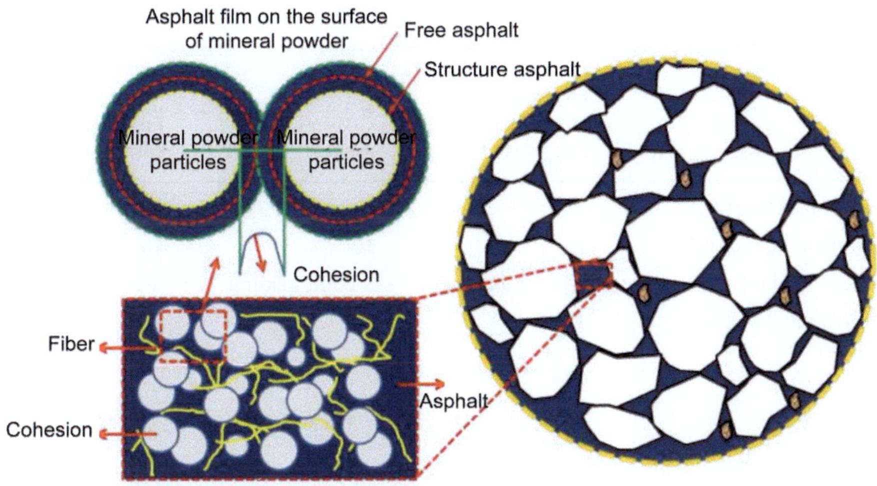

Figure 6.29 Formation mechanism of RHVE-SMA structure.

The pavement functional layer material developed by this technology offers good driving comfort, resistance to rutting, water stability, and fatigue resistance and has significant technical advantages compared to similar materials domestically and internationally.

6.2.3.4 Fatigue Resistance Performance of Steel Bridge Deck Composite Pavement Structure

The pavement design scheme "steel plate + steel mesh + shear studs + lightweight high-strength and high-toughness cementitious engineering composite material + stress absorption functional layer + high-viscosity SMA-13" is adopted. The lightweight high-strength and high-toughness cementitious engineering composite material, stress absorption functional layer, and high-viscosity SMA-13 are sequentially poured to form a composite pavement component. The structural parameters are selected as follows: steel plate (14 mm), steel mesh (diameter ϕ10, mesh size 100 × 100 mm), shear studs (spacing 400 mm, height 45 mm, diameter ϕ16 mm), lightweight high-strength and high-toughness cementitious material (thickness 50 mm), stress absorption functional layer, and high-viscosity SMA-13 (thickness 50 mm). Comparative analysis was conducted with currently available domestic bridge deck pavement materials exhibiting excellent fatigue performance—cast-in-place asphalt concrete and epoxy asphalt concrete—as presented in Table 6.26.

From the table above, it can be seen that although there are slight differences in the test design and index control of different pavement structures, the composite pavement structure mentioned in this paper, which uses the prepared lightweight high-strength and high-toughness cementitious engineering composite material as the concrete paving layer, exhibits excellent fatigue performance, reaching more

Table 6.26 Comparison of fatigue test indicators for different pavement structures.

Index	Composite pavement structure in this paper	Cast-in-place asphalt	Epoxy asphalt concrete
Test temperature (°C)	20	20	20
Loading frequency (Hz)	10	10	10
Loading method	3-point loading	5-point loading	3-point loading
Loading point distance from support (mm)	150	125	150
Strain level (με)	700		
Steel plate plan Dimensions (mm)	550 × 150	700 × 200	380 × 100
Steel plate thickness (mm)	14	14	14
Fatigue life/10,000 times	>1000	>1000	>1000

than 10 million cycles, and achieves the same excellent level of resistance to bending and tensile fatigue as epoxy asphalt concrete and cast-in-place asphalt concrete [59–62].

6.2.4 Construction Technology and Engineering Application of Steel Bridge Deck Pavement

6.2.4.1 Construction Technology

(1) Surface treatment of steel structure mainly includes the following steps:

(1) **Welding shear studs:** Welding points are set using the hanging line method, and the weld points are ground to reveal the metal luster with a grinding machine, and shear studs are welded one by one using arc welding. The welding machine automatically controls the welding parameters, with ceramic porcelain rings assisting in forming. The penetration depth is set according to the thickness of the top plate of the steel box beam and is determined by adjusting the current and welding time, generally controlled within 1–2 mm.
(2) **Sandblasting and rust removal and spray coating with anti-corrosion paint:** Automatic dust-free sandblasting machine for sandblasting and rust removal: used until there are no visible oil films, contaminants, and rust on the steel plate surface, and the steel plate surface shows a clear metal luster; Sa2.5 level, Rz40–80 μm. Anti-corrosion paint is sprayed using high-pressure spraying equipment: the adhesion is greater than or equal to 5 MPa; the dry film thickness is controlled within 75–100 μm; it is required to be smooth, uniform, and free of bubbles and cracks.
(3) **Tying of steel mesh:** LL450, ϕ5.5, cold-rolled ribbed steel mesh with a mesh size of 10 cm × 15 cm is used, with longitudinal reinforcement on top and transverse reinforcement at the bottom, the centerline of the transverse reinforcement is 2.2 cm away from the steel bridge deck panel. Tying is performed at all intersections of the steel mesh and shear studs, and spot welding is used every 3 points to weld the steel mesh and shear studs together. Spacers are set under the steel mesh, with a height of 2 cm.

(2) Preparation of high-strength and high-toughness lightweight aggregate concrete mainly includes the following steps:

(1) **Determination of concrete mix proportion:** After extensive experiments and on-site mixing conditions at the mixing station, a high-toughness lightweight aggregate concrete mix ratio is proposed in Table 6.27 (the recommended mix

Table 6.27 Mix ratio of lightweight aggregate concrete.

Cement	Fly ash	Sand	Lightweight aggregate	Water	Polymer	Steel fiber	Water reducer	Stabilizer MB
kg/m^3							Percentage of binder materials	
440	80	639	510	150	20	78	1.0%	0.05%

ratio, which should be slightly adjusted in actual projects based on the raw material conditions). Lightweight aggregates have a greater quality variability compared to ordinary aggregates, and they need to be pre-wetted before use. To ensure that the concrete used in actual projects has the same workability as the laboratory-mixed concrete, strict control over the raw materials, production process, transportation, and pouring techniques of lightweight aggregate concrete is required [47, 63].

(2) **Pre-treatment of raw materials:** The pre-wetting process of lightweight aggregates requires strict control of the pre-wetting time, and the treated lightweight aggregates should be in a saturated surface-dry state. The water content of sand must be accurately and site determined because lightweight aggregate concrete is particularly sensitive to water; next is the order of material placement, which should be strictly followed according to the sequence in Figure 6.30. The placement of steel fibers is the most critical in this process. To ensure their uniform distribution in a dry state, steel fibers must be added after the dry materials are fully and evenly mixed. The steel fibers should be shaken out of the bag before use and must not be poured into the mixer in heaps.

(3) **Concrete production and construction:** a. *Mixing:* Strictly control the order of material addition and mixing time and accurately weigh cement, aggregates, and water to prevent significant fluctuations in the water–binder ratio. Strengthen the slump test when discharging the mix, as the slump is one of the important indicators for controlling the construction quality of lightweight aggregate concrete and should be strictly controlled. b. *Pouring and vibrating:* Clean the bridge surface before pouring, and pre-wet it with water before proceeding with pouring. During the vibration process, initially level the surface with tools such as a straightedge and steel shovel, then use a flat vibrator to vibrate until the cement-based material surface is paste-like and stops bubbling, finally use a drum roller to level the surface and assign a dedicated person to monitor whether additional screening is needed. If clumping of steel

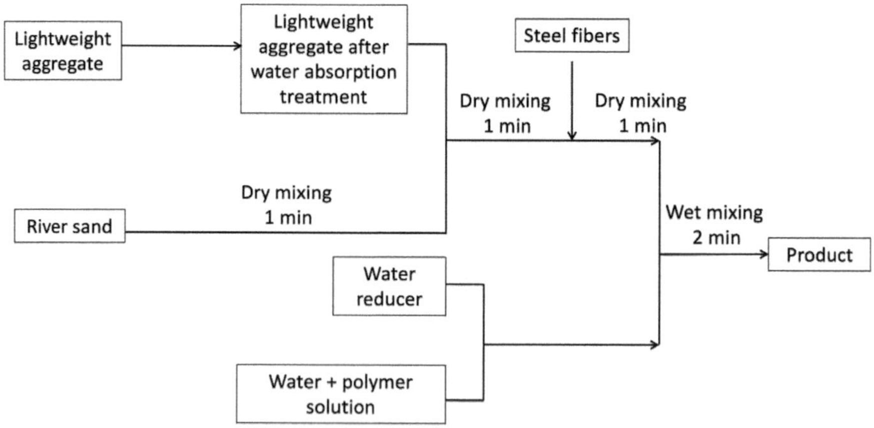

Figure 6.30 Mixing process of high-toughness lightweight aggregate concrete.

fibers is found during the spreading process, it must be promptly torn apart and shaken out by hand, and care should be taken to avoid segregation during spreading. c. *Finishing:* Finishing, also known as fine surface treatment, ensures the flatness of the bridge surface, which is checked with a 3 m straightedge, and areas that do not meet the standards are screened according to the actual situation. d. *Curing:* After the concrete is poured, proper curing should be carried out to prevent early shrinkage cracks in the concrete. Immediately cover the surface with a film after pouring, and after the concrete begins to set, cover the surface with wet jute bags or wet straw mats. At the same time, because steel fiber concrete has a higher early strength, early moist curing should be intensified, with water sprinkled several times a day to maintain a damp state [39, 64–67].

(3) Construction of high-viscosity and high-elasticity stress-absorbing bonding layer

(1) Technical requirements for high-viscosity and high-elasticity asphalt of the stress-absorbing bonding layer: The high-viscosity and high-elasticity asphalt used as the stress-absorbing bonding layer should emphasize the prevention of reflective cracks and interface shear resistance. The requirements for penetration, viscosity, and elastic recovery are relatively high, as shown in Table 6.28.
(2) Technical requirements for the stress-absorbing bonding layer: The technical requirements for the high-viscosity and high-elasticity stress-absorbing bonding layer are shown in Table 6.29.
(3) Construction of the stress-absorbing bonding layer: a. *Concrete slab treatment:* The concrete slab should be leveled and roughened (sandblasting), and the concrete surface should be ensured to be clean and dry. b. *Preparation of raw materials:* The asphalt should be stored at low temperatures before use and rapidly heated to 190–200 °C before use, and it should be directly injected into the asphalt pipe in front of the asphalt and chipped stone spreader. The chipped

Table 6.28 Technical requirements for rubber high-viscosity and high-elasticity modified asphalt used in stress-absorbing bonding layers.

Test content	Units	Specification requirements	Actual control indicators
Penetration (25 °C/100 g/5 s)	0.1 mm	≥40	40–60
Extension (5 cm/min)	cm	≥50(15 °C)	≥50(15 °C) ≥40(5 °C)
Softening point	°C	≥80	≥90
Viscosity at 60 °C	Pa·s	≥20,000	≥60,000
Flash point	°C	≥260	≥260
Elastic recovery/25 °C	%	≥80	≥90
Adhesive toughness/25 °C	N·m	≥20	≥20
Toughness/25 °C	N·m	≥15	≥15

Table 6.29 Technical requirements for the stress-absorbing bonding layer.

Item	Technical requirements	Remarks
Asphalt application rate	1.7–2.0 kg/m^2	—
Chipped stone application rate	Chipped stone of 9.5–13.2 mm, coverage rate of 70–85%	—
Tensile strength	≥0.3 MPa (at room temperature) ≥0.2 MPa (at temperatures ≥35 °C)	Use the American digital display pull-off type adhesion tester.

stone aggregates are heated through the mixing tower, controlling the heating temperature of the chipped stone aggregates (175–190 °C), and pre-mixed with SBS-modified asphalt (around 0.3% oil-to-stone ratio). c. *Spreading process:* The stress-absorbing layer should not spread too quickly during the spreading process, and the spreading width should not exceed 3.5 m. The asphalt application rate is controlled at 1.7–2.0 kg/m^2, and chipped stone with a particle size of 9.5–13.2 mm is spread simultaneously, with the coverage rate controlled at 70%–85%, and the specific spreading quality is based on the indicators determined by the test section. d. *Rolling:* After spreading the chipped stone, roll with a tire roller, 2–3 times; the rolling speed should not be too fast; and the roller should control to roll at a speed close to that of the spreader.

(4) Asphalt mixture construction

(1) Technical requirements for SMA mixture. The comparison of actual construction index control and specification requirements for SMA paving is shown in Table 6.30.

(2) Production and paving of SMA asphalt. a. Mixing and construction temperature control: asphalt heating temperature 165–175 °C, aggregate temperature 180–190 °C, asphalt mixture delivery temperature 170–180 °C, temperature upon arrival at site 160–180 °C, paving temperature 160–170 °C, Initial rolling temperature must not be lower than 160 °C, subsequent rolling temperature must not be lower than 150 °C, and final rolling temperature must not be lower than 110 °C. b. Transportation: Use large material transport vehicles with good thermal insulation measures for transporting the SMA mixture. c. Paving: Due to the relatively low production of high-viscosity SMA asphalt mixture, the paving machine speed should be controlled at 2.0–3.0 m/min. Use a high-frequency, low-amplitude vibration method to balance initial compaction and avoid aggregate breakdown. The paving speed should match the supply speed of the mixer, maintaining a constant paving speed. For small local damaged areas, manual paving is used. d. Compaction: The rolling equipment uses a 10-ton static dual steel-wheel roller, and initial rolling is carried out immediately after the mixture is paved. First, use two dual steel-wheel rollers for initial rolling for 2 times, then use two steel-wheel rollers for subsequent rolling for 2–4 times, and finally use one dual steel-wheel roller for final rolling

Table 6.30 Technical requirements for high-viscosity SMA mixture.

Test item	Units	Specification requirements	Actual control indicators
Marshall test specimen size	mm	ϕ101.6 × (63.5 ± 1.3)	ϕ101.6 × (63.5 ± 1.3)
Number of blows (both sides)	Times	75	75
Void ratio	%	3–4	3–4
Stability	kN	≥6	≥6
Mineral aggregate void ratio	%	≥16.5	≥16.5
Coarse aggregate framework void ratio	—	≤VCADRC	≤VCADRC
Asphalt saturation	%	75–85	75–85
Shellenberger bleeding loss	%	≤0.1	≤0.1
Kentucky flying loss	%	≤15	≤5
Dynamic stability	Times/mm	≥3000	>6000
Residual stability after immersion	%	≥80	≥90
Freeze–thaw splitting strength ratio	%	≥80	≥85
Permeability coefficient	mL/min	≤80	≤80
Low-temperature bending failure strain	μ_ε	≥2500	≥2500

and finishing for 1–2 times to eliminate surface wheel marks and improve the flatness of the paving layer. Traffic can only be opened for the first time when the temperature is below 50 °C.

6.2.4.2 Engineering Applications

The integrated technology described by the author's team has been successfully transformed and applied (Figure 6.31), with a total of more than 1000 spans of steel beams and over 1 million square meters of area. Until now, all paving materials have shown good durability, with no reports of damage, including the Sichuan Hejiang Changjiang Bridge with a steel structure span of 530 m and the Zhi Cheng Changjiang River Bridge, which is a combined road and rail bridge. The longest application, the Wuhan Outer Ring Expressway elevated interchange bridge, has been in use for over 15 years, achieving significant economic and social benefits [68].

The technology application can be divided into three scenarios:

(1) The overall application of the outcome technology in the design and construction of new bridges, including the direct use of paving structure connection technology (Type A) and the multilayer composite structure design of performance transition layer, interface reinforcement layer, and pavement functional layer

Figure 6.31 Application of composite paving structure of steel bridge surface and high-strength and high-toughness concrete.

in the design of new bridges, as well as material preparation and supporting construction technology. Representative projects include the Sichuan Hejiang Changjiang Bridge across the Yangtze River (main span of 530 m) and other projects.

(2) The overall application of the outcome technology in the pavement engineering of new and old bridges, including the use of paving structure connection technology (Type B) and multilayer composite structure design of performance transition layer, interface reinforcement layer, and pavement functional layer, as well as material preparation and supporting construction technology. Representative projects include Wuhan Jin Qiao Avenue (spanning the Beijing-Guangzhou Railway Hankou Station line, with a span of 150 m) and other projects.

(3) The partial application of the outcome technology in the pavement engineering of new and old bridges, including the use of interface reinforcement layer and pavement functional layer material paving technology and the application of concrete bridge surface, high-toughness lightweight aggregate concrete, and high-viscosity high-elasticity modified asphalt in bridge surface repair projects. Representative projects include Wuhan Erqi Cable-Stayed Changjiang Bridge (main span of 2 × 650 m), Wuhan Yingwuzhou Suspension Changjiang Bridge (main span of 2 × 650 m), Wuhan Changjiang Bridge I and II repair, and other projects.

6.2.4.3 Technical Application Effects

(1) The material and structural design scheme of the durable steel box girder bridge deck composite paving layer, which uses "shear keys + steel mesh + high-strength and high-toughness lightweight aggregate concrete" as the bottom layer and anti-slip, noise-reducing asphalt concrete as the top layer can

effectively address the technical challenges of easy displacement, rutting, and cracking that are common with this type of bridge deck paving layer.

(2) High-strength, high-toughness, and durable lightweight aggregate concrete has been developed, which serve as a gradient transition between steel and asphalt layer materials, effectively mitigating the incompatibility issues caused by the significant performance differences between steel and asphalt. The concrete has a compressive strength of 50–70 MPa, flexural strength greater than 7.0 MPa, toughness index η_{30} greater than 30, 90-d drying shrinkage less than 2.8×10^{-4}, impermeability class of P18, freeze–thaw resistance class greater than F300, and fatigue load life greater than 3 million cycles.

(3) The SMA asphalt mixture suitable for the top layer of steel box girder bridge deck paving, with good anti-slip properties and durability, has been prepared using self-developed high-viscosity modified asphalt. It has a construction depth of ≥1.1 mm, dynamic stability at 60 °C of ≥9000 cycles/mm, dynamic stability at 70 °C of ≥5000 cycles/mm, residual stability after immersion of ≥9%, freeze–thaw splitting residual strength ratio (TSR) of ≥90%, bending strain at −10 °C of $\geq 5 \times 10^{-3}$, and water immersion flying loss of ≤15%.

(4) A homogeneity control admixture for lightweight aggregate concrete has been developed, enabling the long-distance pumping construction of 200-m steel fiber high-strength lightweight aggregate concrete; a comprehensive set of design, construction, and quality control technologies for steel box girder bridge deck paving has been established, and the corresponding construction techniques have been formulated.

6.2.4.4 Technical and Economic Comparison

Based on theoretical research, technological development results, and a large number of engineering application practices, a technical and economic comparison analysis is conducted, as follows:

(1) Pavement material technology

From Table 6.31, it can be seen that the C50 lightweight high-toughness concrete developed by this technology has significantly improved in workability compared to ordinary steel fiber concrete, steel fiber lightweight aggregate concrete, and lightweight aggregate concrete. The slump loss of the concrete over time is small, making it suitable for long-distance pumping; the apparent density has been reduced by 32% compared to C50 ordinary steel fiber concrete, meeting the load-bearing requirements for the paving materials of steel box girder bridge decks; the toughness index at 30 d reaches 28, which is more than 20 times higher than that of C50 ordinary lightweight aggregate concrete, and also shows a certain improvement in toughness compared to steel fiber concrete; the drying shrinkage rate is small, and the impermeability class has significantly improved compared to ordinary concrete. It is evident that the developed C50 high-toughness lightweight concrete is very suitable as a bridge deck paving material.

Table 6.31 Comparison of C50 concrete performance.

Technical content indicator	Steel fiber concrete	Polymer concrete	Lightweight aggregate concrete	This technology
Density (kg/m^3)	2500 ± 100	2450 ± 100	1900 ± 50	1900 ± 50
Elastic modulus adjustment range (GPa)	35–40	30–38	22–32	20–35
Toughness index (η_{30})	15–20	5–10	1–1.5	25–28
Fatigue resistance to bending (stress ratio 0.7, 10 Hz)/10,000 times	55	40	30	200
Slump/mm (0 h/1 h)	200/180	200/180	200/140	240/240

Table 6.32 Comparison of mechanical performance, road performance, and durability of paving layer asphalt concrete.

Test indicator	Cast-in-place (GA)	Epoxy (EA)	Ordinary (SMA)	This technology
Maximum tensile stress of the top layer (MPa) (horizontal/longitudinal)	0.41/0.33	0.28/0.22	0.35/0.28	0.10/0.08
Freeze–thaw splitting strength ratio (%)	95	88	90	95
Dynamic stability/times/mm	865	12,000	7120	8854
Texture depth (mm)	0.1	0.32	1.0	1.15
Fatigue resistance to bending (700 με, 10 Hz)/10,000 times	>1000	>1000	510	>1000

(2) Asphalt pavement material technology

From Table 6.32, compared with several commonly used pavement materials at home and abroad, SMA-13 has the deepest texture, reaching 1.15 mm, which provides better anti-slip capabilities; its noise level is between 74 and 78 dB, which is 5–8 dB lower than concrete pavement and 3–4 dB lower than AC, EA, and other asphalt concrete pavements. This indicates that SMA-13, as a bridge deck paving material, has good anti-slip and noise reduction effects, improving the comfort and safety of driving on the bridge surface.

(3) Economic cost comparison

In terms of the performance of the developed materials, the effectiveness in engineering use, construction convenience, as well as engineering quality and comprehensive economy, the technology described in this paper has obvious

Table 6.33 Comparison of service life and cost of different steel box girder bridge paving materials.

Technical content	SMA/GA	Double-layer SMA	Double-layer EA	This technology
Design service life/year (top/bottom layer)	15/15	15/15	15/15	15/50
General common maintenance start year (year)	6–10	2	6–10	>10
Cost (yuan/m^2)	1250	1312(175 × 7.5)	1450	410

technical and economic advantages compared with similar technologies at home and abroad (see Table 6.33).

Compared with the three major existing technologies, the one-time construction cost of the new steel bridge deck paving structure is only higher than that of double-layer SMA and much lower than GA and EA. However, in most cases, SMA begins to deteriorate after about 2 years and requires repeated repairs and maintenance before reaching its designed service life, with the actual construction and maintenance costs far exceeding those of this technology. The improvement in the performance of the new paving material and structure significantly extends its service life, effectively reducing regular repairs and maintenance. Especially, the use of concrete as the bottom material of the paving structure has a designed service life of up to 50 years, which is much longer than the 15 years for other technologies. In this way, when the upper layer of asphalt paving material with a shorter service life needs to be repaired, the bottom layer does not need to be replaced, which can further save maintenance materials and funds. Obviously, due to the good stability and durability of this technology, the long-term comprehensive cost is the lowest among all technologies, only one-third of others, and has significant economic advantages.

(4) Social benefits

(1) The enhancement of the performance of the new paving materials and structures extends their service life, effectively reducing the need for frequent repairs. In particular, the use of concrete as the lower material of the paving structure has a designed service life of over 50 years. In this way, when the upper layer asphalt paving material with a shorter service life needs repair, the lower layer does not need to be replaced, which can further save on maintenance materials and funds. This achievement has significant social benefits in resource conservation and environmental protection.
(2) The improvement in various road functions (noise reduction, anti-slip) and durability of bridge paving materials, without diseases such as slippage, rutting, cracking, and potholes, results in a smooth road surface, improving road quality, enhancing driving comfort and safety, and greatly reducing

traffic accidents. This significantly increases the bridge's capacity for passage, promoting local economic and social development.

(3) The bridge paving technology enhances the waterproof and anti-corrosion protection capabilities of the paving layer for the steel bridge deck; the structural connection of the paving materials with the steel plate can better distribute the road load evenly, improving the structural stability of the bridge deck steel plate and effectively enhancing the safety of the bridge structure.

(4) With the rapid development of China's economic construction, more and more urban roads and highways will adopt steel box girder bridges that have a low self-weight, are conveniently erected, and have great spanning capabilities. The technical achievements have formed a core technology system with independent intellectual property rights for the design, preparation, and composite structure design and assembly of paving layer materials for steel box girder bridge decks, as well as their engineering applications. This lays an important foundation for the paving technology of steel box girder bridges in China and has significant economic and social benefits in promoting the development of China's new steel box girder bridge industry.

6.3 Composite Materials for Structural/Functional Tunnel Concrete

6.3.1 Overview

6.3.1.1 Technical Requirements for Shield Tunnel Structural Material

With China's rapid economic development and increasing urbanization level, land resources are becoming increasingly scarce and large-scale development and utilization of underground space for infrastructure construction is an inevitable trend. Since entering the 21st century, large-scale tunnel projects across rivers and gulfs and connecting islands have been planned and constructed in our country, and these underground projects, compared with the past, are larger in scale, more difficult to construct, have a more hostile environment, and are more technically difficult, and the construction materials used in the projects have also put forward higher requirements and challenges.

Shield tunnel is the main development direction of modern large-scale tunnel engineering technology. The main structure of a shield tunnel is composed of a segment lining structure spliced by concrete segments, concrete lining structure, and cylinder structure (Figure 6.32). Under severe and complex geological conditions with high water pressure, the technical requirements for large-section shield tunnel structure and materials are as follows: (1) The strength of the concrete segment can meet the construction requirements of shield tunneling and segment ring. (2) The tunnel structure has high strength, high impermeability, and durability. (3) The tunnel structure can withstand a variety of complex external forces and accidental damage.

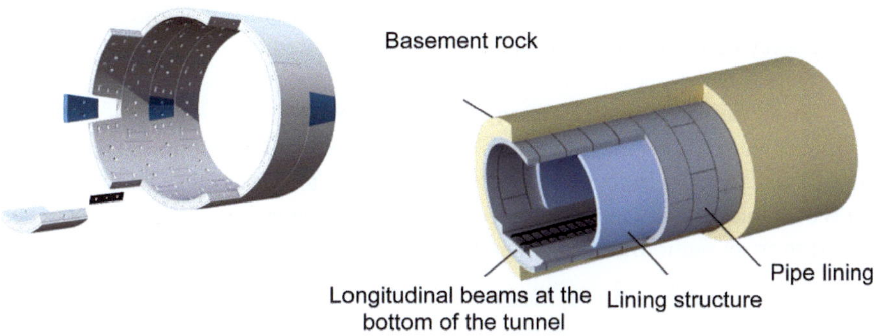

Figure 6.32 Shield tunnel concrete segment, liner, and cylinder structure.

6.3.1.2 Characteristics of Deepwater Large-Diameter Shield Tunnel Structure Materials

Large-diameter (more than 10 m) shield tunnel is a major technology that has been rapidly developed in China's tunnel construction since the new century. The Wuhan Yangtze River Tunnel is China's first deep water large-section shield tunnel. Before the construction of the Wuhan Yangtze River Tunnel, the only large-diameter shield tunnel in China is the one in Shanghai across the Huangpu River, whose stratum is dominated by silty clay, which is very different from the composite, highly permeable stratum that the Wuhan Yangtze River Tunnel traverses, and the foreign engineering experience in similar strata is also scarce. Second, the maximum water pressure of a large-diameter shield tunnel in China is about 0.4 MPa, and there are only a few foreign tunnels reaching the level of 0.6 MPa, such as the Tokyo Bay Undersea Tunnel. Third, the maximum water pressure of a domestic large-diameter shield tunnel is about 0.4 MPa, and there are only a few foreign tunnels with a level of 0.6 MPa, such as the Tokyo Bay Undersea Tunnel, etc. In addition, the longest distance of a large-diameter shield tunnel is about 1400 m domestically and only 2500 m abroad.

The concrete segment of shield tunnels is the key material composing the lining cylinder, and the performance and quality of the segment are crucial, which affect the construction safety of the tunnel and determines the long-term safety of the tunnel. Concrete segments are not only exposed to high-pressure water conditions and harsh service environments but also need to take into account the influence of fire, explosion, seismic, and other incidental factors. Therefore, the concrete segment not only requires structural performance design but also comprehensive research from the perspective of the segment and its lining cylinder structure, to give the segment seepage resistance, corrosion resistance, crack resistance, fire resistance, explosion resistance and other multifunctional resistance, and develop the segment assembly structure design technology and safety enhancement technology. The difficulty of deepwater large-diameter shield tunnel engineering technology lies in the high water pressure, large cross-section and long distance to the shield tunnel main structure strength, stiffness, and impermeability durability and tunnel safety operation put forward higher requirements and challenges. Existing technology can no longer meet the needs of tunnel construction, and new breakthroughs in engineering technology and key construction materials are urgently needed.

6.3.1.3 Major Research and Technology Development

The author's team has carried out the following research and development work and achieved the following innovative results in response to the requirements and difficulties of shield tunnel engineering technology and material performance in deep water, large cross-section, and harsh and complex geological conditions:

(1) Researching and innovating integrated structural/functional design and preparation technologies for shield tunnel concrete. Propose guidelines for the design of tunnel concrete durability and develop technologies for highly impermeable, fire-resistant, burst-resistant, stray-current-resistant and chloride-ion-resistant concrete structures and materials, to improve the ability of tunnels to withstand all kinds of external damage, and to improve the durability and safety of concrete structures.
(2) Research on the layered design technology of tunnel concrete segment structure, optimize the composition structure, form the overall performance of the composite effect, innovate the concrete segment lining structure suitable for such conditions, and develop a new form of "segment lining + non-enclosed lining" superimposed structure, so as to enhance the overall structural strength of shield tunnels, and adapt to the construction of large-diameter and long-distance tunnels.
(3) Research and development of high-performance large-diameter shield tunnel concrete segment production process. Research and development of high-temperature steam curing process and layered structure preparation technology to improve the processing precision and volume stability of large-size concrete segment pieces. Research and development to form from the selection of raw materials, concrete mixing, pouring, vibration, steam curing, and other preparation processes and quality control of a complete set of technology and promote the results of technology in many practical engineering applications.

6.3.2 Structural Design of Tunnel Concrete Segment and Lining

6.3.2.1 Integrated Design of Segment Structure/Function

Existing domestic tunnel segment materials have poor durability, making it difficult to withstand harsh environmental erosion and meet project requirements. In response to the high impermeability and long service life requirements of the concrete segment for the deep water, high-pressure, and large-section project of the Wuhan Yangtze River Tunnel, the method of functional layered gradient design of the concrete segment is proposed based on the principle of functional gradient material design. The functional design is shown in Figure 6.33(a), in the vertical direction along the soil and water pressure, in the order of high impermeable protective layer, structural layer, fire, and blast resistant layer. The structural design of the segment is shown in Figure 6.33(b). In the vertical direction along the soil and

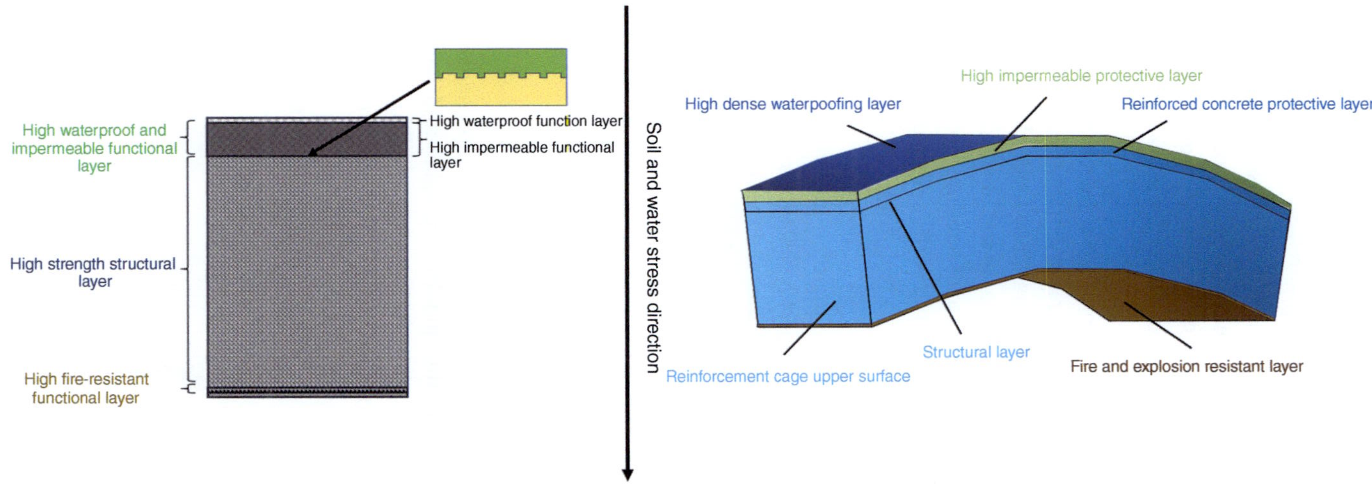

Figure 6.33 Integrated concrete segment functional/structural design and structural modeling.

water pressure, there are five gradient material layers in the order of high-density waterproofing layer, highly impermeable protective layer, reinforced concrete protective layer, high-strength structural layer, and fire-resistant and anti-explosive layer from the top down, and they are combined into three functional layers. Among them, the high-density waterproof layer, high impermeable protective layer, and reinforced concrete protective layer together constitute the concrete segment impermeable functional layer; reinforced concrete protective layer and high-strength structural layer constitute the tunnel segment structural functional layer; and fire-resistant and anti-explosive layer are reinforced to prevent accidental damage to the tunnel.

At the same time, to effectively improve the interfacial bonding performance between the impermeable protective layer and structural layer, the interface between the two functional layers is designed as an embedded folded-line interface, which greatly increases the interface contact area, so that the different materials penetrate each other, occlude, and lap. The interfacial bonding strength, shear strength, etc. can be greatly improved to ensure that the performance of each functional layer of the segment is uniformly transitioned to realize the integrated design of function and structure.

In the specific segment preparation, the total thickness of the segment used in the Wuhan Yangtze River Tunnel Project is 500 mm, of which the thickness of the outer protective layer material is 25–35 mm and that of the structural layer is about 465–475 mm. The lining layer inside the tunnel is evenly brushed with fire-resistant and anti-explosive layer material, thus realizing the multifunctionality of the segment.

6.3.2.2 Optimized Design of Concrete Segment and Lining Structure for Large Diameter Shield Tunnel

(1) Force analysis and parameter optimization.

The beam-spring model was adopted to carry out a comprehensive calculation and analysis (Figure 6.34), to study the quantitative value and law of the soil-water pressure acting on the main structure of the shield tunnel, to find out the interaction characteristics between the main structure of the shield tunnel and the strata, and

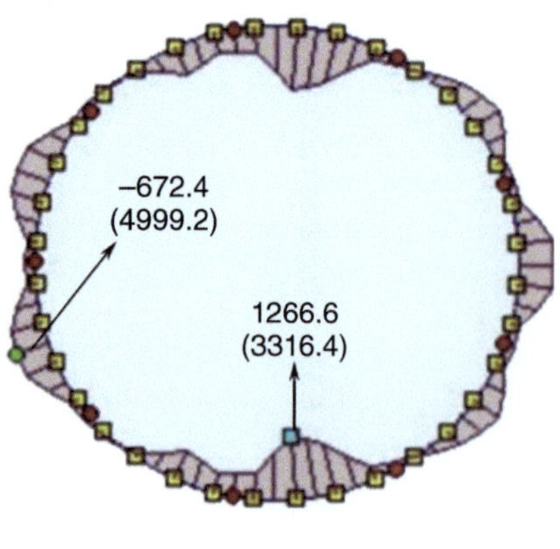

Figure 6.34 Beam spring model.

Figure 6.35 Shell-joint spring model.

to find out the actual state of the shield tunnel and the force characteristics of the main structure of the shield tunnel. The shell-spring model (Figures 6.35 and 6.36) was developed to study the spatial mechanical characteristics of the large-width segment structure, which provides the basis for the overall calculation of the structure. A three-dimensional solid structure finite-element model is used for analysis (Figure 6.37) to study the static mechanical behavior of single-layer assembled segment lining of the trans-river shield tunnel under high water pressure conditions, focusing on the selection of the segment, its thickness, blocking, width, and reinforcement method to carry out a systematic study, and put forward optimized design parameters to meet the requirements of the structural design.

(2) Parameter design of segment

A new structural type of lining ring of "universal wedge ring, 2 m ring width, nine equal parts" is proposed [69], which meets the design and construction requirements of large-diameter shield tunnels. Before the Wuhan Yangtze River Tunnel Project, the maximum ring width of the large-diameter shield tunnel lining ring in China was 1.5 m, and the number of blocks was eight equal parts or 7 + 1 small capping blocks. This kind of structure has the following disadvantages: (1) Due to the many types of lining rings, more steel mold plates are needed, which is not economical; (2) due to the smaller ring width, the longer the total length of segment joints, the more difficult to guarantee the waterproofing effect; (3) the way of lining ring blocking is

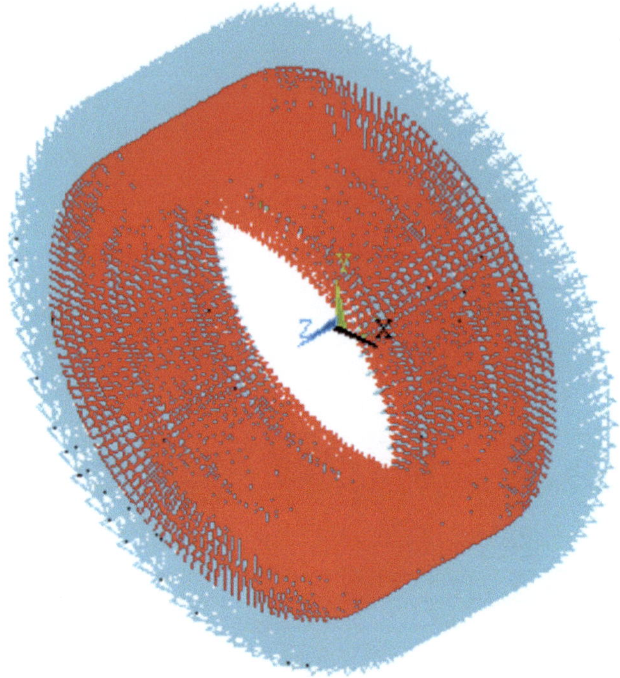

Figure 6.36 Overall schematic of loads and structure.

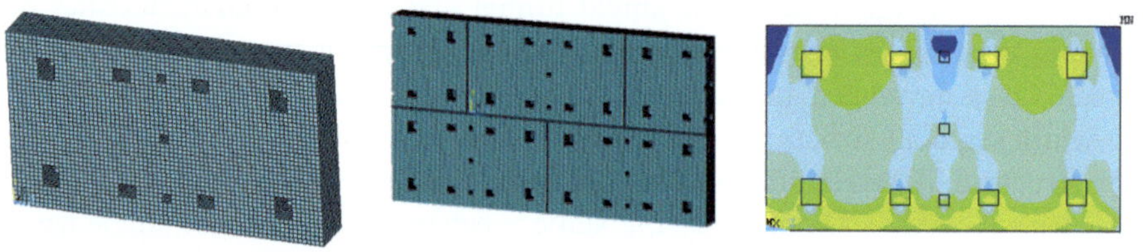

Figure 6.37 3D solid structure finite-element modeling.

directly related to the structural stiffness, and the irrational blocking will increase the structural deformation and waterproofing, and at the same time, it is inconvenient to put it together. Combined with the characteristics of the Wuhan Yangtze River Tunnel, the study proposes to adopt the new lining ring structure of a "universal wedge ring, 2 m ring width, nine equal parts." Based on the above optimized design parameters, the lining ring is divided into pieces for research and design.

(1) **Lining Ring Split Blocks:** Five generalized wedge-shaped segment basic assembly methods were studied. Table 6.34 shows that, irrespective of using staggered-seam assembling or through-seam assembling, in general, the overall stiffness of the 9-equivalent way is large, which is favorable to the deformation control and waterproofing of the structure and is especially suitable for high water pressure and strong permeable strata. At the same time, the increase

Table 6.34 Analysis of structural internal forces and deformations for different tube segmentation schemes (basic assembling method).

No	Chunking	Maximum deformation at a single point (mm)	Maximum positive bending moment (kN·m)	Maximum positive bending moment corresponding to the axial force (kN)	Maximum negative bending moment (kN·m)	Maximum negative bending moment corresponding to the axial force (kN)
FK1	7 + 1	11.815	570.46	1867.6	−326.47	2687
FK2	8 Equal parts	11.876	409.46	2035.7	−181.42	2619.8
FK3	8 + 1	13.555	415.96	2039.7	−181.82	2625.3
FK4	9 Equal parts	11.777	596.5	1909.6	−321.98	2652.3
FK5	9 + 1	11.777	595.6	1992.4	−318.05	2724.3

Table 6.35 Structural internal force and deformation analysis results for different segment thicknesses.

No	Chunking	Maximum deformation at a single point (mm)	Maximum positive bending moment (kN·m)	Maximum positive bending moment corresponding to the axial force (kN)	Maximum negative bending moment (kN·m)	Maximum negative bending moment corresponding to the axial force (kN)	No.
H1	0.45	9 equal	12.571	982.88	3907.8	−503.09	5153.3
H2	0.48	parts	12.082	1106.7	3854.9	−580.34	5313.7
H3	0.50	Ring	11.779	1193.2	3819.1	−644.04	5304.6
H4	0.55	width 2 m	11.41	1443.3	3709.8	−861.06	5539.1

in structural internal force is not too large, and the amount of reinforcement required increases very little, so it is recommended to use the 9-equivalent universal wedge ring program.

(2) **Segment Thickness:** Compare the four segment thickness programs of C50 reinforced concrete flat slabs with a ring width of 2.0 m and 9 equal sections. Select a typical section of the segment thickness program to calculate the maximum deformation of a single point, maximum positive and negative bending moment, and corresponding axial force. From Table 6.35, with the increase in segment thickness, the maximum deformation of the segment ring decreases and the bending moment increases. Considering that the tunnel is in a geological environment of high water pressure and strong water permeability and stress state of the segment structure under different thicknesses, it is recommended that the thickness of the segment be 0.50 m.

(3) **Segment Ring Width:** The selection of segment ring width needs to consider the structural stress, waterproofing, shield machine manufacturing technology, line curve conditions, and other factors. In the past, the maximum ring width of domestic large-diameter shield tunnels did not exceed 1.5 m. Five ring width programs, namely 1.0, 1.2, 1.5, 1.8, and 2.0 m, were studied and compared from the angles of stress and waterproofing. The adoption of larger ring width segments demonstrates that the maximum positive and negative bending moments at the ring width center are significantly reduced compared to those at the ring width edges. From the perspectives of structural optimization and construction cost reduction, employing larger ring widths proves advantageous. However, if the ring width is selected excessively large, it becomes detrimental not only to segment prefabrication, transportation, and assembly during construction, but also substantially increases the manufacturing complexity of shield equipment (Figure 6.38). Therefore, based on analogous overseas shield tunnel engineering projects, a segment ring width of 2.0 m is recommended.

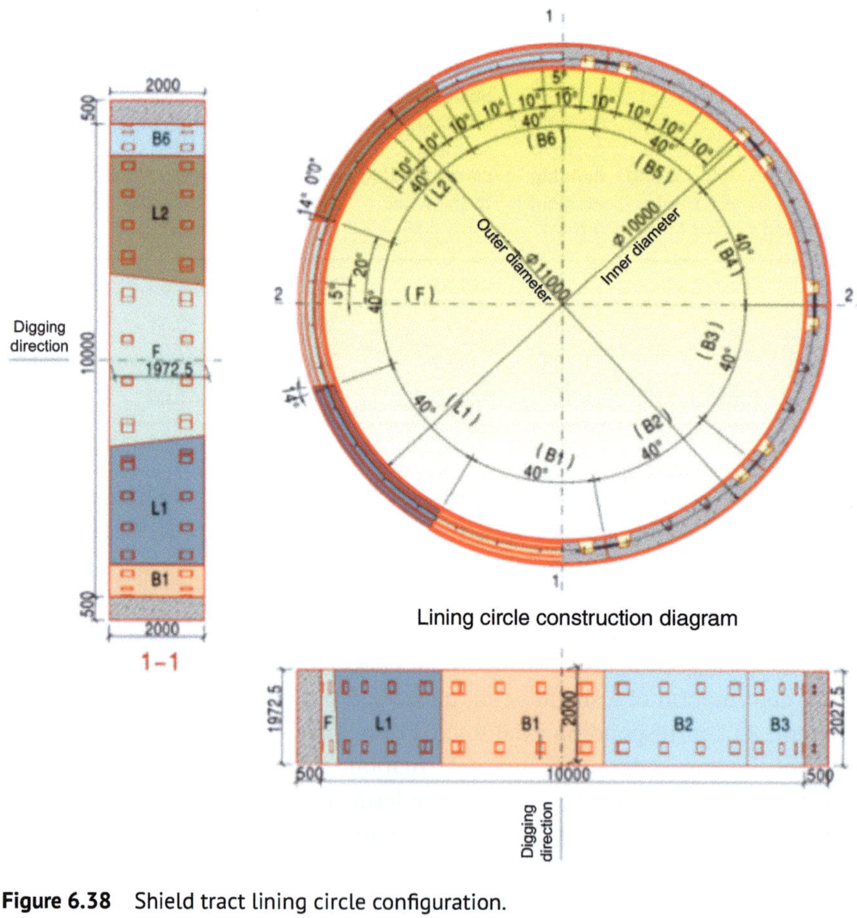

Figure 6.38 Shield tract lining circle configuration.

6.3.2.3 Assembly Design of Segment and Lining Structure

The innovative proposal of the stacked structure of non-closed lining (secondary lining) and segment ensures that the structural design is compatible with the unfavorable geological conditions and river conditions with huge scouring and siltation amplitude and improves the long-term safety of the tunnel structure [70]. Prior to this project, there were only two structural types of shield tunnels: single-layer segment lining and segment lining + fully enclosed inner lining. Single-layer segment lining is the most common shield tunnel structure, but it has low stiffness and is difficult to cope with the highly permeable tunnel environment. Segment lining + fully enclosed inner lining has high stiffness and good waterproofing, but it is uneconomical because of the need to increase the tunnel diameter. Combined with the Wuhan Yangtze River Tunnel conditions, the study proposes the use of non-closed lining and segment lining superposition structure, that is the use of circular tunnel lateral and bottom of the surplus space to set up a non-closed lining, lining and segment lining through the reinforcement connection for the superposition of the structure, the structure not only to strengthen the overall rigidity of the structure but also to completely prevent the tunnel below the center line of the circle due to local seepage and the occurrence of the " segment surge" phenomenon, to ensure that the foundation of the ground, and to ensure that the tunnel is fully enclosed. This structure can not only strengthen the overall rigidity of the structure but also completely prevent the "segment surge" phenomenon from occurring due to local leakage below the circular center line of the tunnel and ensure the foundation stability. The three lining schemes are analyzed and compared (Figures 6.39–6.41).

(1) Single-layer segment lining scheme (Scheme 1).

The tunnel is lined with a single-layer reinforced concrete segment lining (Figure 6.39), and all loads during the construction and operation periods are carried by the segment lining structure. The inner diameter of the tunnel is 10.0 m, the outer diameter is 11.0 m, and the segment is 0.5 m thick.

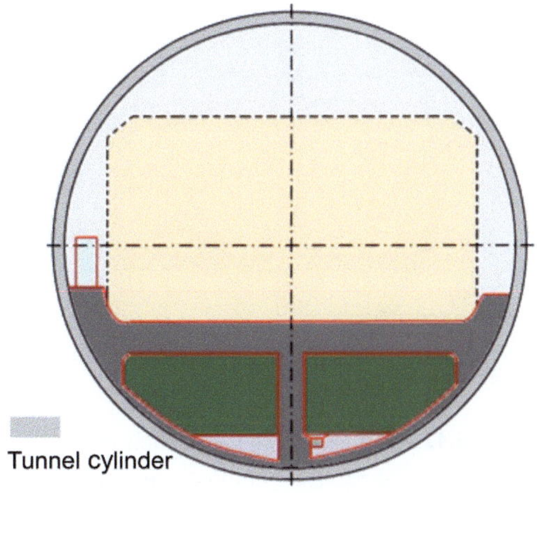

Figure 6.39 Multiple segment lining.

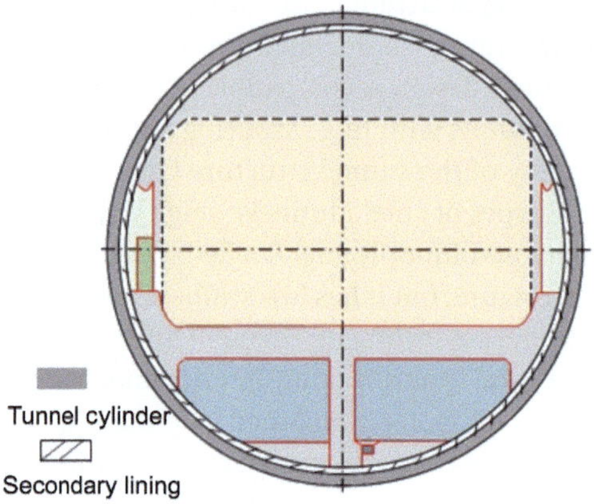

Figure 6.40 Segment lining + fully enclosed inner lining.

(2) Double-layer lining scheme with segment lining and fully closed lining (Scheme 2).

This scheme consists of precast reinforced concrete segment lining and full-ring cast-in-situ reinforced concrete secondary lining, with no waterproofing slabs between the inner and outer linings to form a stacked structure (Figure 6.40). The outer segment lining carries the external soil and water loads during the construction period, and the external soil and water pressure change loads during the operation period are carried by both the inner and outer linings. The inner diameter of the segment lining is 10.4 m, the outer diameter is 11.4 m, the thickness of the segment is 0.5 m; the thickness of the inner lining is 0.2 m, and the inner diameter of the tunnel clearance is 10.0 m.

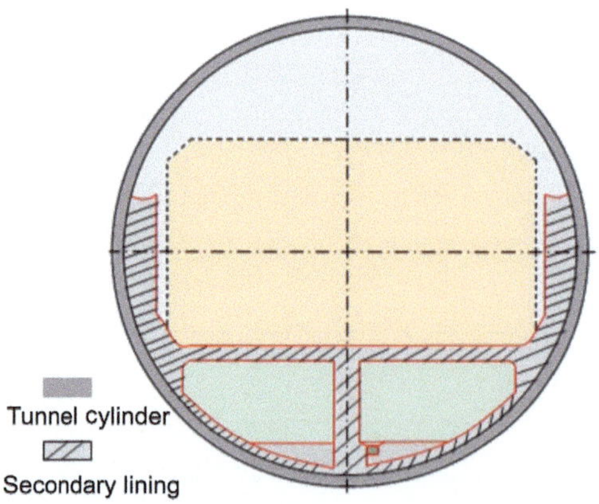

Figure 6.41 Segment lining + unenclosed interior lining.

(3) Double Layer Lining with Segment Lining and Unenclosed Lining (Scheme 3).

Based on Scheme 1, this Scheme utilizes the surplus space on both sides of the travel lanes and at the bottom of the tunnel to set up cast-in-place reinforced concrete, which is used together with the lane slabs to form an unenclosed lining structure, thus forming a double lining structure with a segment lining and an unenclosed lining (Figure 6.41). No flashing is provided between the unenclosed liner and segment lining, but appropriate anchors are required to form a stacked structure. The outer segment lining carries the external soil and water loads during the construction period, and the external soil and water pressure change loads during the operation period are carried by both the inner and outer linings. The inner diameter of the segment lining is 10.0 m, the outer diameter is 11.0 m, and the thickness of the segment is 0.5 m; the thickness of the unenclosed lining is 0.2 m at the thinnest point (at the bottom of the tunnel) and 0.5 m at the thickest point (at the side wall part of the tunnel), and the thickness of the lane slabs and supporting walls is 0.4 m.

Full-ring double lining has the highest cost due to the largest excavation diameter and largest amount of lining work. The non-enclosed liner double-layer lining has the same excavation diameter as the single-layer segment lining but has a larger lining volume and therefore has the middle cost. Based on preliminary estimates, compared with the single-layer segment lining, non-closed liner double lining structure per linear meter increased cost by about 3000 yuan and the whole ring double lining per linear meter increased cost by about 10,000 yuan.

In summary, compared to the single-layer segment lining structural scheme, the segment lining + non-enclosed inner lining double-layer lining scheme maintains equivalent construction difficulty while requiring only modest increases in cost and construction duration. However, it achieves substantial improvements in waterproofing performance and structural stability, exhibits reduced lateral deformation under riverbed scour and siltation variations, and benefits the stability of internal facilities and installations. When compared to the full-ring double-layer lining scheme, it offers reduced construction difficulty, lower cost, shorter construction duration, and comparable lateral structural deformation performance under riverbed scour and siltation changes, while effectively addressing waterproofing performance and long-term structural stability issues at critical locations including the tunnel invert and sidewalls. Therefore, the segment lining + non-enclosed inner lining double-layer lining structural scheme is recommended for adoption.

6.3.3 Properties of Concrete Segment

6.3.3.1 Concrete Segment Materials

(1) Proposed high-performance segment concrete design principles and methods.

Proposed high-performance segment concrete design principles with durability as the main design index, good physical and mechanical properties, and volumetric stability, taking into account the economy and rationality [71]; the use of the self-developed expert system for concrete safety, according to the specific characteristics of Wuhan Tunnel Project optimized the design of the segment concrete, segment concrete arrives at the following indexes: a, strength grade: 28 d compressive strength >60.0 MPa. b, creep coefficient <2.0. c, impermeability grade P18. d, drying shrinkage value: 28 d drying shrinkage value $< 2.0 \times 10^{-4}$, 90 d drying shrinkage value $< 2.5 \times 10^{-4}$. e, alkali content ≤ 2.5 kg/m^3. f, no alkali activity of aggregate. g, Cl^- diffusion coefficient $\leq 8 \times 10^{-13}$ m^2/s. h, alkali content ≤ 2.5 kg/m^3. h, alkali content ≤ 2.5 kg/m^3. f, no alkali activity of aggregate. g, Cl^- diffusion coefficient $\leq 8 \times 10^{-13}$ m^2/s. h, anti-sulfate erosion coefficient ≥ 1.4. i, rapid carbonization test: 28 d carbonation depth < 1 mm.

(2) The effects of steam curing on the properties of tunnel shield concrete were investigated.

(1) Extending the static curing time promotes the physical and mechanical properties of high-strength concrete in the later stage; slow warming is desirable, and both extending the constant temperature time and increasing the constant temperature improve early strength [72]. (2) Effects of steam curing on the volume stability of tunnel shield concrete: Extending the pre-curing period improves the volume stability of tunnel segment concrete. Excessively rapid heating rates cause coarse crystallization of hydration products and their uneven distribution within the concrete, resulting in deteriorated pore structure and increased harmful porosity. Among the four steam curing parameters, heating rate and constant temperature significantly influence concrete volume stability and should be prioritized for control. High-energy delayed expansion agents for concrete have been developed to control concrete volume stability [73]. (3) Effect of steam curing on the durability of tunnel shield concrete. Steam curing has a significant effect on the durability performance of concrete. Prolonging the curing time can improve the degree of cement hydration, increase the cement hydration products, and reduce the adverse effects caused by heating deformation; slowing down the heating rate can reduce the concrete temperature gradient, improve the uniformity of the distribution of hydration products, and reduce the internal defects of concrete; shortening the isothermal temperature time and lowering the isothermal temperature can alleviate the decrease in $Ca(OH)_2$ content in the concrete, reduce the expansion of bubbles inside the concrete, rupture and connectivity caused by the number of convenient channels for CO_2 intrusion into concrete, reduce the number of harmful pores, and improve the carbonation and frost resistance of concrete [74–78]. (4) The effect of steam curing on the internal humidity and autogenous-shrinkage of tunnel shield concrete is studied, and the relationship between internal humidity, autogenousautogenous shrinkage, and resistivity of steam-cured high-strength concrete is investigated, and a technical method is proposed to introduce humidity-regulating porous materials to compensate for the loss of internal humidity in the concrete, to reduce the autogenous-shrinkage of the concrete, and to control the volumetric stability of the concrete; for simple and accurate characterization of internal moisture and autogenous-shrinkage patterns in concrete, detecting electrical resistivity is proposed. [79–81].

(3) Durability design of concrete under the joint action of stray current and chloride ions.

Under the action of stray current, the compressive strength of concrete will suffer a certain degree of loss, and with the extension of the action of time and aggravation, the self-corrosion potential and corrosion current of built-in reinforcing steel bars in the concrete increase. Different mixing ratios of concrete in the resistance to stray current have a large difference in the impact of the use of the supplementary mineral admixture technology method can improve the resistance of concrete under the action of stray current resistance to infiltration. Studies have shown that the composite admixture of fly ash and slag powder will significantly reduce the effect of stray current action [82].

For subway tunnel concrete, higher design requirements for ion permeation resistance must be adopted. The dense structure design of concrete is realized through the introduction of mineral admixtures, functional particles, and aggregate particle design to meet the requirement of high permeability resistance of concrete [83]. Through the design of high chloride binding capacity of concrete, the binding capacity of hardened cement paste for chloride ions is improved, so as to realize the design and preparation of highly durable concrete that is resistant to chloride erosion and stray current [84, 85].

6.3.3.2 Highly Impermeable Protective Layer Material

Since the protective layer material needs to meet the requirements of high permeability resistance of the segment, therefore, from the consideration of increasing the compactness of concrete, the meso-defect interface transition zone of concrete is strengthened and the pore structure of concrete is refined as a means of preparation, and meso-defect interface transition zone-free cement-based material (MIF) is prepared. The design principle of MIF material is mainly to eliminate the interfacial transition zone in concrete through the selection of raw materials, elimination of coarse aggregate components, strict control of the particle size of the concrete skeleton components, and dense filling of ultra-fine particles. Thus, the performance of concrete is optimized, especially to improve its interfacial transport properties, isolate the rapid channels of ion migration and solution penetration, greatly enhance its impermeability and material compactness, and ultimately significantly extend the service life of the structure [86].

The main material composition of MIF: 42.5R and above grade ordinary portland cement or portland cement (C) 40%–70%, modified reinforced dense padding components (MRP) 30%–50%, shrinkage-reducing and anti-cracking components (SRC) 2%–4%, and main skeleton fine granular material (1.0–1.5 times the amount of cement). Modified reinforced compact filling component (MRP) consists of high-silica nano-sized powder material, low-calcium and high-silica-aluminum derivative ultra-fine powder material, and water-reducing dispersant, which has the effect of compact filling and water-reducing enhancement. Shrinkage reduction and anti-cracking component (SRC) consist of shrinkage reduction material and organic mixed fiber (volume doping 0.1%–0.2%), through the reduction of capillary liquid surface tension and fiber enhancement and other shrinkage reduction and anti-cracking effect. The main skeleton fine-grained material is siliceous particles with a particle size range of 0.2–0.8 mm [87, 88].

Table 6.36 Cl^- diffusion coefficients and lifetime predictions.

Name of material	Thickness of protective layer (mm)	Cl^- diffusion coefficient (10^{-13} m^2/s)	Projected life (a)
MIF concrete (FGM design)	25	2.08	686.7
Ordinary concrete (single-story design)	50	14.3	52.0
Stobelt Tunnel Segment, Denmark	50	6.00	212.5

The impermeability of the protective layer material was evaluated by testing the chloride ion permeability and quantified by the diffusion coefficient of chloride ions in the concrete. Table 6.36 shows the comparison between the permeability resistance of MIF concrete and normal concrete (Cl^- diffusion coefficient and life prediction). From Table 6.36, the introduction of MIF material as the outer protective layer material of the concrete segment can greatly improve the permeability resistance of the concrete segment and extend the service life of the structure by more than 10 times compared with the ordinary reinforced concrete segment, and the result reaches the leading level of the international projects of the same kind [89].

6.3.3.3 Fireproofing Functional Layer Materials

(1) Fireproofing of functional material coatings.

The fire-resistance limit test tested concrete materials subjected to simulated fire-induced high-temperature effect and resistance to high-temperature damage effect. Fire-resistance limit is the time to ensure that no significant damage occurs to the reinforcement at 50 mm of the protective layer of concrete under the condition that the target temperature is 800 °C, i.e., the time when the temperature at the thickness of the protective layer of the concrete specimen reaches 250 °C. The specimens with brushed fire-resistant and blast-resistant layer coatings and untreated specimens were tested, and the test results are shown in Table 6.37. The study shows that the structural layer of concrete by brushing fire-resistant anti-explosive layer coating, its fire resistance performance is greatly improved, the fire-resistant time limit from 18 to 85 min, 4.72 times more than the untreated specimens. Therefore, the functional material fire- and explosion-resistant coating can effectively improve the fire- and explosion-resistant performance of the lining layer of the segment,

Table 6.37 Fire and explosion resistance of the FGCS lining layer.

Sample	Fire resistance time limit (min)	Proportion
Structural layer concrete	18	1.00
Structural layer concrete + 13 mm coating	85	4.72

Note: The concrete specimens were subjected to heat on six sides, and therefore, 13 mm of fire and blast-resistant paint was applied to all six sides of the specimens.

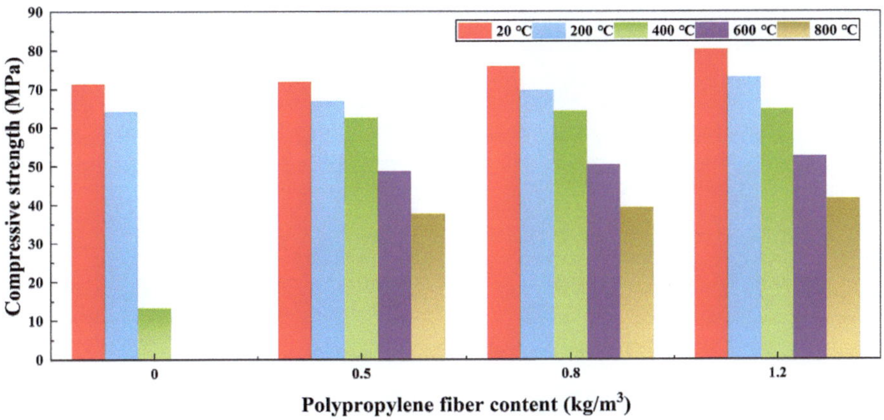

Figure 6.42 Comparison of residual strength of concrete after high temperature for different polypropylene fiber dosages.

which, when used in actual projects, will extend the performance of the underground engineering tunnel segment, reflecting the multifunctional integration of functionally graded concrete segment (FGCS).

(2) Polypropylene fiber toughening and fire prevention.

When the tunnel suffers from fire, the concrete segment is subjected to high temperature; when the temperature exceeds the melting point of polypropylene fibers of 165 °C, polypropylene fibers volatilize, leaving a pore equivalent to the volume of the fibers in the concrete, so that the water vapor and heat generated inside the concrete is smoothly discharged, avoiding the bursting of the segment, thus improving the fire-resistant performance of the concrete segment, and at the same time improving the cracking resistance of the concrete segment. In this project, polypropylene hollow fibers were mixed in the inner 5 cm of the segment to toughen and fireproof the concrete of the segment, the toughness index I5 of the segment concrete mixed with 0.8 kg/m^3 fibers was 2.1, and the number of initial cracking impact resistance was 116, which was 2.1 and 2.8 times of the ordinary concrete, respectively, and effectively solved the problem of the large hand hole of the large-size segment that was prone to cracking [90]. The residual compressive strength after being subjected to 800 °C (30 min) was still more than 50% of the remaining strength, whereas ordinary concrete had burst at 600 °C and had no strength (Figures 6.42 and 6.43).

6.3.3.4 Influence of Episodic Factors on the Properties of Segment Concrete

(1) Analysis of concrete bursting resistance after high temperature.

When the concrete prepared with polypropylene fiber (PF) and steel fiber (HPF) was subjected to high temperature at 800 °C, cracks could be clearly seen on the surface of the specimen, and PF even cracked, loosened, and collapsed. While HPF has only some cracks after high temperature at 800 °C, the damage is not obvious, as shown in Figures 6.44 and 6.45 control.

Figure 6.43 Morphology of concrete after thermal treatment.

Figure 6.44 Morphology of PF after 800 °C.

Figure 6.45 Morphology of HPF after 800 °C.

Figure 6.44 represents the obvious cracks on the surface of PF after the action of high temperature at 800 °C, and there is also the phenomenon of spalling, resulting in its low residual strength. Figure 6.45 shows the state of concrete exposed to high temperature of 800 °C after mixing steel fiber, which can be seen from the photo that there are only some small cracks on its surface and keep its good integrity, so the residual strength of HPF is much larger than PF. The reason for the above phenomenon is that the melting point of polypropylene fiber is 165 °C; when the temperature exceeds the melting point, PF will volatilize and escape, then the role played by polypropylene fiber on the strength will disappear, especially the fiber volatilization will introduce a certain number of pores in the concrete, so it is not good for the strength of concrete. In the range of 200–400 °C, due to the action of high temperature, the internal vapor pressure increases dramatically, forming pores in the concrete just to play a function of uniform heat dissipation, enhancing the ability of concrete to resist the action of high temperature. When the temperature rises to a certain degree, only steel fibers are left in the concrete to play a toughening role, steel wire type of steel fiber bond to the concrete to enhance the tensile strength of the concrete, so the high temperature at 800 °C after the role of mixed fiber concrete specimens helps maintain good integrity and only a small number of cracks, while the PF concrete will have a relatively large number of cracks and result in bursting causing damage. However, the above appearances are much better than in ordinary concrete; untreated fiber concrete, due to the absence of PF volatilization and formation of pores, cannot alleviate the increasing internal pressure due to water evaporation, so ordinary concrete at 400 °C above the explosion of the phenomenon is very easy to appear, as shown in Figure 6.46.

Figure 6.46 Morphology of ordinary concrete after 400 °C.

(2) Damage analysis of tunnel lining structure under the action of explosive load.

At the moment of the explosion of explosives, it will produce a high temperature of several thousand degrees and a high pressure of tens of thousands of MPa, forming an explosive shockwave of several kilometers per second and an explosive energy of thousands of joules, causing serious damage to the surrounding structures. Once an explosion occurs in a shield tunnel, the damage to the lining structure is also more serious due to the closed environment, where the high-temperature and high-pressure gases and explosion shockwaves cannot be evacuated in time.

Figure 6.47 shows the shield tunnel damage map under the contact blast load at the upper edge of the middle of the traveled way plate, and the working conditions I to IV represent the gradual increase of the radius of the TNT bar explosive package.

In the explosion of case one of the lane board and lining structure, damage is very small. Case two, due to the explosion of the explosive package to produce a large explosive energy and burst pressure directly on the lining structure, so the lane board and explosive package contact parts quickly causing damage, and in the middle of the lane board to form an explosive cavity, while the cavity outside the lane board is less damaged. The damage difference in case three is very large, not only in the middle of the roadway plate to form a large explosion cavity but also the lining structure of the left and right sides of the gang near the area of serious damage and the top plate near the area of the formation of a lot of radial cracks. In case four relative to case three, damage is more obvious, the roadway plate formed by the explosion cavity is larger, the roadway plate above the lining structure of radial cracks is more intensive, and the explosion is formed by the cavity so that the explosion shockwave is in the overall lining structure between the propagation of the lining structure, resulting in the roadway plate below the formation of the lining structure of a number of small cracks.

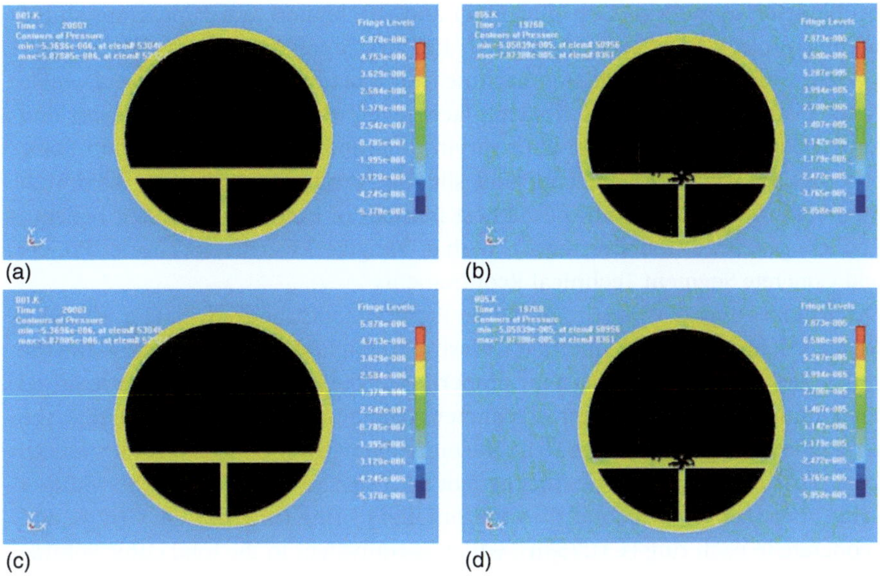

Figure 6.47 Damage to lining structure under contact blast loading: (a) Operating Condition 1, (b) Operating Condition 2, (c) Operating Condition 3, and (d) Operating Condition 4.

As can be seen from Case one to Case four, the explosive energy generated by the explosion of TNT strip packages is directly transferred to the lining structure. Package and the lane plate contact parts in the explosion shockwave, explosive energy and high-temperature and high-pressure explosive gas under the joint action of high radial and tangential compressive stress, and compressive stress are much greater than the dynamic compressive strength of the lane plate, so the contact parts are strongly compressed and crushing damage occurs and soon in the lane plate in the middle of the formation of a cavity. At the same time, the lining structure to withstand the explosion shockwave, explosive energy, and high temperature and high-pressure explosive gas together, resulting in the lining of concrete by radial compressive stress at the same time in the tangential direction by the tensile stress, and the concrete material is a brittle material; its tensile strength is very low. Therefore, when the tangential tensile stress value is greater than the tensile strength of concrete, the lining structure concrete that is pulled off, which in the lining structure of the inner edge of the radial produced a lot of radial cracks, the lining structure caused serious damage.

6.3.4 Preparation and Engineering Application of Functional Composite Concrete Segment

6.3.4.1 Overview of the Wuhan Yangtze River Cross Tunnel Project

The Wuhan Yangtze River Cross Tunnel Project is the first over-the-river traffic tunnel project on the Yangtze River, designed as a two-hole bidirectional four-lane tunnel project. The distinctive characteristics of this tunnel project include:

large-diameter tunneling (11m diameter) and the highest water pressure shield tunneling project (maximum 0.64MPa); the first application of large-diameter composite slurry balance shield machine construction with the longest single advance distance (2540m); the first full-face penetration through sand layers and heterogeneous composite strata with varying soft and hard geological conditions; the world's greatest riverbed scour and siltation variation amplitude (maximum annual variation of 20m) and the highest number of buildings tunneled beneath (54 buildings with 5 floors or more). Wuhan Yangtze River Cross Tunnel Project Shield Concrete Segment Technical Requirements

The Wuhan Yangtze River Cross Tunnel Project adopts the composite mud-water balanced shield construction method with strong adaptability to the geology, and the waterproofing and seepage-resistant properties of the shield segment are required to be extremely high. The reinforced concrete segment adopts C50S15 concrete, the inner diameter of the segment is 10 m, the outer diameter is 11 m, and the arc length of each piece is about 4 m. The thickness of the ring sheet is 500 mm, the width of the ring sheet is 2000 mm, and the total number of rings is about 2533 rings, and the concrete in each ring is 31.75 m^3, which is equivalent to the total cubic volume of concrete that is about 80,423 m^3. Each ring is divided into 9 pieces, including 1 capping piece, 2 neighboring pieces, and 6 standard pieces, and the pieces are equipped with concave and convex grooves in the longitudinal direction of the ring. The Wuhan Yangtze River Cross Tunnel Project has the highest requirements for segment performance among similar tunnel projects at home and abroad.

Concrete segment performance requirements are shown in Table 6.38 to ensure the precision of engineering construction after the production of the concrete segment; the production precision of concrete and volume deformation requirements are shown in Table 6.39. The design parameters of the functional concrete segment are shown in Table 6.40.

Table 6.38 Mechanical and durability indicators for concrete.

No	Control items	Design indicators
1	Compressive strength	C50
2	Impermeability rating	P12
3	Cl^- diffusion coefficient of concrete	$\leq 8 \times 10^{-13}\ m^2/s$

Table 6.39 Segment production accuracy requirements.

Project	Precision requirement (mm)	Project	Precision requirement (mm)
Width	±0.5	Thickness	+3, −1
Arc length	±1	Bolt hole diameter and location	±1

Table 6.40 Design parameters for functional concrete segments.

Functional layer	Material		Thicknesses	Performance
Layer 1	Highly dense waterproofing layer	Highly efficient osmotic crystalline waterproof material	0.8–1.5 mm. Choose 1.2 mm	Cl^- diffusion coefficient $\leq 0.8 \times 10^{-13}$ m^2/s
Layer 2	Protective layer with high impermeability	Transition zone-free cement-based material (MIF)	1.0–3.0 cm. Choose 2.0 cm	Cl^- diffusion coefficient $\leq 1 \times 10^{-13}$ m^2/s
Layer 3	Reinforced concrete protective layer	C50 concrete (highly active mineral admixtures such as fly ash or slag powder)	2.0–4.0 cm. Choose 3.0 cm	Cl^- diffusion coefficient $\leq 15 \times 10^{-13}$ m^2/s
Layer 4	High-strength structural layer	C50 concrete (highly active mineral admixtures such as fly ash or slag powder)	42.0–48.0 cm. Choose 45.0 cm.	Cl^- diffusion coefficient $\leq 15 \times 10^{-13}$ m^2/s
Layer 5	Flameproof and anti-explosive layer	Tunnel fireproof coating	10.0–16.0 mm. Choose 12.0 mm	Fire resistance limit >3.5 h

6.3.4.2 Optimized Design of Segment Concrete for Shield Tunnels

Prefabricated shield segment has strict requirements for durability and dimensional accuracy and appearance such as impermeability, shrinkage, and alkali aggregate reaction resistance, etc. In the case of meeting the specifications and strength requirements, the following technical route is determined: minimize the amount of cement and use a large number of admixtures. Admixtures can not only reduce the heat of hydration in the early stage of cement hydration but also reduce the possibility of cracks caused by the changes in temperature and volume of the concrete itself, as well as improve the volume stability of concrete. Moreover, it can reduce the alkali content in concrete and prevent the occurrence of alkali aggregate reaction, and it can also utilize the microaggregate filling effect of mineral admixture to make concrete further dense, which is conducive to improving the impermeability and durability of concrete, and at the same time, it can effectively reduce the cost of concrete [91, 92].

The challenges and key points in the mix design of shield segment concrete lie in: achieving good thixotropic properties in concrete with relatively low slump during casting; ensuring minimal aggregate settlement and reduced segregation during vibratory molding; enabling the concrete to rapidly lose workability after placement

Table 6.41 Fly ash segment concrete proportions.

No	Cement	fly ash	Sand	Gravel	Water	Additive	Strength
			(kg/m³)				(MPa)
1	360	120	692	1128	150	7.2	65.4
2	380	100	692	1128	150	7.2	66.7
3	400	80	692	1128	150	7.2	67.5

Table 6.42 Slag powder segment concrete mix ratios.

No	Cement	Fly ash	Sand	Gravel	Water	Additive	Strength
			(kg/m³)				(MPa)
4	330	130	698	1140	152	6.9	64.8
5	360	100	698	1140	152	6.9	65.6
6	390	70	698	1140	152	6.9	67.6

to form initial structure for easy finishing; achieving high early-age strength to meet the requirement of two-cycle turnover within 24 h. Concrete demolding strength varies according to different requirements of the lifting process, traditional lifting needs to be ≥25 MPa, vacuum lifting process needs to reach about 15 MPa, to meet the alkali content $\leq 2.5\ kg/m^3$, 28-d drying and shrinkage value $<2.0\times10^{-4}$, 90-d drying and shrinkage value $<2.5\times10^{-4}$, and Cl^- diffusion coefficient $\leq 8\times10^{-13}\ m^2/s$.

The dense stacking design method is used to design high-strength segment concrete for the Wuhan Yangtze River Tunnel. Based on the strength and durability requirement setting, YD42.5R grade ordinary portland cement, YL Thermal Power Plant I grade fly ash, Bahe river sand with fineness modulus of 2.6, 5–25 continuous graded gravel, and HWMD150 modified naphthalene highly efficient water-reducing agent were selected. The water–binder ratio is in the range of 0.28–0.35, the total amount of cementitious materials is 450–500 kg/m^3, and the sand rate is 35%–40%. The proportion of using fly ash and slag powder as an admixture is shown in Tables 6.41 and 6.42, respectively.

The seepage resistance and volume stability performance of the abovementioned concrete were tested, and the Cl^- diffusion coefficient was $7.2\times10^{-13}\ m^2/s$, the 28 d drying shrinkage value was 1.75×10^{-4}, and the 90-d drying shrinkage value was 2.4×10^{-4}, which met the design requirements of the segment.

6.3.4.3 Preparation of Tunnel Shield Sheets

(1) Factory production process of the segment.

The production of shield tunnel functional composite concrete segment FGCS is carried out in strict accordance with the pre-established production process, the concrete is added in the order of the two ends of the mold first and then the middle and poured according to the order of the structural layer and protective layer, so as to realize the vibration of the layered vibration. The structural layer is added to the appropriate amount of concrete and then covered with an interface treatment cover, and then the protective layer of MIF material is poured and vibrated into shape using flat plate vibration. The production process of FGCS is shown in Figure 6.48.

(2) Construction preparation

(1) **Mold Assembly:** Thoroughly clean the residues on the inner and outer surfaces of the mold and evenly apply the mold release agent. According to the order of the end template, side template and bottom template will be fixed with bolts installed, from the middle position to the two ends of the order to tighten; prohibit the reverse order of operation, so as not to cause deformation of the mold.
(2) **Mold Tuning:** After assembling the mold, its width, arc length, and hand hole position measurement, unqualified for timely adjustment, must reach the mold limited tolerance range to ensure accuracy. Detection methods are as follows: the use of 0–2100 mm range of the inner diameter to detect the width of the steel mold, the error is +0.2–0.4 mm; the use of 0–5 m range of steel tape measure to detect the arc length of the bottom plate of the steel mold, the error is ±1 mm.
(3) Reinforcing the steel skeleton into the mold and the installation of embedded parts with a four-point hook will be reinforcing the steel skeleton according to the specifications of the mold number into the mold. The lifting process must be smooth, not to make the steel skeleton and mold collide. The head of the screw must be inserted into the mold hole of the hand hole holder to prevent the connection from being not tight enough to cause gaps, resulting in slurry leakage. Check whether the accessories are placed completely and firmly as required. Check whether the reinforcing steel skeleton protective layer pads are correctly placed to ensure that the protective layer of the main reinforcement is 50 mm on the outside and 40 mm on the inside.

(3) Concrete pouring for the structural layer.

Concrete is placed in the order of the two ends of the mold first and then in the middle, and the concrete is poured in batches, thus realizing layered vibration. When the concrete is added to the appropriate amount of cover interface treatment cover (Figure 6.49), use the manual insertion vibration mode and vibrating rod vibration molding.

(4) High impermeability protection layer concrete casting.

High impermeability protection layer adopts cementitious material MIF without a fine interface transition zone, and the extension degree is controlled within 140 ± 10 mm. After testing, MIF material steam curing 12 h demolding strength≥15 MPa, 7 d out of the pool when the compressive strength>35 MPa, and 28 d strength≥60 MPa; Cl^- diffusion coefficient $\leq 0.8 \times 10^{-13} m^2 s$, impermeability grade ≥ S40.

High impermeable protective layer concrete proportion: cement 686 kg/m^3, modified enhanced dense filling component M1 is 51.5 kg/m^3 (5%), M2 is 257.7 kg/m^3

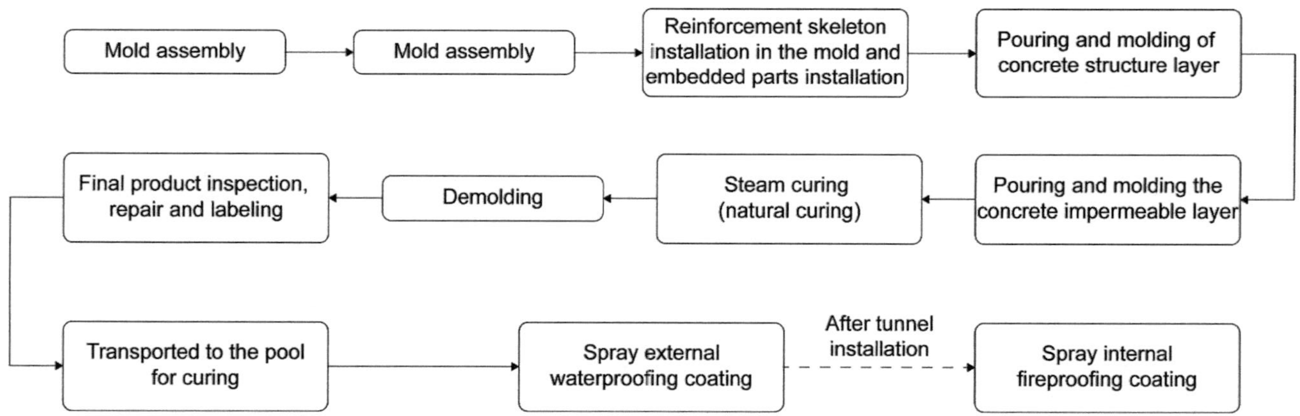

Figure 6.48 Process flow of production of functional composite concrete segments for shield tunnels.

Figure 6.49 FGCS interface treatment cover plate.

Figure 6.50 Effect of embossing on concrete surfaces.

(25%), M3 is 30.9 kg/m^3 (3.0%). Shrinkage anti-cracking component S1 is 20.6 kg/m^3 (2.0%), S2 is 1.82 kg/m^3, set ratio 1.33, and water–binder ratio 0.24. About 20 min after the structural layer is poured and vibrated, the well-mixed protective layer concrete is poured on the top of the structural layer, and the embossing effect is shown in Figure 6.50. Vibratory rods are used for point-by-point insertion vibration, with each vibration point controlled within 10-20 seconds. After achieving vibration compaction, the vibratory rods must be slowly withdrawn, followed by covering with the top slab.

(5) Concrete plastering.

The time to open the top plate is generally about 45 min after the concrete is poured, and the specific time depends on the temperature and concrete setting. (1) Rough plastering. Use an aluminum alloy pressure ruler; scrape off excess concrete (or fill the depression), so that the concrete surface is smooth. (2) Middle plastering. After the concrete surface is watered down, use a gray spoon to smooth out the surface of the segment. (3) Fine plastering. When there is a slight flat indentation in the concrete by lightly pressing with fingers, use a long spoon to smoothen the surface, and strive to make the surface bright without gray spoon marks. Spray concrete anti-cracking maintenance liquid M1500 on the surface of concrete about 1 h after pouring.

(6) Steam curing.

Steam curing is employed to enhance the demoulding strength of concrete and reduce the curing time. After the initial setting of the concrete, the top plate is closed, and a tightly sealed, airtight curing enclosure is placed around the mold to prior to steam curing. After the concrete cools down, the same-condition molded concrete specimens will be tested in the same period. When the strength reaches 15 MPa or more, start demolding. Demolding sequence: loosen the grouting hole fixing screw, open the mold side template, open the mold end plate, attach the spreader to the segment, vibration demolding. The curing system is shown in Tablc 6.43; under the premise of meeting the strength of concrete demolding, the curing parameters such as resting time can be adjusted appropriately according to climate change, and the speed of heating and cooling should not be changed in general.

(7) Spraying waterproofing coating.

After 7 d of water curing following segment demolding, the outer waterproof coating is spray-applied. 1) To ensure the waterproofing function of the concrete substrate, the surface must be clean and free from oil stains, dust, and other debris.

Table 6.43 Steam curing regimes for functionally graded concrete segments.

	Steam curing	
Item	**Spring, Summer, Fall**	**Winter**
Standstill time (h)	3	5
Heating up time (h)	3	5
Temperature rise rate (°C/h)	10–15	10–15
Constant temperature time (h)	4	3
Maximum steam curing temperature (°C/h)	40–45	55–60
Cooling time (h)	2	3
Cooling rate (°C /h)	<10	<10
Temperature difference from outside during demolding (°C)	<20	<20

No water accumulation shall occur within 48 h after completion of the waterproof coating application. 2) The waterproof coating consumption rate should be at least 1.5 kg/m^2. 3) Construction employs spray application, with the spray nozzle positioned close to the coating surface to ensure the mortar can penetrate into surface micropores or microcracks. Upon completion of the first waterproof coating layer, when finger pressure leaves no impression, the second waterproof coating layer can be applied after 4 h. If the surface becomes excessively dry, water spraying should be applied for moisture curing.

6.3.4.4 Segment Quality Inspection

During the segment production process, in addition to conducting compressive strength, impermeability, and durability testing on raw materials and concrete specimens as required by design specifications and standards, the following additional inspections are required.

(1) Appearance dimension quality standards for individual pieces.

Each piece of a functional composite concrete segment should be examined for appearance, size, and quality standards as shown in Table 6.44. Vernier calipers are used to measure the width and thickness of the functional composite concrete segment, using a steel tape measure the arc length of the concrete segment, with a nylon line to check the functional composite concrete segment twisted deformation.

(2) Three-ring assembling inspection of segment.

The three-ring assembly test is mainly applied to check the precision and overall matching of the segment mold; through the three-ring assembly, one can find out whether the production precision of each piece of mold meets the requirements.

Table 6.44 Functional gradient concrete segment piece single block appearance and dimensional quality standard.

No	Items		Tolerance (mm)	Measurement frequency		Measurement methods
				Range	Point count	
1	Overall dimensions	Width	±0.5	Each	3 points each on the inside and outside	Dial caliper
		Chord length Arc length	±1.0	Each	3	Tape rule
		Thickness	+3.0, −1.0	Each	3	Dial caliper
2	Screw hole diameter and position		±1.0	Each	3	Dial caliper
3	Strength of cementitious composite		≥Design strength grade	—	—	—
4	Seepage resistance of cementitious composite		≥Design impermeability grade	—	—	—

Figure 6.51 Three-ring assembly inspection.

The three-ring assembly should be carried out during the trial production of the segment, and the three-ring assembly is also a necessary condition for transferring from trial production to normal production. The test result of the three-ring assembling performance shows that the assembling performance is good and fully meets the requirements of the project (Figure 6.51).

(3) Single piece of segment piece impermeability test.

Individual segment impermeability testing is conducted on a specialized impermeability test platform after the segments reach the required age (Figure 6.52). The impermeability performance of segments is tested by applying water pressure to the outer curved surface of the segments. To a certain extent, individual segment impermeability testing serves as an evaluation of the operational standards in segment production. The testing criterion requires maintaining constant water pressure of 0.8MPa for 4 h, with the seepage line not exceeding 1/5 of the segment thickness.

Through the above inspection test and strict control of the quality process of each procedure, one can ensure the production of the segment meets the design and specification requirements, as shown in Table 6.45.

The results show that, compared with the ordinary segment, the functional composite concrete segment shows more excellent impermeability performance, 0.8 MPa constant pressure for 4 h with no leakage, and better capability to meet the tunnel high-pressure water-rich environment of high impermeability requirements. Through on-site inspection, the quality, precision, and durability indexes of the autoclaved high-strength concrete segment produced by this research result have reached and exceeded the design indexes, and the appearance of the segment is

Figure 6.52 Seepage resistance test for a single segment.

Table 6.45 Segment impermeability properties.

Test	Standard value	Results	
		Ordinary segment	FGCS
Impermeability	Constant pressure at 0.8 MPa for 4 h, penetration depth not more than 50 mm	Constant pressure at 0.8 MPa for 4 h, a line of infiltration, infiltration depth of 15 mm	Constant pressure at 0.8 MPa for 4 h, no infiltration, depth of infiltration 0 mm

in full compliance with the requirements of shield construction, which is used in actual shield tunneling projects to achieve good results.

6.3.4.5 Engineering Application

Application benefits: By adopting the project's pioneering new structural system optimization technology for large-diameter shield tunnel segments and linings, the inner main reinforcement and outer reinforcement of the segments are adjusted to save steel and fabrication and installation costs. Reducing the purchase of segment templates saves costs and realizes the world's largest class, structurally and functionally integrated segment with a diameter of more than 15 m and an accuracy of 0.3 mm in large-scale production. The synergistic improvement of permeability, deformation resistance, and precision of the segment greatly improves the overall structural rigidity and permeability resistance of the segment assembled into a ring, thus changing the form of the tunnel lining structure and creating the first semi-closed lining structure that is more economical and more efficient in construction, with a reduction of more than 15% in the project cost (Figures 6.53–6.55).

Figure 6.53 Concrete segment production.

Figure 6.54 Tunnel segment lining structure.

All of the developed technologies have been applied to the Wuhan Yangtze River Tunnel, as well as Wuhan's successive subway lines 2 (the first Yangtze River subway tunnel), 8 (the largest single-hole, two-lane Yangtze River subway tunnel), and 7 (the world's first public railway concavity tunnel with a super-large diameter in the Yangtze River). As well as popularized and applied in Guangzhou-Shenzhen Shiziyang Tunnel (the world's first underwater high-speed railway tunnel with a design speed of 350 km per hour) and many other large-scale tunneling projects (Figures 6.56–6.58), it has provided important technical support for the construction

Figure 6.55 Tunnel interior lining structure.

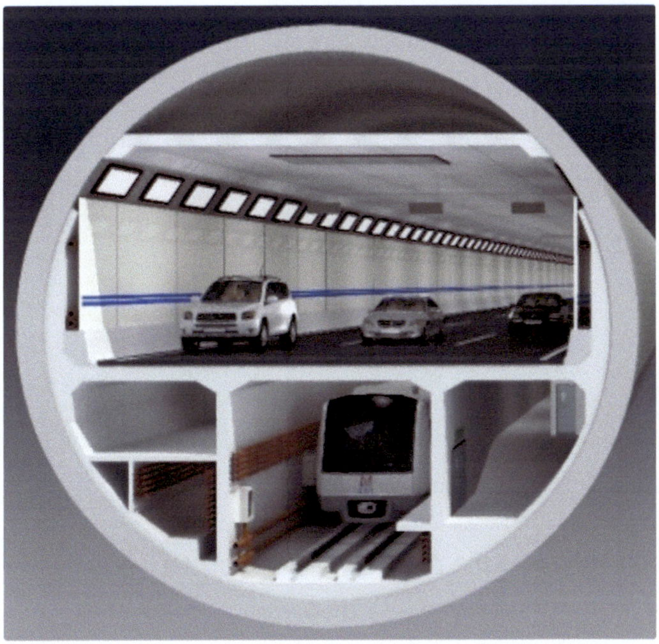

Figure 6.56 Sanyang Road Yangtze river public railway tunnel.

of super-large-diameter shield tunnels under complex and severe working conditions. Through the precise selection of raw materials, performance improvement, and structural optimization, the accumulated direct economic benefits have exceeded one billion yuan, and the social benefits are remarkable.

Figure 6.57 Wuhan metro line 8 tunneling project (one hole and two lines).

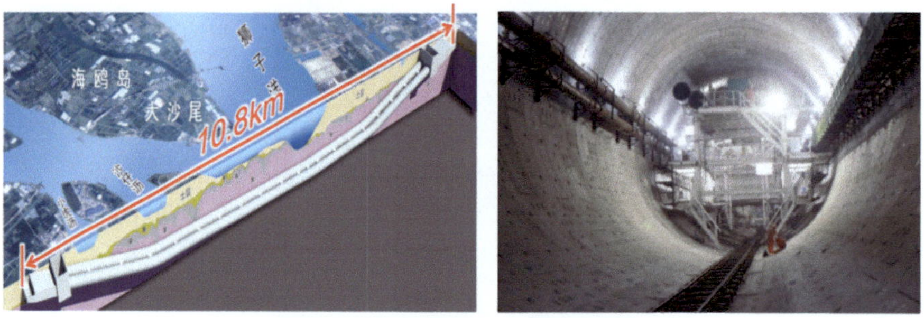

Figure 6.58 High-speed Railway Shiziyang tunnel project.

6.4 High-strength Lightweight Aggregate Concrete

6.4.1 Overview

6.4.1.1 High-strength Lightweight Aggregate Concrete

High-strength lightweight aggregate concrete (HSLC) stands out for its remarkable characteristics, including superior strength, reduced weight, consistent volume stability, and enhanced durability [93–97]. This innovative concrete manages to diminish the self-weight of structures by over 20% without compromising on strength, distinguishing itself as a pivotal solution for load-bearing constructions. Its unique attributes render it particularly advantageous for specialized engineering endeavors and structural designs where minimizing weight is crucial, such as in skyscrapers, large-span bridges, and offshore projects within contemporary large-scale developments. Consequently, HSLC has emerged as a critical focus for research and innovation in the realm of lightweight aggregate concrete. The traditional limitations of conventional concrete, characterized by its considerable weight and relatively lower specific strength, have long constrained its application scope in structural engineering and the optimization of concrete's material properties. In this context, HSLC represents a strategic advancement toward achieving a lighter, stronger, and more efficient concrete, marking a significant evolution in the material's development toward high performance and sustainability. This approach not only addresses previous constraints but also opens up new possibilities for the future of construction and engineering disciplines [66, 98, 99].

6.4.1.2 The Main Problems of Lightweight Aggregate Concrete used for Structures

The deployment of lightweight aggregate concrete in large-scale structural engineering endeavors has historically been met with challenges, hindering its breakthroughs in this domain. The primary obstacles can be articulated as follows: Initially, the inherent structure of lightweight aggregates, characterized by a multitude of pores, predisposes them to the formation of crack defects during the firing phase. This defect is the main cause of reduced strength and poor performance of lightweight aggregates and directly affects the interface between the aggregate and cement matrix, the more fragile component of concrete. This results in inadequate strength and performance of the concrete to meet the stringent requirements of structural engineering applications. Furthermore, when compared with traditional concrete, lightweight aggregate concrete exhibits notable discrepancies in characteristics such as elasticity modulus, creep behavior, brittleness, and shrinkage. To effectively integrate it into structural engineering projects, it is imperative to refine design methodologies, standards, and technical guidelines that resonate with its unique properties. This ensures the integrity and safety of concrete components and structures while optimizing the inherent benefits of lightweight aggregate concrete. The third challenge revolves around the lightweight and porous attributes of lightweight aggregates, which predispose them to floatation and uneven distribution during the concrete mixing process. Managing the water content within aggregates and mixture becomes increasingly challenging during the pumping of concrete construction. This issue complicates pumping operations and leads to inconsistent concrete quality. Such predicaments significantly limit the utilization of lightweight aggregate concrete in structural engineering, necessitating innovative solutions to overcome these barriers.

The issues highlighted previously have predominantly confined the use of lightweight aggregates to the production of concrete blocks, wall panels, and similar applications. While there have been instances of lightweight aggregate concrete being employed in structural engineering internationally, the prevalent solution to the aforementioned challenges involves the adoption of high-performance lightweight aggregates alongside specialized concrete preparation and construction techniques. These approaches are characterized by stringent demands on the quality of raw materials, elevated energy consumption, intricate processing stages, and considerable expenses. Furthermore, these methods grapple with issues related to inconsistent quality and adaptability, which hinder their widespread adoption and implementation. In China, there is a notable absence of systematic research concerning the deployment of lightweight aggregate concrete within large-scale structural engineering projects, underscoring a significant gap in the field's development and application in the country.

6.4.1.3 Main Research and Technology Development

In response to the outlined scenario with an emphasis on the application of lightweight aggregate concrete within the realm of structural engineering, the research group led by the author undertook an array of research and development initiatives, culminating in pioneering findings:

(1) An in-depth exploration was conducted to identify the ideal synergy between the strength of lightweight aggregates, cement matrix, and overall density of concrete. This exploration extended to analyzing the brittleness mechanisms and developing toughening strategies for high-strength lightweight aggregate concrete. Furthermore, an innovative technology was devised to enhance the interface structure between lightweight aggregates and hardened cement paste. This led to the formulation of advanced design and modification techniques for high-performance lightweight aggregate concrete, aligning with the stringent specifications for strength, density, and performance requisite for structural engineering applications.
(2) Leveraging the intrinsic properties of lightweight aggregate concrete and evolutionary dynamics of its composite material structure, the team adopted a gradient design strategy to innovate a crack prevention technology. This technology is specifically tailored for the anchorage ends of prestressed components and integrated structure of steel plates (or box girders) with lightweight aggregate concrete. Through this endeavor, the team has mastered crucial design and fabrication techniques for lightweight aggregate concrete structures and components, thereby addressing the critical issue of crack formation in such concrete structures.
(3) The research further delved into the fundamental factors and principles influencing the uniformity of lightweight aggregate concrete. In doing so, control technologies for maintaining homogeneity, along with methods for detecting and evaluating lightweight aggregate concrete mixtures, were developed. A concentrated effort was made to surmount the significant technical challenges associated with the pumping construction of lightweight aggregate concrete. The team has acquired proficiency in essential construction technologies for the deployment of lightweight aggregate concrete in large-scale structural engineering projects, thereby facilitating its broader application in significant construction endeavors.

6.4.2 Key Technologies for the Design and Preparation of High-strength Lightweight Aggregate Concrete

6.4.2.1 Strength Design of Lightweight Aggregate Concrete

(1) The relationship between lightweight aggregate structure, density, and water absorption rate

The interrelation between the structural composition of lightweight aggregates and their physical properties, specifically density and water absorption capacity, is governed by intrinsic material properties and the aggregates' pore structure. The density of these aggregates is a function of both the constituent material and spatial arrangement of pores within the aggregate. In instances where lightweight aggregates are fabricated from identical materials, a direct correlation is observed where an increase in structural porosity inversely affects the aggregate's density. Conversely, the rate of water absorption is contingent upon both the porosity and interconnectivity of the pores within the aggregate. This study introduces the utilization of an internal pore connectivity coefficient, denoted as "ϕ" (as per Eq. (6.15)), to quantitatively assess the extent of pore interconnection within lightweight aggregates. It is suggested that, for aggregates possessing equivalent densities, an augmented "ϕ" value signifies enhanced pore interconnectivity, thereby elevating the rate of water absorption. This absorption rate escalates progressively with prolonged saturation until nearing a vacuum-saturated state. Comparative analysis, as demonstrated in Table 6.46, reveals that lightweight aggregate A exhibits a connectivity rate of merely 61%, in stark contrast to the 84% connectivity rate of lightweight aggregate B, elucidating the disparity in their respective water absorption kinetics over time.

$$\phi = \frac{w_{24\,\mathrm{h}}}{w_{\mathrm{vac}}} \times 100\% \tag{6.15}$$

where $w_{24\,\mathrm{h}}$ is the 24 h water absorption of lightweight aggregate and w_{vac} is the 24 h vacuum saturation rate of lightweight aggregate.

(2) Matching Design of Concrete Strength and Density

Density and strength are important indicators for the design of lightweight aggregate concrete. The study obtained the relationship between strength and density of lightweight aggregate concrete shows that for each density level of concrete, they are matched with the best strength range (Figure 6.59). The higher the density class of concrete, the greater the strength design range. When designing and preparing

Table 6.46 Water absorption characteristics of lightweight aggregates and their relationship with structural pore connectivity.

Lightweight aggregates	Water absorption (%)	Vacuum water absorption	Connectivity	Bulk density
Compressive strength (MPa)	5 min	30 min	60 min	24 h
A	4.1	4.1	4.3	4.3
B	10.0	12.2	14.1	19.4

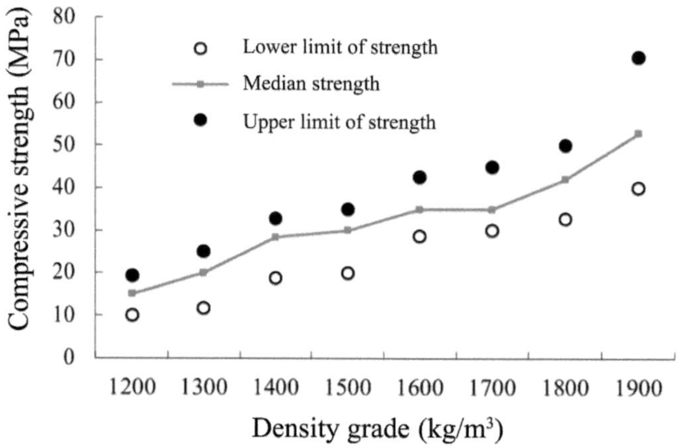

Figure 6.59 Matching relationship between strength and density of lightweight aggregate concrete.

lightweight aggregate concrete, it is important to match the strength and density of the concrete.

(3) Strength-matching design of lightweight aggregate, hardened cement paste, and concrete

Concrete is a multiphase composite material composed of hardened cement paste, aggregate, and its interface transition zone. The strength of each phase of hardened cement paste, aggregate, and interfacial transition zone and its matching determine the strength of concrete. A strength-matching design method is proposed by studying and analyzing the relationship between the strengths of lightweight aggregate, hardened cement paste, and concrete [100–102].

(1) Strength-matching Design of Light Aggregate Concrete and Hardened cement paste.

According to the relative magnitude of the elastic modulus of the two phases of aggregate and hardened cement paste in concrete, concrete can be divided into two categories: composite hard base materials ($E_a > E_m$) and composite soft base materials ($E_a < E_m$). The stress distribution and strain magnitude of the internal phases composed of the two types of materials are different when they are subjected to external load.

The strength-matching relationship between concrete and hardened cement paste was obtained from the study (Figure 6.60). In the OA section of the figure, the strength of the concrete is low, and the lightweight aggregate has suffered damage in the BC section. According to the theory of soft base composites, to prepare a high-strength concrete, the strength of the hardened cement paste must be much higher than the designed concrete strength. Although the purpose can be achieved by methods such as high mixing of cement, it will adversely affect the durability of concrete. We believe that, in the AB section of the figure, the strength of hardened cement paste and concrete have a better match, and the relationship between them

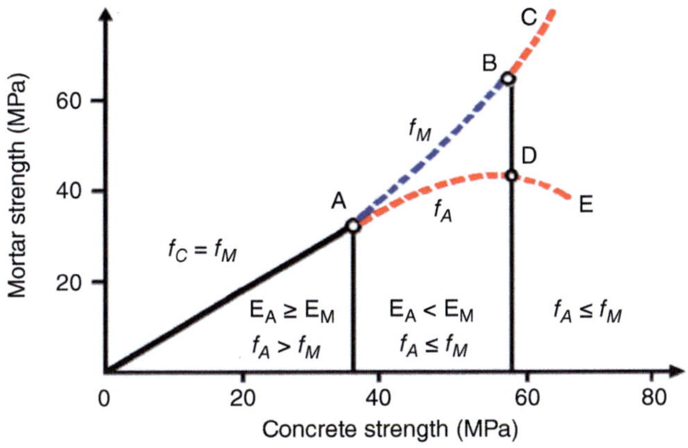

Figure 6.60 Matching relationship between lightweight aggregate concrete and mortar strengths.

obtained from the study is shown in Eq. (6.16):

$$f_m = ae^{bf_c} f_{ma} < f_m < f_{mb} \tag{6.16}$$

where f_m is the mortar strength in MPa, f_c is the concrete strength in MPa, a, b is constant, f_{ma} is the strength of hardened cement paste corresponding to point A in Figure 6.60 measured in MPa, and f_{mb} is the hardened cement paste strength corresponding to point B in Figure 6.60 measured in MPa.

(2) Strength matching design of lightweight aggregate and hardened cement paste

The cylinder compression strength and modulus of elasticity of lightweight aggregate are low, and in the preparation of high-strength concrete, it is necessary to use hardened cement paste whose strength matches the strength of the aggregate. According to the study of the strength relationship between lightweight aggregate concrete and hardened cement paste, for a certain strength of fa lightweight aggregate, there exists a strength interval of hardened cement paste that matches with f_a.

The upper limit of hardened cement paste strength, f_{mb}, was determined by regression analysis in relation to f_a (Eq. (6.17)):

$$f_{mb} = ce^{df_a} \tag{6.17}$$

where f_{mb} is the upper limit of hardened cement paste strength in MPa, f_a is the lightweight aggregate cylinder compressive strength in MPa, and c and d are constants.

In the design and preparation of high-strength lightweight aggregate concrete, the lightweight aggregate type and hardened cement paste composition can be optimized according to the design requirements of concrete strength and the matching relationship of Eqs. (6.18) and (6.19), so that the strengths of lightweight aggregate, hardened cement paste and concrete can be matched with each other, so as to economically and reasonably prepare high-strength concrete.

6.4.2.2 Structural Strengthening Technology for Lightweight Aggregate-Hardened Cement Paste Interface

According to the strength design method of lightweight aggregate concrete, high-strength lightweight aggregate with cylinder compression strength of 7.0 MPa, matched with hardened cement paste with a strength of 60–70 MPa, can theoretically be prepared to reach the strength of LC50-LC60 high-strength concrete. However, in practice, it is difficult to prepare LC50-LC60 high-strength concrete in practical application due to a large number of original defects on the surface of lightweight aggregate during the preparation process of cooling and crushing, which leads to the destruction of lightweight aggregate at a lower level of loading stress. To address this problem, we have developed a technology to strengthen the interface structure between lightweight aggregate and hardened cement paste [100, 103].

(1) Technical principles

The design establishes a mechanism of interfacial interaction between cementite and lightweight aggregate, so that the interaction between lightweight aggregate and hardened cement paste can be generated; through the interfacial reinforcement process, the interface between lightweight aggregate and hardened cement paste can form a structurally dense high-strength lightweight aggregate structure, which improves the stress mode of the lightweight aggregate, and plays a reinforcing role for the lightweight aggregate.

(2) Interface strengthening process

The lightweight aggregate-hardened cement paste interface is reinforced using the process shown in Figure 6.61, which consists of the following two main steps.

a) **Pre-wetting of Lightweight Aggregates:** pre-wetting of lightweight aggregates by spraying or soaking to a degree of 20% or more.
b) **Enhancer:** mixed powder consisting of ultra-fine inorganic powder materials such as silica fume, fly ash, slag powder, etc., with an average specific surface area greater than 500 m^2/kg.

(3) Technical effects

The high-strength lightweight aggregate with a cylinder compression strength of 7.5 MPa was selected, and the effects of the three techniques of no reinforcing agent, direct mixing of the reinforcing agent with cement, and interfacial reinforcement

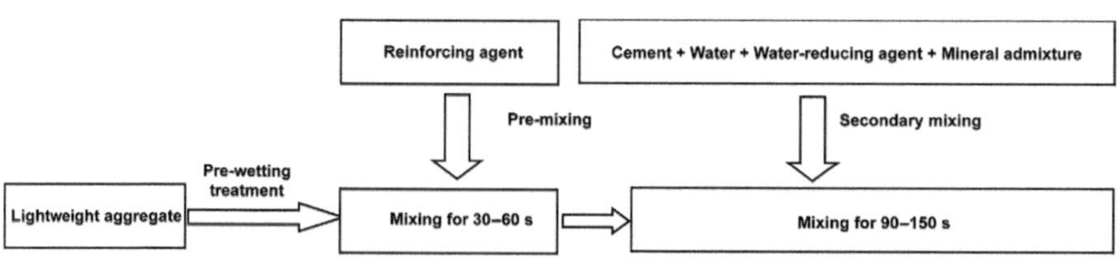

Figure 6.61 Lightweight aggregate and hardened cement paste interface strengthening technology process.

Table 6.47 Comparison of enhancement effect of different processes on lightweight aggregate concrete.

Number	Enhancer B	Process	Slump (cm)	Spread (cm)	Compressive strength (MPa)	
					7 d	28 d
A	No	Normal mixing process	18	55	46.5	60.0
B	Yes	Normal mixing process	24	65	44.0	64.0
C	Yes	Interface strengthening process	22	64	56.2	72.1

Note: The water–binder ratio is 0.30, the amount of cementitious material is 530 kg/m^3, the dosage of fly ash is 10%, and the sand rate is 0.40.

Table 6.48 Influence of interfacial strengthening techniques on the impermeability of lightweight aggregate concrete.

Number	Enhancer (%)		Conductivity (C)	Chloride permeability coefficient ($\times 10^{-13}$ m^2/s)
	Powder A	Powder B		
A	—	—	2495	14.9
C1	10	10	742	6.2
C2	10	5	414	4.6
C3	—	10	1467	9.8

Note: The water–binder ratio is 0.30, the amount of cementitious material is 530 kg/m^3, the dosage of fly ash is 10%, and the sand rate is 0.40.

were compared (Table 6.47). It can be seen that after the application of this technology, the strength of concrete was significantly improved, the permeability was significantly reduced, and the chloride ion permeability coefficient could reach a level lower than 5 × 10–13 m^2/s (Table 6.48).

Figure 6.62 compares the Si/Ca ratio in the interface zone between lightweight aggregate and hardened cement paste prepared by the interfacial reinforcing technology and ordinary process, indicating that the Si/Ca ratio in the interface zone is significantly increased by the pozzolanic reaction using interfacial reinforcing technology [101]. The hydrated calcium silicate gel generated by the reaction between the reinforcing agent and $Ca(OH)_2$ fills up the gaps in the hardened cement paste in the interface zone to make the structure of the hardened cement paste in the interface zone denser, and this effect makes the microhardness of the interface area between lightweight aggregate and hardened cement paste increase greatly (Figure 6.63). Through this effect, the aggregate surface layer to form a dense, high-strength hardened cement paste structure, plays a role in hindering the expansion of cracks. The combined effect of the above results in a significant increase in the strength and permeability of lightweight aggregate concrete.

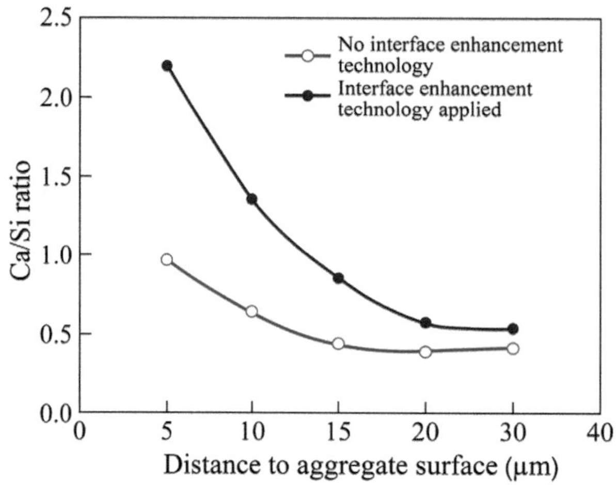

Figure 6.62 Si/Ca ratio in the interfacial region.

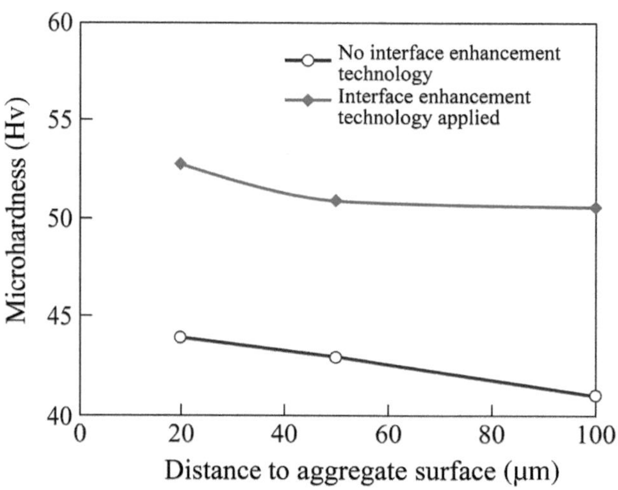

Figure 6.63 Interfacial microhardness.

6.4.2.3 Structural Strengthening Technology for Lightweight Aggregate-Hardened Cement Paste Interface

Pretreatment of lightweight aggregate before use is a special feature different from the ordinary concrete production process. A proper pretreatment process can improve the workability of concrete; improve the mechanical properties, volume stability, and durability performance of concrete; but if not properly controlled, it will also produce some negative effects, so the pretreatment process is very critical [63]. The principle of pre-wetting treatment of lightweight aggregate is as follows:

(1) For lightweight aggregate with water absorption less than 8% in 30 min, it can be used directly for concrete preparation without pre-wetting treatment. To prevent lightweight aggregates from absorbing water during mixing, which would reduce concrete workability, a certain amount of additional water must be added during

concrete mixing. The amount of additional water (W) is calculated according to Eq. (6.18):

$$W = Q \times R \tag{6.18}$$

where Q is the amount of lightweight aggregate per square of concrete in kg/m^3 and R is the water absorption of lightweight aggregate for 30 min presented in %.

For a 30-min water absorption rate higher than 8% of the lightweight aggregate, it should be pre-wet treated. Pre-wetting treatment should first grasp the water absorption characteristics of lightweight aggregate, drawing water absorption-water-saturated time relationship graph, and then according to the technical requirements of construction and environmental conditions to determine the optimal degree of pre-saturation (calculated in accordance with Eq. (6.19)) control range, and then according to the relationship between lightweight aggregate absorption-water-saturated time to determine the optimal pre-saturation time "D":

$$D = \frac{w_t}{w_{\text{vac}}} \times 100 \tag{6.19}$$

where w_t is the water absorption of light aggregate saturated with water for time t, presented in % and w_{vac} is the vacuum water saturation rate of light aggregate saturated with water for time t, presented in%.

According to the construction conditions, the pre-wetting treatment method can be spraying, soaking, heat treatment, vacuum saturated with water, and so on. After pre-wetting treatment of the lightweight aggregate surface attached to a certain amount of water, this part of the water added to the concrete will have a certain impact on the performance of the concrete; the method of draining or drying must be used to make the surface reach the saturated surface dry state.

6.4.2.4 Homogeneity Control Technology for Lightweight Aggregate Concrete Mixes

The constituent materials in concrete mixtures—aggregates, cement binder materials, and water—exhibit certain density differences. During mixing, external segregation phenomena occur where materials with lower density tend to float upward while those with higher density tend to settle downward. The intensity of this external segregation effect influences concrete homogeneity. The density of light aggregates is smaller than that of hardened cement paste, and light aggregates are prone to an uplift phenomenon in the concrete mix. Because the density difference between lightweight aggregate and cement paste is much larger than that between ordinary crushed stone aggregate and cement paste, the outer layering effect of lightweight aggregate concrete is stronger; therefore, if effective homogeneity control technology is not adopted in the preparation of lightweight aggregate concrete, a large number of lightweight aggregates will be floating in the middle and upper layers of the concrete structure, which results in the unevenness of the concrete structure, thus affecting the physical and mechanical properties of the concrete and its durability.

Therefore, homogeneity control is one of the core technologies for the preparation of high-strength lightweight aggregate concrete.

(1) Lightweight aggregate in the concrete mix movement law

Lightweight aggregate in the concrete mix movement speed is an important factor affecting the homogeneity of concrete, the use of hydrodynamic principles of analysis of the lightweight aggregate in the concrete mix particles in the equation of motion, and deduction of the lightweight aggregate in the concrete mix uplift velocity "v" expression (Eq. (6.20)):

$$v = \frac{Kr^2 g(\rho - \rho_c)}{\eta} \tag{6.20}$$

where ρ is the apparent density of the lightweight aggregate, g/cm^3; ρ_c is the apparent density of cement paste, g/cm^3; r is the particle radius, cm; η is the coefficient of cohesion of concrete, Pa·s; v is the velocity of movement of light aggregates, cm/s; and K is the influence factor, the size of which is related to the light aggregate surface properties, material, and other factors.

The main factors affecting the homogeneity of lightweight aggregate concrete mixes were thus obtained in the following order: particle size of lightweight aggregate, viscosity of cement paste, and density difference between cement paste and lightweight aggregate.

(2) Development of key components for homogeneity control and homogeneity control technology

The systematic research on the rheological properties of lightweight aggregate concrete shows that the plastic viscosity of the cement paste should be controlled between 1.5 and 2.5 Pa·s, and the yield stress should be controlled between 4 and 10 Pa for the formulation of high-flow lightweight aggregate concrete with excellent homogeneity [104–107]. The authors used the technical method of material compounding, compounding viscosity increasing, water retaining, and density-regulating components and developed a homogeneity controlling component stabilizer MB, whose technical effects are shown in Table 6.49.

Table 6.49 Effectiveness of MB.

			Slump (cm)		Spread (cm)		Pressurized 6.0 MPa bleeding (ml)	
Number	MB (%)	Stratification (%)	Initial	90 min	Initial	90 min	V10	V140
0	0	14.6	23	16	58	36	8.0	34
1	0.1	12.4	24	22	60	55	5.0	22
2	0.3	5.8	24	22	58	56	4.0	20
3	0.5	4.6	22	18	55	53	3.5	19

Note: The degree of stratification is an indicator of the homogeneity of concrete, which is tested by special test equipment. The smaller the FCD, the better the homogeneity of concrete.

As can be seen from the table, the use of stabilizer MB can significantly improve the homogeneity of freshly mixed lightweight aggregate concrete, improve the water retention properties of concrete under pumping pressure, reduce the time loss of slump of lightweight aggregate concrete, and play an important role in the mixing and pumping of high-performance lightweight aggregate concrete. In addition, the effects of concrete water–cement ratio, sand rate, lightweight aggregate particle size, and mineral admixture on the homogeneity of lightweight aggregate concrete were systematically studied, and the key technical parameters for the preparation of high-performance lightweight aggregate concrete were determined. By optimizing the concrete preparation parameters and adjusting the composition and admixture of MB, the homogeneity of lightweight aggregate concrete of different strengths and densities and with different flow degrees can be effectively controlled.

6.4.3 High-performance Lightweight Aggregate Concrete Performance Design and Modification Technology

6.4.3.1 Brittleness Mechanism and Toughening Technology of High-strength Lightweight Aggregate Concrete

(1) Brittle mechanism of lightweight aggregate concrete

Since the modulus of elasticity of lightweight aggregate is generally smaller than that of hardened cement paste, under axial compressive load, tensile stresses are generated at the upper and lower parts of the aggregate and compressive stresses are generated at the sides. Under tension-compression biaxial stress conditions (tensile stress vertically and compressive stress horizontally), the failure of lightweight aggregate concrete initiates with tensile cracks forming in the cement paste matrix. These cracks subsequently propagate into the aggregates. Due to the inherently low strength and high brittleness of lightweight aggregates, cracks rapidly penetrate through the aggregates. This results in relatively low-energy dissipation during the entire failure process, leading to higher brittleness in lightweight aggregate concrete compared to normal aggregate concrete of equivalent strength grade. Moreover, this brittle characteristic becomes more pronounced with increasing concrete strength [97].

(2) Toughening techniques for high-strength lightweight aggregate concrete

Toughening of high-strength lightweight aggregate concrete is mainly carried out by the following techniques:

(1) Polymer toughening technology for lightweight aggregates and concrete matrix: The low cylinder compressive strength of lightweight aggregates, outstanding brittleness of the material, and poor ability to impede crack extension are important factors affecting the brittleness of concrete. The toughening modification of lightweight aggregates is achieved by utilizing the principle of polymer demulsification and film formation within the aggregates, thereby enhancing their intrinsic resistance to crack propagation [108–112]. The polymer used was modified polymer emulsion (JBX). First, the lightweight aggregate was poured into a vacuumed container (vacuum degree 0.1 MPa), and then the prepared JBX

Table 6.50 Toughness of concrete with polymer pretreated lightweight aggregates.

Preparation method	28 d compressive strength (MPa)	28 d flexural strength (MPa)	Flexural compression ratio
Base concrete	49	3.1	0.063
Lightweight aggregate treated with JBX	50	3.8	0.076
Matrix doped with 20% JBX, lightweight aggregate not treated with polymer	52	4.2	0.081
Lightweight aggregates treated with polymer, matrix mixed with 20% JBX	56	4.8	0.086

Note: Cementitious material 500 kg/m^3, fly ash dosage 15%, water-binder ratio 0.3, and sand rate 40%.

emulsion with a certain concentration was injected into the container, and then the lightweight aggregate was taken out after 1 min and air dried to the saturated surface -dry state and then used for the preparation of concrete. Meanwhile, the synergistic effect of concrete matrix polymer toughening and lightweight aggregate polymer toughening technology was further investigated, and the technical effects are shown in Table 6.50.

After the lightweight aggregate is toughened and modified, the polymer can enter into the open pores inside the lightweight aggregate under vacuum, and after the lightweight aggregate is dried, the polymer breaks into a film in the internal pore wall of the lightweight aggregate and laps into a network, which can play a role in hindering the propagation of cracks when cracks expend to the lightweight aggregate. Therefore, the toughness of lightweight aggregate concrete can be improved after the treatment of lightweight aggregate with JBX. Mixed with 20% JBX; the toughness of concrete has a relatively significant improvement compared with that of control concrete; in the concrete matrix mixed with 20% JBX, while lightweight aggregate pretreated with JBX, the polymer network can be formed throughout the entire concrete, and its toughening effect reaches the best.

Figures 6.64 and 6.65 show the comparison of the effect of polymer toughening. As can be seen from the figures, the densification of the hydration products in the interface zone between the lightweight aggregate and hardened cement paste is effectively improved after the polymer toughening treatment.

(2) Steel fiber and polymer fiber hybrid toughening technology

A systematic study was conducted on the toughening effects of steel fibers and polymer fibers in high-strength lightweight aggregate concrete (Table 6.51) The results demonstrate that the incorporation of steel fibers significantly enhances concrete toughness. At a steel fiber dosage of 1.0%, the flexural-to-compressive strength ratio reaches 0.096, and the flexural toughness index increases by over

Figure 6.64 Unmodified interfacial hardened cement paste structure.

Figure 6.65 Modified interfacial hardened cement paste structure.

15 times compared to plain concrete. However, steel fibers increase the density of lightweight aggregate concrete. Polymer fibers exhibit less pronounced toughening effects than steel fibers but have a minimal impact on concrete density. The combined use of steel and polymer fibers achieves a 14–16 time improvement in flexural toughness while only marginally increasing density compared to plain concrete [48, 67, 112]. This hybrid approach reduces steel fiber consumption

Table 6.51 Effect of mixed fibers on mechanical properties of lightweight aggregate concrete.

Number	Volumetric dosage of steel fibers (%)	Polyacrylonitrile (kg/m³)	28 d compressive strength (MPa)	28 d flexural strength (MPa)	Flexural compression ratio	Fracture toughness index	Densities (kg/m³)
0#	—	—	49	3.1	0.063	1.26	1887
16#	1.0	—	57	5.5	0.096	20.4	1970
20#	—	1.0	53	4.3	0.081	1.70	1890
27#	0.6	1	58	5.6	0.097	21.0	1925

without compromising performance, hereby lowering material costs and preserving the lightweight advantage of lightweight aggregate concrete while maintaining high strength.

6.4.3.2 Modulus of Elasticity, Creep, and Drying Shrinkage Modification of Lightweight Aggregate Concrete

The modulus of elasticity, creep, and drying shrinkage of concrete are important parameters for structural design, and they have a significant impact on the structural performance of concrete. The modulus of elasticity of lightweight aggregates is generally less than 20 GPa, which results in the preparation of concrete with a low modulus of elasticity (generally less than 30 GPa) and large drying shrinkage and creep. To address this problem, research on modification techniques for modulus of elasticity, creep, and drying shrinkage of lightweight aggregate concrete has been carried out [104–107, 113]:

(1) Normal aggregate and lightweight aggregate composite technology

The modulus of elasticity of concrete has a certain correlation with dry shrinkage and creep volume deformation characteristics. The improvement of elastic modulus is conducive to suppressing the shrinkage and creep of concrete. The elastic modulus of normal aggregate is large, and its introduction into lightweight aggregate concrete can increase the elastic modulus of concrete, but its apparent density is large, and its introduction into concrete will cause an increase in the density of concrete and make the workability of concrete worse. Therefore, the technical key lies in the control of the introduction amount of normal aggregate and design of particle grading.

(1) Design of normal aggregate introduction: The relationship between normal aggregate introduction (V_{NA}) and concrete modulus of elasticity (E) was obtained from the study (Eq. (6.21)):

$$E = \frac{E_{\mathrm{NA}} \cdot E_{\mathrm{LC}}}{[E_{\mathrm{NA}} - V_{\mathrm{NA}}(E_{\mathrm{NA}} - E_{\mathrm{LC}})]} \tag{6.21}$$

where V_{NA} is the volume content of normal aggregate in concrete presented in %, E_{NA} is the elastic modulus of normal aggregate presented in MPa, and E_{LC} is the elastic modulus of high-strength lightweight aggregate concrete presented in MPa.

Considering the effect of the composite use of normal and lightweight aggregates on the homogeneity of concrete, the relationship equation between V_{NA} and the homogeneity of concrete was obtained (Eq. (6.22)). In order to ensure that the concrete has good homogeneity (FCD $\in$ [5%–10%]), the V_{NA} should be controlled at 20%–30%.

$$FCD = 0.180\, {V_{NA}}^2 - 0.202\, V_{NA} + 0.103\; R^2 = 0.980 \tag{6.22}$$

where FCD is the degree of stratification of concrete presented in % and V_{NA} is the volume content of normal aggregate in concrete presented in %.

(2) Aggregate grading design: theoretical analysis and experimental research, lightweight aggregates and ordinary aggregates should be controlled in the same particle size interval or close to the same particle size interval and as far as possible to ensure that the volume of ordinary aggregates remain unchanged under the common aggregates to increase the number of particles, that is, to reduce the average particle size of ordinary aggregates particles. The effect of composite technology of normal aggregate and light aggregate is shown in Table 6.52.

(2) Optimized design of cementitious materials

The study of the effect of single and composite action of mineral admixtures such as fly ash and silica fume found that both the strength and modulus of elasticity of lightweight aggregate concrete can be significantly improved by composite addition of fly ash and silica fume and the drying shrinkage and creep of concrete can be reduced (Table 6.53). Considering that the composite use of normal and lightweight aggregates increases the density of concrete, the effect of over-substitution of sand with fly ash on the density of concrete was investigated (Table 6.54).

From the table, it can be seen that the use of fly ash to replace sand can make the concrete maintain a high modulus of elasticity while reducing the apparent density of concrete, and the greater the rate of fly ash to replace sand, the more obvious the effect.

Table 6.52 Effect of normal aggregate introduction on the modulus of elasticity of concrete.

Number	V_{NA} (%)	28 d compressive strength (MPa)	28 d tensile strength (MPa)	Apparent density (kg/m^3)	Modulus of elasticity (GPa)
1	0	55.4	3.5	1900	26.2
2	20	57.4	3.7	2080	29.7
3	30	59.8	3.8	2160	32.9

Note: Cementitious material was 530 kg/m^3, fly ash was 10%, water-binder ratio was 0.3, and sand rate was 40%.

Table 6.53 Effect of mineral admixtures on concrete properties.

Number	Mineral admixture (%)		28 d compressive strength (MPa)	Modulus of elasticity (GPa)
	F ash	Silica fume		
A	—	—	55.0	26.2
B	10	—	55.4	27.5
C	10	5	64.1	28.8

Note: The amount of cementitious material was 530 kg·m^{-3}, fly ash was 10%, and water–binder ratio was 0.30.

Table 6.54 Effect of fly ash replacement of sand on concrete properties.

Number	Fly ash substitution rate for sand (%)	Sand rate (%)	Apparent density (kg/m^3)	28 d compressive strength (MPa)	Modulus of elasticity (GPa)
FA15	0	40	2080	57.4	29.7
DS05	5	38	2050	54.8	28.7
DS10	10	36	2020	55.7	30.1
DS20	20	32	1980	56.7	30.5

Note: The V_{NA} was 20%, amount of cementitious material was 530 kg/m^3, fly ash was 15%, and water–binder ratio was 0.30.

(3) Fiber induced volumetric stabilization technology

The study compares the effect of fiber on volumetric deformation, creep, and modulus of elasticity of lightweight aggregate concrete; the results show that the addition of polymer fiber can reduce the plastic shrinkage of lightweight aggregate concrete, and the higher the modulus of elasticity of the fiber, the more pronounced the inhibition of the volumetric shrinkage of lightweight aggregate concrete; the 28 d shrinkage lightweight aggregate concrete mixed with 1% polyacrylonitrile fiber (modulus of elasticity of 17,100 MPa) was 2.8×10^{-4}; the 28 d dry shrinkage of lightweight aggregate concrete with 1% steel fiber (modulus of elasticity 200,000 MPa) added by volume rate was 2.4×10^{-4}.

The modulus of elasticity, dry shrinkage, and creep coefficient of concrete before and after technical modification were compared by the combined application of the above modification techniques (Tables 6.55 and 6.56).

6.4.4 Design and Preparation Techniques for Lightweight Aggregate Concrete Structures and Elements

6.4.4.1 Stress–strain Relationship of High-strength Lightweight Aggregate Concrete

The stress–strain relationship of concrete is fundamental for structural design, and the stress–strain relationship of high-strength lightweight aggregate concrete

Table 6.55 Effectiveness of modification techniques for lightweight aggregate concrete.

Number	Polymers (kg/m^3)	Fibers (kg/m^3)	Admixture (kg/m^3)	V_{NA} (%)	Modulus of elasticity (GPa)	Creep factor	28 d drying shrinkage/10^{-4}
1	0	0	0	0	26	1.2	4.0
2	60	0	0	0	24	1.2	3.6
3	0	50	0	0	26	1.2	3.0
4	0	0	70	0	28	1.1	2.8
5	60	50	70	0	30	1.1	2.4
6	0	0	0	30	32	1.0	2.2
7	0	0	70	30	35	1.0	2.0

Table 6.56 Comparison table of properties of lightweight aggregate concrete before and after modification.

Concrete	Performances	Modulus of elasticity (GPa)	Creep factor	28 d drying shrinkage/10^{-4}
General concrete	Range of indicators	30–40	1.0–1.2	2.0–2.4
Lightweight aggregate concrete	Range of indicators before modification	20–30	1.2–1.4	>4.0
	Range of indicators after modification	12–35	1.0–1.2	2.0–4.0

(Eqs. (6.25) and (6.26)) was obtained by using a rigid testing machine loaded in a strain-controlled mode:

$$y = ax + bx^2 + cx^3 \ (x \leq 1) \tag{6.23}$$

$$y = \frac{x}{d(x-1)^2 + ex} \ (x \geq 1) \tag{6.24}$$

where x was $\frac{\varepsilon}{\varepsilon_c}$; y was $\frac{\sigma}{f_c^*}$; ε_c was peak strain; f_c^* was peak stress; a, b, c, d, e a was constant with expressions $a = 1.2$, $b = 0.6$, $c = -0.8$, $d = 15$, and $e = 1$ for LC40 lightweight aggregate concrete and $a = 1.1$, $b = 0.8$, $c = -0.9$, $d = 20$, and $e = 1$ for LC50 lightweight aggregate concrete.

6.4.4.2 Properties of High-strength Lightweight Aggregate Concrete Members

The member is the basic unit of the structure, and the design method of high-strength lightweight aggregate concrete members and its performance evolution rule have been mastered by systematically carrying out the performance

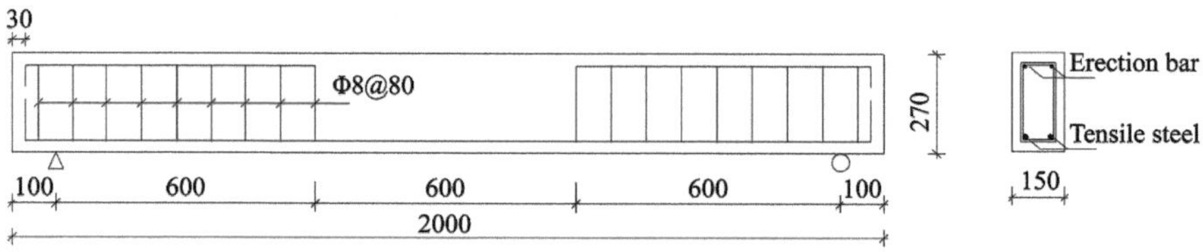

Figure 6.66 Structural design drawings of flexural members (in mm).

experiments of three types of typical members, namely flexural, shear, and prestressed [61, 94, 98, 114, 115].

(1) High-strength reinforced lightweight aggregate concrete flexural members

The structural design of lightweight aggregate concrete flexural members is shown in Figure 6.66, and the effects of reinforcement rate and concrete strength class on the positive section capacity and deflection of the members were investigated separately (Table 6.57).

As can be seen from the table, the measured loads of the members at the yield stage are higher than the theoretical values, indicating that the adopted design method for the flexural capacity of the positive section is safe and reliable. The flexural capacity of lightweight aggregate concrete members is slightly higher than that of ordinary concrete under the same strength class condition, and its magnitude increases with the increase of reinforcement rate and the increase of concrete strength. It is shown that the deflection of lightweight aggregate concrete beams is greater than that of plain concrete beams under the same reinforcement rate and same strength class. However, the maximum deflection under the service load still meets the normal service requirements specified in the code. Under the condition of the same reinforcement rate and same concrete strength, the crack distribution of lightweight aggregate concrete members is more uniform than that of ordinary concrete members, with smaller spacing and finer width, which indicates that lightweight aggregate concrete members have excellent crack resistance.

Based on the results of the study, the formula for calculating the cracking moment M_{cr} for rectangular-section beams of high-strength lightweight aggregate concrete was modified. For high-strength lightweight aggregate concrete (48.5 MPa $\leq f_{cu} \leq$ 72.1 MPa) rectangular cross-section beams,

$$M_{cr} = 1.50\xi W_0 f_{tk} \tag{6.25}$$

where W_0 was the elastic resistance moment of the section at the tension edge; ξ was the section height impact factor, $\xi = 0.7 + 120/h$ (when $h \leq 400$ mm, take $h = 400$; when $h \geq 1600$ mm, take $h = 1600$ mm); and f_{tk} was tensile strength of concrete, MPa.

(2) High-strength reinforced lightweight aggregate concrete shear members

The structural design diagrams of the shear members are shown in Figure 6.67, and the effects of shear-to-span ratio, hoop ratio, the presence or absence of web

Table 6.57 Study on the performance of high-strength reinforced lightweight aggregate concrete flexural members.

Number	Compressive strength of concrete (MPa)	Reinforcement rat (%)	Cracking stage load		Yield stage load		Measured ultimate load (kN)
			Theoretical value (kN)	Measure value (kN)	Theoretical value (kN)	Measure value (kN)	
LC60-10	72.1	0.45	32.00	29.17	55.53	56.29	76.46
LC60-16		1.16	33.46	31.25	110.77	125.54	157.29
LC60-16-16		2.56	34.60	33.93	191.61	202.14	224.29
LC60-16-20		3.37	35.20	37.50	243.62	260.71	282.14
LC50-16	60.9	1.16	27.88	28.16	110.11	120.42	150.42
LC40-16	48.5	1.16	24.96	25.00	109.01	116.25	132.92
NC50-16	54.7	1.16	29.12	29.20	107.72	117.08	142.08

Note: LC is lightweight aggregate concrete and NC is normal concrete.

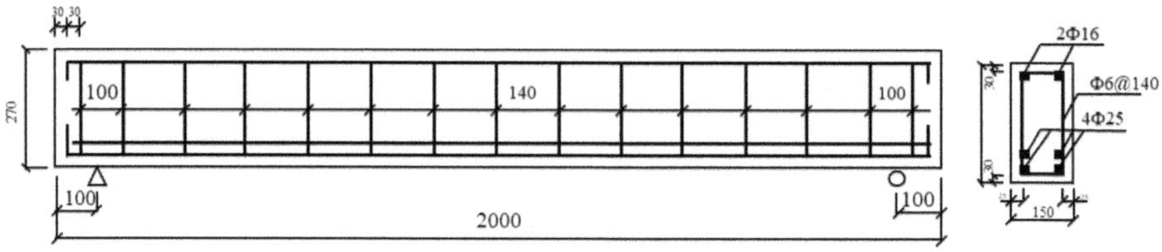

Figure 6.67 Structural design drawings of shear-resisting members (in mm).

reinforcement, and concrete strength grade on the shear capacity and deflection of the members were investigated separately (Tables 6.58 and 6.59).

It is found that the ultimate shear capacity of high-strength lightweight aggregate concrete members under the same shear span ratio is smaller than that of ordinary concrete members, and its ultimate shear capacity decreases with the increase in shear span ratio and improves with the increase in concrete strength or hoop ratio; the deflection of the members increases with the increase in shear span ratio, and the concrete strength and hoop ratio have less effect on deflection; the damage form of lightweight aggregate concrete members is the same as for ordinary concrete members, and it shows obvious brittleness in shear damage, and the damaged surface is smoother; for the beam without web, the member shows obvious brittleness in damage, and the damage form is diagonal tensile damage. It is found that the shear capacity formula specified in the "Lightweight Aggregate Concrete Structural Design Regulations" (JGJ12-99) is safe and applicable for estimating the shear capacity of high-strength lightweight aggregate beams with web reinforcement.

(3) Prestressed lightweight aggregate concrete member

Design LC40 lightweight aggregate concrete prestressed member was designed (Figure 6.68). The member is 20 m long, and the prestressing reinforcement is a high-strength, low-relaxation strand with standard strength = 1860 MPa, modulus of elasticity $E = 1.95 \times 105$ MPa, and tensile stress=0.73Ryb. Structural design loads are standardized to exceed Class 20 for automobiles and Class 120 for trailers. Research shows that in accordance with the normal use of the state of calculation of the maximum load (218 kN) under the action of maximum deflection in the span of the member for 14.03 mm, the maximum deflection and bearing capacity meet the "highway bridge specification" (JTGD62-2004) design requirements. There is no visible crack at the bottom of the member during the loading process, and the residual deformation of the member cross-section in the middle of the span after unloading is 0.26 mm, and the ratio of it to the total deformation is between 2% and 4%, which meets the specification requirements and the design of the load capacity. During loading and unloading, the prestressed lightweight aggregate concrete members showed good elastic deformation performance (Figure 6.69).

6.4.4.3 Structural Gradient Design for High-strength Lightweight Aggregate Concrete

The technical characteristics of high-strength lightweight and high-strength concrete are very suitable for large-span structural projects, but its brittleness, poorer

Table 6.58 Test program for the performance of high-strength reinforced lightweight aggregate concrete flexural members.

Number	Effective height of section (mm)	Shear-span ratio	Main reinforcement (mm)	Reinforcement ratio of main reinforcement μ (%)	Compression steel (mm)	Hoop	Rebar rate μ_k (%)
LC60-1	200	1	4ϕ25	6.55	2ϕ16	ϕ6@140	0.27
LC60-2	200	2	4ϕ25	6.55	2ϕ16	ϕ6@140	0.27
LC60-3	200	3	4ϕ25	6.55	2ϕ16	ϕ6@140	0.27
LC60-4	200	4	4ϕ25	6.55	2ϕ16	ϕ6@140	0.27
NC50-3	200	3	4ϕ25	6.55	2ϕ16	ϕ6@140	0.27
LC50-3	200	3	4ϕ25	6.55	2ϕ16	ϕ6@140	0.27
LC40-3	200	3	4ϕ25	6.55	2ϕ16	ϕ6@140	0.27
LC60-120	200	3	4ϕ25	6.55	2ϕ16	ϕ6@120	0.31
LC60-100	200	3	4ϕ25	6.55	2ϕ16	ϕ6@100	0.38
LC60-N	226.5	3	2ϕ25	2.89	—	—	—
LC50-N	226.5	3	2ϕ25	2.89	—	—	—

Note: LC is lightweight aggregate concrete beam with web reinforcement; 1, 2, 3, 4 refer to shear span ratios of 1, 2, 3, and 4, respectively; LC-100 and LC-120 refer to hoop spacing of 100 mm and 120 mm, respectively, while the rest of the beams have hoop spacing of 140 mm; NC50-3 is plain concrete beam with web reinforcement and a shear span ratio of 3; and LC60-N and LC50-N are lightweight aggregate concrete beams without web reinforcement. LC60-N and LC50-N are lightweight aggregate concrete beams without web reinforcement.

Table 6.59 Main results of shear member performance tests.

Number	f_{cu} (MPa)	Mid-span moment at beam cracking (kN·m)	Oblique crack opening shear (kN)	Critical diagonal crack shear (kN)	Ultimate shear (kN)	Forms of damage
LC60-1	73.1	19.5	98	247	345.9	Bevel
LC60-2	73.2	32.4	87	132	250.1	Shear
LC60-3	69.9	52.2	81	122	194.3	Shear
LC60-4	68.5	48.0	60	113	149.0	Inclined pull
NC50-3	54.7	48.0	80	100	199.0	Shear
LC50-3	60.9	37.2	62	114	181.1	Shear
LC40-3	48.5	45.0	75	97	171.0	Shear
LC60-120	68.8	53.4	89	126	211.5	Shear
LC60-100	67.3	51.6	86	124	219.7	Shear
LC60-N	65.6	40.8	60	68	103.6	Inclined pull
LC50-N	60.9	38.1	56	65	101.3	Inclined pull

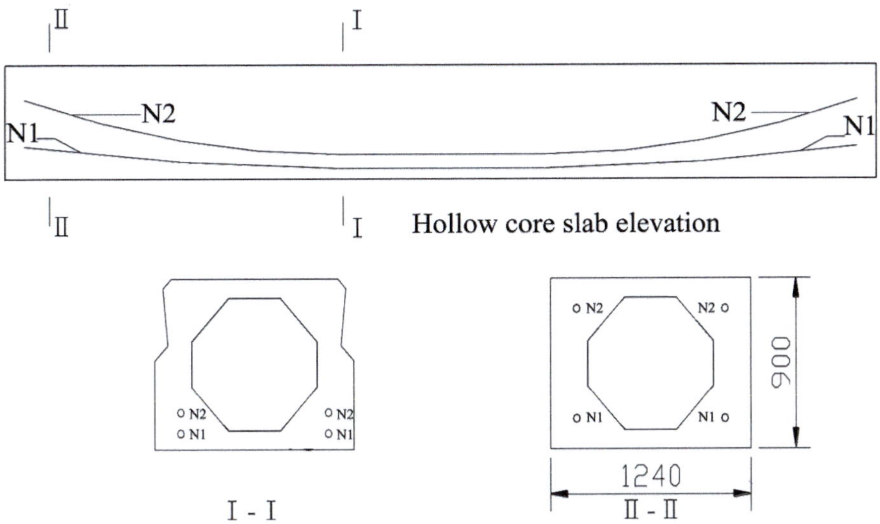

Figure 6.68 Design of prestressed lightweight aggregate concrete members.

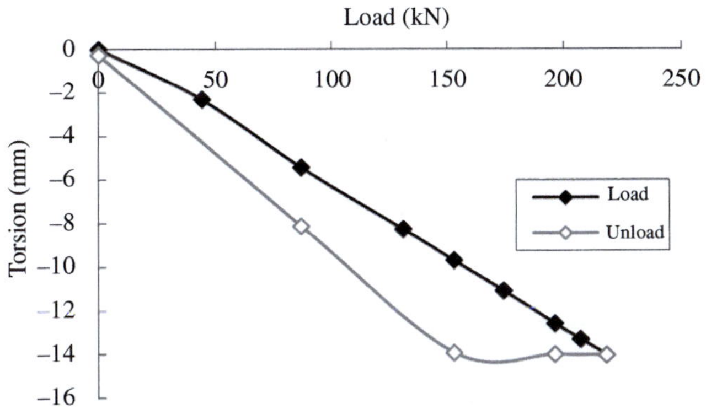

Figure 6.69 Load-deflection relationships for prestressed concrete members.

shear bearing capacity than ordinary concrete, small elastic modulus, and large shrinkage and deformation make its use in large-span structural projects have some adverse effects.

In order to solve the technical problems of the application of lightweight aggregate concrete in large-span structures, we proposed the method of structural gradient design [116]. The principle is to fully utilize the designable characteristics of concrete density and elastic modulus by employing concretes with distinct properties along the structural length, thereby achieving gradient performance variations. This approach optimizes structural stress distribution, enhances span capacity, and improves service performance (Figure 6.70).

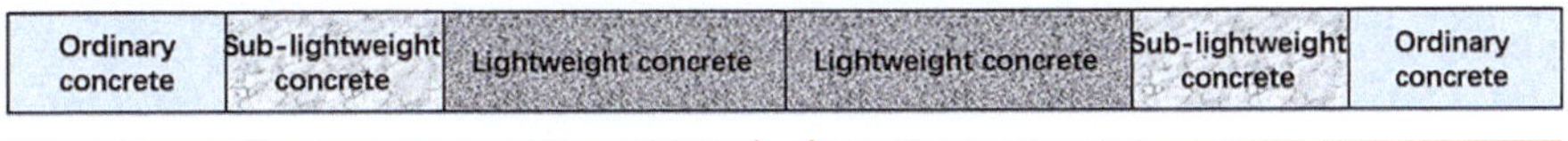

Figure 6.70 Structural gradient design principle.

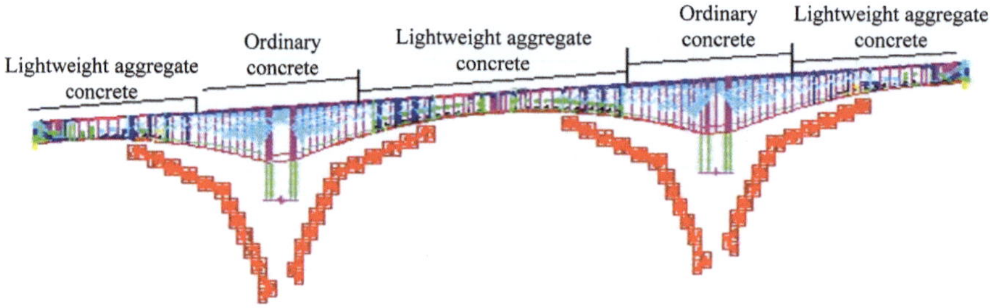

Figure 6.71 Material gradient design technique for large-span structures.

Take the large-span continuous rigid bridge girder as an example. According to the design experience of this type of ordinary concrete bridge, the box girder pier top section, the main girder at the 1/8 cross section of the bridge span is the most stress-concentrated region of the whole bridge as well as having the most complicated stress condition. Through the finite-element calculation simulation analysis, applying the design method of structural gradient, we designed the main bridge with 85 + 150 + 85 m large-span continuous rigid bridge girder (Figure 6.71), the side span support cast-in-place section of 8.8 m, top of the pier of 15 m, and the two sides of each of the five 2.5 m section of a total of 40 m using C50 ordinary concrete; the middle of the span and side spans used the LC60 lightweight aggregate concrete. By comparing with the technical scheme of using normal concrete or lightweight aggregate concrete for the whole bridge, it is considered that this design scheme has better safety and economy.

6.4.4.4 Anti-cracking Technologies for Anchor End of Prestressed Lightweight Aggregate Concrete Members

The anchored ends of concrete members are susceptible to cracking during prestressing tensioning due to insufficient localized bearing pressure. Compared with ordinary concrete of the same strength grade, lightweight aggregate concrete has high brittleness and low tensile strength, so lightweight aggregate concrete members are more prone to end cracking during prestressing tensioning. In response to this problem, an anti-cracking technique for the anchorage end of prestressed lightweight aggregate concrete members was invented [117].

(1) Finite-element modeling of prestressed lightweight aggregate concrete members.

According to the designed dimensions of the prestressed lightweight aggregate concrete members, a reliable finite element model of the prestressed lightweight aggregate concrete members was established by using the ANSYS system using the lightweight aggregate concrete material properties parameters (Figure 6.72).

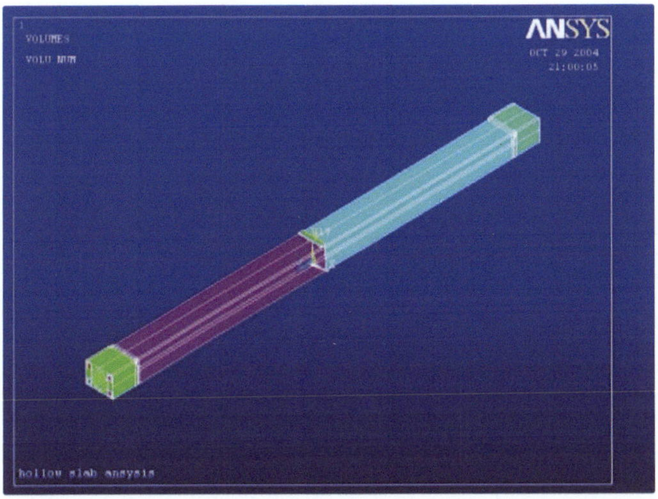

Figure 6.72 Finite-element modeling of prestressed lightweight aggregate concrete members.

(2) Stress analysis of the anchorage end of the prestressed member

The stress distribution of its anchorage end during prestressing tensioning and loading was analyzed using a finite-element model (Figure 6.73), and it was found that the maximum stress in the concrete of the anchorage area during tensioning was 24.0–33.8 MPa, which might exceed the design value of the axial compressive strength of the LC40 lightweight aggregate concrete and would result in cracking of the end of the prestressed member.

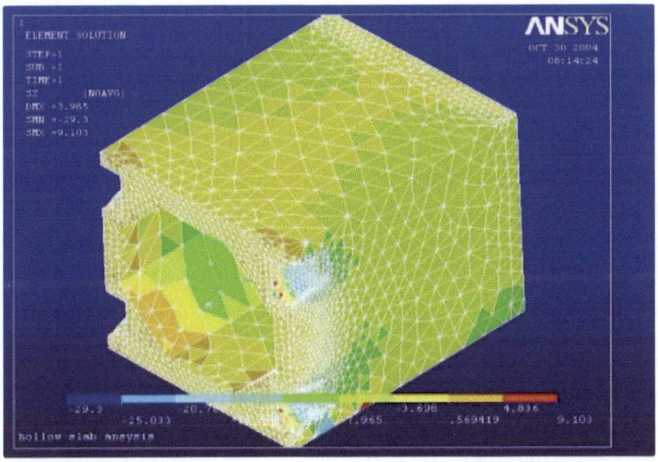

Figure 6.73 Cloud diagram of anchorage end of prestressed lightweight aggregate concrete member.

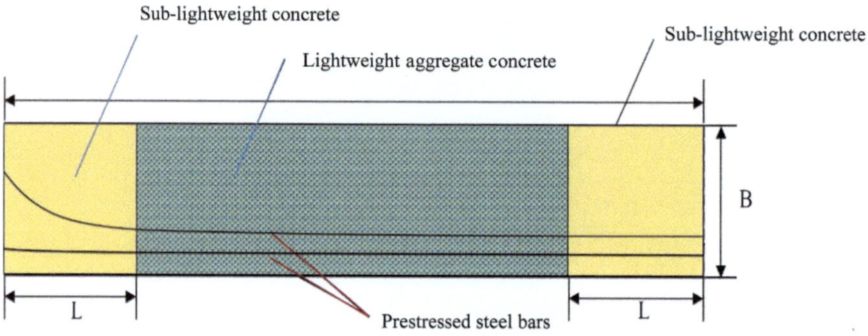

Figure 6.74 Structural gradient design of prestressed concrete members.

(3) Anti-cracking technology at the end of prestressed concrete members

Applying the structural gradient design method, combined with the lightweight aggregate concrete toughening and modification technology, the design method for anti-cracking at the end of prestressed lightweight aggregate concrete members shown in Figure 6.74 has been invented. High-strength lightweight aggregate concrete is used in the middle section of the member, and high-strength sub-lightweight concrete with higher tensile strength and good crack resistance is used in the end section of the member. The length L of the end high-strength sub-lightweight concrete is selected according to Eq. (6.26):

$$L = (0.8 - 1.2)B \tag{6.26}$$

where L is the length of the sub-lightweight concrete used at the end of the member, presented in m and B is the height of the end of the member, presented in m. The longer the member, the upper limit of L is taken, and the shorter the member, the lower limit of L is taken.

The finite-element simulation analysis was used to obtain that the optimal length of sub-lightweight concrete used at the end of the 20 m prestressed lightweight aggregate concrete member was 1 m. Due to the use of high-toughness and high-crack resistance sub-lightweight concrete at the end of the prestressed member, the stress distribution in the anchorage area was significantly improved, the concentrated stress at the end was reduced, and the cracking phenomenon at the end of the prestressed member in the process of prestressing was prevented effectively.

6.4.5 Construction and Application Technology of High-strength Lightweight Aggregate Concrete

6.4.5.1 Evaluation Method for Homogeneity of Lightweight Aggregate Concrete

Previously, the homogeneity of concrete is generally judged by observing the working condition of the concrete mix, which has the disadvantage of large subjectivity.

The density of lightweight aggregate is less than cement paste, and it has the tendency of spontaneous upward movement in freshly mixed concrete, and the uplift of lightweight aggregate will cause it to be unevenly distributed inside the concrete, resulting in segregation that the upper part of the concrete structure is enriched with lightweight aggregate. Aiming at the problem that the uplift of lightweight aggregate is easy to occur in the construction of lightweight aggregate concrete and the lack of a scientific and reasonable homogeneity evaluation method before, we invented a test method and device that can quantitatively evaluate the homogeneity of lightweight aggregate concrete [118].

(1) Homogeneity evaluation indicator – stratification
Stratification is the degree of non-uniformity in the distribution of lightweight aggregates in the concrete mix. The higher the degree of stratification, the worse the homogeneity of the concrete. Conversely, the better the homogeneity of the concrete.
(2) Stratification test method and apparatus
A stratification test method and apparatus were designed and developed. The test apparatus for the stratification degree consists of an S-type flow meter and a set of five $\phi 10 \times 10$ cm or $\phi 20 \times 20$ cm cylindrical steel molds (Figure 6.75). The concrete was mixed and filled into the S-type flowmeter, through which the concrete flowed into the device, and then vibrated on a concrete vibrating table for 30 s. The lightweight aggregates were screened out of each layer separately using the water rinsing method, dried, and then weighed (to the nearest 0.1 g).

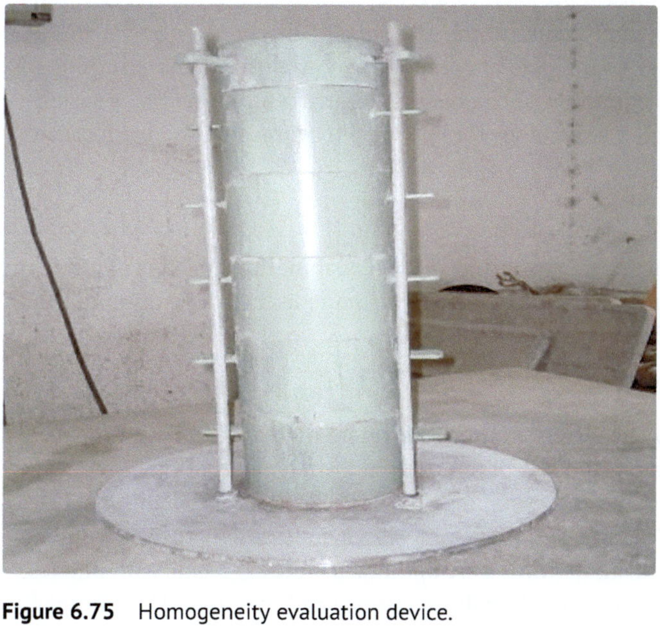

Figure 6.75 Homogeneity evaluation device.

(3) Evaluation method of stratification degree
The flowability of concrete can be evaluated according to the time of concrete flow in the S-flow meter. According to the weight data of lightweight aggregate in each layer obtained from the test, the stratification degree of concrete can be calculated using Eq. (6.27).
Tests show that, for the slump less than 180 mm concrete, stratification degree *FCD* is less than 5% when the concrete generally has good homogeneity; for the slump requirement is greater than 180 mm, the stratification degree is 5–10% when the concrete generally has good homogeneity:

$$FCD = \frac{\sqrt{\sum_{i=1}^{n} (G_i - \bar{G})^2}}{n-1} \times 100\% \tag{6.27}$$

where *FCD* is the stratification degree, %; G_i is the weight of lightweight aggregate in each layer, g; $\bar{G}$ is the average of the weights of the light aggregates in layers 1-n, g; and n is the number of layers, $n = 2, 3, 4, 5$.
This method provides quantitative indexes for the evaluation of homogeneity and quality control of lightweight aggregate concrete during testing and construction. A workability evaluation system of lightweight aggregate concrete is established by combining the stratification degree evaluation method with slump and spread tests. The engineering test proves that the evaluation system is simple, applicable, and effective.

6.4.5.2 Ultra-high, Ultra-long Distance Homogeneous Pumping Technology

The characteristics of lightweight aggregate concrete determine that its pumping has considerable technical difficulties, and the smaller the density or the higher the strength of lightweight aggregate concrete, the more difficult the pumping of concrete. Through a large number of engineering experimental research, a systematic homogeneous pumping technology for lightweight aggregate concrete has been formed [65, 66, 119–121]:

(1) Material performance design

The initial slump of concrete is controlled at 180–240 mm, the spread is controlled at 450–600 mm, the 90 min slump loss through time is less than 60 mm, and the stratification degree of the mix is 5.0%–10.0%.

(2) Preferred raw materials

Lightweight aggregate with low water absorption under high pressure was preferred. Determine the pretreatment process of lightweight aggregate according to the size of lightweight aggregate water absorption as well as construction conditions. Control the maximum particle size of lightweight aggregate less than 16 mm. Cement, water reducer, and mineral admixtures were optimized. Fine aggregate with fineness modulus greater than 2.6 or more of medium-coarse sand was used. Use pumps to deliver the agent.

(3) Optimization of Concrete Proportioning

Optimization of technical parameters for the preparation of concrete such as water-binder ratio, binder-aggregate ratio, sand rate, water reducer, pumping agent admixture, etc.

(4) The pumping process mainly includes the following steps:

(1) Pumping machine and pumping pipe selection.
   The pump pressure and displacement adjustable pump were selected according to the pumping distance. Choose the diameter of a 150–200 mm pump pipe; when the pumping distance is higher, a larger diameter pump pipe is more conducive to pumping.
(2) Pump pipe erection method.
   The pump pipe layout should minimize bends, especially 90° elbows, to reduce concrete pumping resistance. Due to significant changes in the workability of concrete at the pump outlet—where lightweight aggregates are prone to excessive water absorption leading to blockages—the straight-line distance at the pump outlet should be appropriately extended.
(3) Pump before the wet pipe method.
   Before pumping concrete, wet the pump pipe with water, and then lubricate the pump pipe with cement mortar mixed with a certain proportion of fly ash and water reducer.
(4) Stabilized Pressure and Medium-Displacement Continuous Pumping Method
   During initial pumping, control the pump to operate at medium displacement or medium pressure, maintaining a slow, steady, and reversible working state. For normal pumping, appropriately increase pressure and displacement while ensuring uninterrupted concrete supply to minimize pump stoppages. During long-distance or ultra-high pumping, gradually adjust pressure from low to high levels.

(5) Pumping process control

Monitor and adjust initial slump, flow spread, segregation resistance, and slump loss over time to ensure optimal pumpability. Implement secondary mixing of concrete at the pump outlet to mitigate segregation caused by prolonged pumping. Strictly observe pump pressure fluctuations; abnormal pressure spikes may indicate blockages, requiring immediate anti-pump operations (2–3 reverse strokes) to clear the pipeline. Test pumped concrete for density and compressive strength, ensuring density increase $\leq$10–20 kg/m^3, and the loss of strength at 10% or less.

6.4.5.3 Construction Critical Processes

To address the characteristics of lightweight aggregate concrete (LAC)—such as **greater variability in aggregate performance** (requiring pre-treatment), **prone to segregation during transport and vibration**, and **heightened sensitivity to raw material stability and construction process fluctuations**—a **systematic and comprehensive study** was conducted on LAC construction techniques. This encompassed construction processes (mixing, transportation, pumping, vibration, and curing), site organization (equipment selection/layout, construction planning), quality control (process monitoring and mitigation of influencing factors). Key technologies developed include **admixture feeding method, stabilized pressure and medium-displacement continuous pumping, and high-frequency, low-amplitude, dense-spacing vibration.**

(1) Admixture feeding method For pre-wetted and non-pre-wetted lightweight aggregates, the feeding order of admixtures is different: If the lightweight aggregates are pre-wetted, the admixture can be added together with other raw materials. If the lightweight aggregates are not pre-wetted before use, the admixture should be added last.

(2) High-frequency low-amplitude dense-spacing vibration method The high-frequency low-amplitude dense-spacing vibration method specifically includes insertion of vibration compact, high-frequency low amplitude vibration, vibration time, and vibration point dense-spacing control vibration process, the vibration process can make the concrete compact without segregation.

(3) Moisture preservation and heat preservation process When the exposed surface of the concrete is large or in hot climatic conditions or wind speed is high, continuous moist curing (such as wet sacks or spray wet plastic film) should be adopted. If the concrete construction is completed within 3–7 d encountered in negative-temperature weather, it is necessary to use heat preservation and anti-freezing curing measures.

6.4.5.4 Engineering Applications

(1) Overview of technology application

Through systematic research and technological innovation, the author's team has summarized and integrated complete sets of technologies and formed a complete system of independent intellectual property rights through the promotion and application of a large number of structural projects. The team has mastered the complete set of technologies for the preparation of high-strength and high-performance lightweight aggregate concrete and design and construction of structural engineering. By using the high-quality lightweight aggregate prepared by this technology or by using ordinary lightweight aggregate through the reinforcing and toughening modification technology provided by this technology, it can be stably produced to meet the requirements of concrete for all kinds of structural engineering with the strength grades ranging from LC20 to LC60, and the densities within the range of 1200–1950 kg/m^3, so that lightweight aggregate concrete is applied to all kinds of structural engineering buildings. The lightweight aggregate concrete has been applied to various structural engineering buildings and road pavement structures. It was the first time to design and construct the whole high-rise building structure (29-storey) Yichang Riverside International Garden. It has constructed the super large-area structural floor slabs – the floor slabs of the Wuhan Tianhe Airport Terminal Building Phase II Project reached 120,000m^2 in a single floor; the bridge deck structure of large-span bridges of Hanjiang Bridge of the Beijing-Zhuhai

Wuhan Qintai Grand Theater
(Floor and decorative frame structure)

Wuhan Tianhe Airport Phase II Project
(Terminal floor structure)

Wuhan Securities Building
(Structure with floor plates above 68 levels)

Yichang Binjiang International Garden Building (All lightweight aggregate concrete structure)

Splendid Yangtze River Tower
(Ultrahigh elevation pumping)

Figure 6.76 Typical high-strength lightweight aggregate concrete works.

Expressway (span of 180 m × 24 m); the girder structure of the bridge of the motorway – the Tuanshan River Bridge of the Hubei Xiaoxiang Expressway (span of 80 m); the structure of the bridge of the motorway. Structure (span 80 m), ultra-high-rise (330 m) steel building floor slab combination structure-Wuhan Securities Building 64–68 floor slabs; and 10 steel box girder bridge deck combination pavement structure projects-Wuhan Central Ring Road West Ring Section Viaduct Steel Box Girder Bridge Deck Pavement Layer and Wuhan Roundabout Highway East–West Lake Steel Box Girder Bridge, etc. Realization of lightweight aggregate concrete at a horizontal distance of more than 200 m – Wuhan Jinxiu Yangtze River Building Project – and vertical height is more than 180 m – Wuhan Qintai Grand Theatre project of ultra-long and ultra-high pumping construction. Some of the projects are shown in Figure 6.76.

The outstanding features of high-strength lightweight aggregate structural concrete technology are as follows:

(1) Wide range of strength and density regulation, strength regulation range LC7.5-LC60, density regulation range 1200–2200 grade.
(2) Good homogeneity of concrete, excellent workability, and longtime of workability maintenance and regulation.
(3) Good pumpability of concrete, the designed LC7.5-LC60, 1200–2200 grade concrete can be constructed by pumping technology. Concrete can be constructed

by pumping technology, and the horizontal pumping distance reaches 300 m and the vertical distance reaches more than 180 m in the actual engineering application.

(4) Excellent durability, and the chlorine-ion impermeability coefficient is less than $5 \times 10^{-13}\ m^2/s$.
(5) Simple construction process, and lightweight aggregates cannot be pre-wetted.
(6) Outstanding economic benefits. Through the application of lightweight aggregate concrete in structural engineering, the structural deadweight is reduced, the height of the structure is increased under the condition of no-foundation treatment, the cost of reinforcement is saved, and the area of the building used is increased.
(7) The social benefits are remarkable, the use of natural aggregates is saved due to the use of lightweight aggregate, and a large amount of lightweight aggregates prepared by silt and fly ash is applied, so that silt is effectively utilized to improve the urban environment, and the mining of natural aggregates is reduced at the same time. Mining of natural aggregates is also reduced.

(2) Application Examples

High-strength lightweight aggregate concrete has been successfully used in the Jingzhu Expressway Wuhan Caidian Hanjiang Bridge Project and Yichang Binjiang International Garden Project.

(1) Wuhan Caidian Hanjiang Bridge Project for Beijing-Zhuhai Expressway
The Caidian Hanjiang Bridge is a national key project for the Beijing-Zhuhai (Beijing-Zhuhai) Expressway and Wuhan Outer Ring Road, with a total length of 1607 m, the main span is a pre-stressed box-reinforced concrete structure, and the maximum span of the bridge is 400 m, of which the main span is 180 m and the spans of the two sides are both 110 m (Figure 6.77). Due to design changes, more than 600 t of weight reduction was required, and it was decided to use about 8 cm of lightweight aggregate concrete as the bridge deck structure high-performance lightweight aggregate concrete of $1200\ m^3$, which weighs about 2280 t. Compared with ordinary concrete, the weight reduction is more than 600 t and meets the design requirements.

Figure 6.77 Lightweight aggregate structural engineering and lightweight aggregate bridge deck structure construction for the Caidian Hanjiang Bridge.

Table 6.60 Construction mix ratio of pumped lightweight aggregate concrete for the Caidian Hanjiang Bridge.

Water	Cement	HCF	Lightweight aggregate	Granule	DLNT fibers	FDN (%)	Slump (cm)	Extensibility (cm)
			kg/m^3					
160	477	73	506	619	1.0–1.2	0.8	18–22	55–60

The design requirements of the project are as follows: a. The dry apparent density of concrete is less than 1900 kg/m^3. b. The strength grade reaches above LC40 and the flexural strength is more than 6.0 MPa, in addition to better seismic, cracking, and abrasion resistance, as well as smaller drying shrinkage. c. It is required to be constructed by pumping with a pumping distance of more than 200 m horizontally and 50–70 m vertically.

According to the above technical requirements, it is determined that the raw materials are as follows: a. HX525# cement. b. FDN9001 high-efficiency water-reducing agent. c. DLNT fiber, the length of which is 6 mm. d. Crushed shale grains. The strength standard is 40 MPa, the barrel compression strength is more than 7.0 MPa, the grading of the grains is 5–20 mm continuous grading, and the average value of the stacking density is 795 kg/m^3. e. Secondary fly ash; the construction ratio determined after trial mixing is shown in Table 6.60.

The developed fiber-reinforced lightweight aggregate concrete can fully meet the pumping construction requirements: The pump pressure before the piston is 5.5–8.0 MPa, the slump of concrete before pumping is 18 cm, and it is 16 cm after pumping. The loss of the slump of concrete is small. The 7-d compressive strength of fresh concrete is 45.5 MPa, and it reaches 58 MPa in 28 d; after pumping, the 7-d compressive strength of concrete is 44.0 MPa, and 28-d compressive strength of 55.0 MPa or more. Concrete 28-d flexural strength is 6.8 MPa. Impermeability performance is of P12 or more, the surface of the concrete did not find cracks, the durability of the concrete has been significantly improved, so that the bridge deck is more robust and durable.

Yichang Riverside International Garden Project The Yichang Riverside International Garden Project has a total building height of 101.1 m, 29 floors, with a total floor area of 36,197.9m^2, and all of the podiums on the ground floor (commercial) above the fourth floor are built with lightweight aggregate concrete. This project is the first record of using lightweight aggregate concrete in the structural system (simple body, shear wall, beam, column, and slab) in China, and at the same time, it has realized the high-strength lightweight aggregate concrete pumping technology for a high-rise with a vertical height of 101 m. The lightweight aggregate used in the project was produced by Yichang Baozhu Ceramic Granule Development Co., Ltd., and the author's team provided technical guidance for the enterprise's research and development of high-strength lightweight aggregate, while the China Academy of Architectural Design and Research (CAADR) provided technical guidance for the construction of this architectural structure project.

References

1 Hu, S. and Ding, Q. (2007). *Steel Tube Confined Concrete*, 1–12. Beijing: China Communications Press (in Chinese).

2 Hu, S., Ding, Q., Lu, L. et al. (2007, May 23). A method for fine control of steel pipe concrete expansion. CN 200510019476. (in Chinese).

3 Hu, S., Wu, J., Yang, W. et al. (2010). Relationship between autogenous deformation and internal relative humidity of high-strength expansive concrete. *Journal of Wuhan University of Technology, Materials Science Edition* 25 (3): 504–508.

4 Hu, S. and Li, Y. (1999). Research on the hydration, hardening mechanism, and microstructure of high performance expansive concrete. *Cement and Concrete Research* 29 (7): 1013–1017.

5 Li, Y. and Hu, S. (2001). The microstructure of the interfacial transition zone between steel and cement paste. *Cement and Concrete Research* 31 (3): 385–388.

6 Lu, L., He, Y., Li, Y. et al. (2003). The microstructure of the interfacial transition zone on steel tube-high strength expansive concrete. *Journal of Huazhong University of Science and Technology (Nature Science Edition)* 31 (11): 89–91. (in Chinese).

7 Li, Y., Hu, S., Ding, Q. et al. (2001). Properties of ettringite type expansive agent. *Journal of Wuhan University of Technology-Materials Science Edition* 16 (4): 53–56.

8 Hu, S., Guan, X., and Ding, Q. (2002). Research on optimizing components of microfine high-performance composite cementitious materials. *Cement and Concrete Research* 32 (12): 1871–1875.

9 Lu, Z., Li, S., and Chen, Y. (2001). Experiment study on self-stress of high-strength, low-heat and micro-expansion concrete-filled steel tube-all databases. *Journal of Wuhan University of Technology-Materials Science Edition* 16 (2): 83–86.

10 Hu, S., Ding, Q., Wang, H. et al. (2007, May 09). A kind of high energy delayed expansion agent for closed concrete and its preparation method. CN 200510019477. (in Chinese).

11 Ding, Q., Mou, Y., Li, C. et al. (2015, November 25). A kind of C30 self-compacting micro-expansion steel pipe concrete and its preparation method. CN 201510500483. (in Chinese).

12 Hu, S., He, Y., and Lu, L. (2006, 3). Water-releasing factor technology to modify the internal relative humidity of concrete and its application. *Journal of Railway Science and Engineering* 11–14. (in Chinese).

13 Hu, S., Ding, Q., Wang, H. et al. (2007, February 14). Highly active shrinkage-compensating mineral admixture and its preparation method. CN 200510018820. (in Chinese).

14 Hu, S., Ding, Q., He, Y. et al. (2005, December 28). Steel pipe concrete with continuous and stable expansion and its construction process. CN 200410012910. (in Chinese).

15 Li, Y., Hu, S., and Ding, Q. (2000). Study on the expansive concrete filled steel tube and its application. *Journal of Shandong Institute of Building Materials* 14 (3): 189–192. (in Chinese).

16 Hu, S., Ding, Q., Zou, D. et al. (2007, February 14). Expansion-designable high-strength steel pipe concrete and its preparation method. CN 200510019475. (in Chinese).

17 Hu, S., Ding, Q., Peng, Y. et al. (2007, July 04). Slow-setting, high-efficiency water-reducing and plastic-retaining additives for steel pipe concrete. CN 200510019478. (in Chinese).

18 Peng, Y., Ding, Q., Lu, L. et al. (2006). Preparation of pumped c50 expensive steel tube-confined concrete. *Journal of Wuhan University of Technology* 28 (3): 39–42. (in Chinese).

19 Hu, S., Ding, Q., He, Y. et al. (2007, July 04). A construction process of steel pipe concrete. CN 200510018794.5. (in Chinese).

20 Ding, Q., Hu, S., and Guan, B. (2001). Application of large-diameter and long-span micro-expansive pumping concrete filled steel tube arch bridge. *Journal of Wuhan University of Technology-Materials Science Edition* 16 (4): 73–76.

21 Li, Y. and Zhang, L. (2005). The investigation and application of self-compacting expensive concrete filled steel tube. *Journal of Beijing University of Technology* 31 (5): 496–499. (in Chinese).

22 Tan, L., Ding, Q., and He, Y. (2005). Design and construction of micro-expansion steel pipe concrete for Yonghe bridge in Nanning. *Overseas Building Materials Technology* 26 (4): 14–15. (in Chinese).

23 Ding, Q., Chen, J., Li, Y. et al. (2001). Mix design and construction technology of concrete filled steel tube of the Longtan river bridge. *Journal of Building Materials* 4 (3): 265–269. (in Chinese).

24 Li, Y., Ding, Q., Hu, S. et al. (2000). Study on the mechanical properties of expansive concrete filled steel tube and its expansion model. *Journal of Wuhan University of Technology* 22 (6): 25–28. (in Chinese).

25 Liu, M., Yuan, W., and Ren, F. (2003). Study on the safety evaluation for long-span CFST arch bridge based on fuzzy integrated evaluation. *Journal of Wuhan University of Technology* 25 (5): 33–36. (in Chinese).

26 Ding, Q., Li, Y., Hu, S. et al. (2000). Study on the non-vibrated expansive high strength concrete. *Journal of Wuhan University of Technology* 22 (6): 1–4. (in Chinese).

27 Ren, F., Liu, M., Yuan, W. et al. (2003). Study on intelligent system of safety evaluation for CFST arch bridge. *Journal of Wuhan University of Technology* 25 (6): 28–30. (in Chinese).

28 Liu, M. and He, Z. (2004). Safety evaluation for CFST arch bridge based on uncertain analytic hierarchy process method. 26 (6): 33–35. (in Chinese).

29 Liu, M. and Yuan, W. (2004). Research on safety assessment of long-span concrete-filled steel tube arch bridge based on fuzzy-neural network. *China Journal of Highway and Transport* 17 (4): 55–58. (in Chinese).

30 Ren, Z., Hu, S., and Ding, Q. (2010). Research on the effect of solar radiation model on temperature field of concrete-filled steel tube pier. *Engineering Mechanics* 27 (4): 246–250, 256. (in Chinese).

31 Ren, Z. and Hu, S. (2012). Plane strain analytical solutions to concrete-filled steel tube under axisymmetric variable temperature. *Journal of Huazhong University of Science and Technology. Nature Science* 40 (8): 34–38. (in Chinese).

32 Liang, M., Liu, M., and Feng, Z. (2001). Study of ultrasonic detection and evaluation for the deficiency recognition of concrete-filled steel tube. *Journal of Wuhan University of Technology* 23 (12): 44–46. (in Chinese).

33 Wang, F., Hu, S., Ding, Q. et al. (2005). Research on high performance lightweight concrete and its application in old-bridge deck rebuilding. *Journal of Highway and Transportation Research and Development* 22 (3): 86–88+100. (in Chinese).

34 Ding, Q., Qiao, L., Zhao, M. et al. (2015, April 29) A bridge deck paving method. CN 201510027419. (in Chinese).

35 Ding, Q., Li, Q., Huang, S. et al. (2010, March 03). A paving method for steel box girder bridge deck paving layer. CN 200910063292. (in Chinese).

36 Hu, S., Lin, H., Ding, Q. et al. (2005, June 08). A paving method of steel bridge deck combined layer. CN 200410061407. (in Chinese).

37 Liu, M., Cao, Y., and Ding, Q. (2008). Study on extracting bridge dynamic characteristics from the dynamic response of a passing vehicle. *Journal of HUST (Urban Science Edition)* 25 (4): 21–25. (in Chinese).

38 Ding, Q., Wang, F., Huang, S. et al. (2002). Material design of bridge pavement. *Journal of Wuhan University of Technology* 24 (4): 55–58. (in Chinese).

39 Ding, Q., Zhang, F., Lin, Q. et al. (2006). Research and application of lightweight concrete steel bridge deck pavement. *Sino-Foreign Highway* 26 (4): 188–190. (in Chinese).

40 Hu, S., Xie, X., Ding, Q. et al. (2009, February 18). A paving method for a large-span steel box girder bridge deck anti-thrusting combination structure. CN 200810197027. (in Chinese).

41 Hu, S., Zhang, H., Lin, Q. et al. (2005, December 21). A paving method to prevent asphalt concrete from pushing away on steel box girder bridge decks. CN 200510019039. (in Chinese).

42 Hu, S., Lu, J., Zhao, M. et al. (2015). Research of the performance of embedded cold-paved antiskid wear layer materials. *Journal of Wuhan University of Technology* 37 (3): 1–7. (in Chinese).

43 Huang, X., Ding, Q., and Hu, S. (2009). A new study on deck surfacing anti-slipping materials for steel-box girder bridge. *Journal of Building Materials* 12 (3): 276–280. (in Chinese).

44 Wang, F., Hu, S., Ding, Q. et al. (2001). Research on high-performance ultra-thin bridge deck paving layer materials. *Highway* 7: 105–108. (in Chinese).

45 Ding, Q., Hu, S., Xie, X. et al. (2008, August 06). Preparation method of anti-slip and wear-resistant lightweight steel box girder bridge deck pavement layer. CN 200810046968. (in Chinese).

46 Mou, Y., Ding, Q., Huang, X. et al. (2013, March 20). High toughness low shrinkage crack-resistant concrete for bridge deck paving and its preparation method. CN 201210566338. (in Chinese).

47 Ding, Q., Huang, X., Ma, P. et al. (2007). Design of high strength semi-lightweight concrete and application in steel bridge deck pavement. *Construction Technology* 36 (12): 64–66. (in Chinese).

48 Zhang, F., Ding, Q., Lin, Q. et al. Research on steel fibre reinforcement lightweight aggregate concrete for deck pavement of steel box bridge. *Concrete* 2005 (11): 46–49. (in Chinese).

49 Ding, Q., Wang, X., and Hu, S. (2009). A study on properties of lightweight aggregate concrete of steel box girder bridge pavement. *Highway* 4: 60–65. (in Chinese).

50 Ding, Q., Li, Q., Huang, S. et al. (2010, January 20). A preparation method of waterproof bonded stress-absorbing bridge deck paving material. CN 200910063308. (in Chinese).

51 Han, H., Huang, S., Ding, Q. et al. (2005). Study on the performance of OGFC-containing high viscosity asphalt. *Journal of Wuhan University of Technology* 27 (3): 41–43. (in Chinese).

52 Liu, X., Ding, Q., Yao, Y. et al. (2007). SMA pavement using high-viscosity modified asphalt in steel box girder bridge. *Highway* 9: 97–100. (in Chinese).

53 Hu, S., Li, Q., Sun, Z. et al. (2010). Development and performance study of highly viscous and highly elastic modified asphalt based on waterproof bonded stress absorption layer. *Highway* 2: 134–137. (in Chinese).

54 Ding, Q., Lu, C., Xu, B. et al. (2013, December 18). A cementitious composite material for steel box girder bridge deck and its preparation and paving. CN 201310419625. (in Chinese).

55 Ding, Q., Shen, F., Huang, S. et al. (2011, September 07). A kind of water-based epoxy modified emulsified asphalt concrete for steel bridge deck and its preparation method. CN 201110003603. (in Chinese).

56 Hu, S., Huang, S., Zhang, H. et al. (2004). Study on the design and performance of the open-graded Asphalt friction course. *Journal of Wuhan University of Technology* 26 (8): 23–25. (in Chinese).

57 Ding, Q., Xu, M., Huang, S. et al. (2010). Assessment and application on noise-damping performance of antiskid and fire-retarding asphalt pavement. *Journal of Wuhan University of Technology (Transportation Science & Engineering)* 34 (5): 1044–1048. (in Chinese).

58 Hu, S., Ding, Q., Lu, L. et al. (2004, December 22). Black lightweight bridge deck paving layer concrete. CN 03128045. (in Chinese).

59 Ding, Q., Hu, S., and Huang, X. (2009, December 02). A resin-infused asphalt concrete steel box girder bridge deck combined structure paving method. CN 200710053025. (in Chinese).

60 Liu, M., Li, O., Ding, Q. et al. (2006). Flexural performance of reinforced high-strength lightweight concrete beams. *Journal of Huazhong University of Science and Technology (Natural Science Edition)* 34 (10): 100–103. (in Chinese).

61 Liu, M., Yin, H., Ding, Q. et al. (2008). Calculation of the shear resistance of the reinforced high-strength lightweight concrete beams. *Journal of Huazhong University of Science and Technology (Natural Science Edition)* 36 (3): 46–49. (in Chinese).

62 Mou, Y., Ding, Q., Zhou, X. et al. (2012). Dynamic characteristics analysis of extroversion CFST arch bridge with single rib. *Journal of Wuhan University of Technology (Transportation Science & Engineering)* 36 (5): 984–987+991. (in Chinese).

63 Hu, S., Wang, F., Ding, Q. et al. (2002). Influence of water absorption and pre-wetted method of lightweight aggregate on the workability of concrete. *Journal of Huazhong University of Science and Technology (Urban Science Edition)* 2: 1–4. (in Chinese).

64 Ding, Q., Guan, L., Mou, Y. et al. (2012, June 12). Bridge deck paving concrete prepared with slag aggregate and its production method. CN 201310121722. (in Chinese).

65 Ding, Q., Zhang, Y., Wang, F. et al. (2001). Study on the pumping high-strength light-weight aggregate concrete. 23 (9): 4–6. (in Chinese).

66 Ding, Q., Zhang, Y., Wang, F. et al. Pumping technology of high-strength light weight aggregate concrete on bridge construction. *Concrete* 2002 (1): 58–60. (in Chinese).

67 Zhang, C. and Ding, Q. (2011). Preparation and application of steel fiber light aggregate and self-compacting concrete in the pavement of steel bridge deck. *Concrete* 3: 59–62. (in Chinese).

68 Ding, Q., Wang, X., and Lin, Q. (2008). Design and construction technology Hancai steel box-girder bridge deck surfacing material. *Concrete* 12: 108–112. (in Chinese).

69 Xiao, M., Deng, Z., Pei, L. et al. (2011, March 23). A shield tunnel lining ring with 7+1 blocking method. 201020235113. (in Chinese).

70 Xiao, M., Deng, Z., Lu, Z. et al. (2011, August 17). A shield tunnel lining structure with non-enclosed secondary lining. CN 201110111644. (in Chinese).

71 Ma, B., Wang, X., Hu, S. et al. (2006, June 21). Shield tunnel lining pipe sheet and its preparation method. CN 200510120533. (in Chinese).

72 Peng, B., Ding, Q., and Hu, S. (2007). Factors affecting the strength of concrete in tube sheets of shield tunnels. *Building Technique Development* 34 (11): 25–26. (in Chinese).

73 Peng, B., Ding, Q., Wang, H. et al. (2007). Advances in key technology of steam-cured cement and concrete products. *Overseas Building Materials Technology* 28 (3): 21–24. (in Chinese).

74 Tian, Y., Peng, B., Ding, Q. et al. (2009). Influence of steam—curing regimes on the carbonation of high strength concrete. *Journal of Building Materials* 12 (6): 720–723. (in Chinese).

75 Tian, Y., Peng, B., Ding, Q. et al. Carbonation properties of autoclaved high strength concrete and its prediction model. *Journal of Wuhan University of Technology* 2009 (20): 34–38. (in Chinese).

76 Tian, Y., Li, W., Peng, B. et al. (2010). Influence of steam' curing regimes on the freezing-thawing resistance of high strength concrete. *Journal of Building Materials* 13 (4): 515–519. (in Chinese).

77 Tian, Y., Peng, B., Ding, Q. et al. (2010). Effect of curing parameters on the sulphate attack resistance of high strength concrete. *China Concrete and Cement Products* 3: 1–4. (in Chinese).

78 Peng, B., Hu, S., Ding, Q. et al. (2007). Influence of steam-curing regimes on the chloride ingress of high strength concrete. *Journal of Wuhan University of Technology* 29 (5): 27–30. (in Chinese).

79 Hu, S., Ma, J., Ding, Q. et al. (2008). Study oil anti-crack concrete in Yangtze river tunnel of Wuhan. *Construction Technology* 37 (12): 1–3. (in Chinese).

80 Nie, S., Zhang, W., Hu, S. et al. (2018). Improving the fluid transport properties of heat-cured concrete by internal curing. *Construction and Building Materials* 168: 522–531.

81 Shen, P., Lu, L., He, Y. et al. (2018). Experimental investigation on the autogenous shrinkage of steam cured ultra-high performance concrete. *Construction and Building Materials* 162: 512–522.

82 Hu, S., Jin, Y., Xie, X. et al. (2009, August 03). High impermeable concrete admixture for shield tube sheet and its preparation method. CN 200710051238. (in Chinese).

83 Nie, S., Hu, S., Wang, F. et al. (2016). Internal curing – a suitable method for improving the performance of heat-cured concrete. *Construction and Building Materials* 122: 294–301.

84 Hu, S., Geng, J., and Ding, Q. (2008). The characters of cement hardened paste with mineral admixtures binding chloride ion under the condition of stray current interfering. *Journal of Huazhong University of Science and Technology (Natural Science Edition)* 36 (3): 32–34. (in Chinese).

85 Ding, Q., Wu, X., and Geng, J. (2008). Experimental study of binding capability of chloride ion under stray current in cement hardened paste admixed with fly ash. *Journal of Building Materials* 11 (1): 80–83. (in Chinese).

86 Wang, H., Tao, W., Du, X. et al. (2013). Effects of CCCW on properties of cement-based material in different mediums. *China Building Waterproofing* 1: 4–7. (in Chinese).

87 Lu, L., Hu, S., Ding, Q. et al. (2003). Research on high-performance crack-resistant and anti-seepage additives. *Water-Proof Materials and Construction* (11): 4–7. (in Chinese).

88 Lu, L., Hu, S., and Ding, Q. (2003). Preparation of high performance chemical admixture and its influence on properties of concrete. *Bulletin of the Chinese Ceramic Society* 4: 16–20. (in Chinese).

89 Wang, X., Ma, B., and Zhang, A. (2009). Transport property and service life prediction of gradient structural concrete. *Materials Reports* 23 (20): 78–81. (in Chinese).

90 Ding, Q., Zeng, B., Peng, B. et al. (2006). Factors and prevention measure of cracking of big hand hole segment. *Concrete* 7: 59–60+68. (in Chinese).

91 Huang, X., Hu, S., Wang, F. et al. (2019). The effect of supplementary cementitious materials on the permeability of chloride in steam cured high-ferrite Portland cement concrete. *Construction and Building Materials* 197: 99–106.

92 Ding, Q., Geng, J., and Hu, S. (2010). The influence of mineral admixtures on concrete anti-chloride ion permeability with stray current. 50-Year Teaching and Research Anniversary of Professor Sun Wei.

93 Hu, S. and Wang, F. (2006). *Lightweight Aggregate Concrete*. Beijing: Chemical Industry Press (in Chinese).

94 Ye, J. (2005). *Optimised Design and Performance Study of High Strength Lightweight Aggregate Concrete Elements*. Wuhan: Wuhan University of Technology (in Chinese).

95 Ding, Q. (2006). *Research and Application of High Strength Sub-lightweight Concrete*. Wuhan: Wuhan University of Technology (in Chinese).

96 Zhang, L. (2007). *Structural Characteristics of High Strength Lightweight Aggregate Concrete Continuous Rigid Frame Bridge*. Wuhan: Wuhan University of Technology (in Chinese).

97 Jiang, C. (2010). *Research on Brittleness Mechanism and Modification of High Strength Lightweight Concrete*. Wuhan: Wuhan University of Technology (in Chinese).

98 Liu, M., Yin, H., Ding, Q. et al. (2008). Experimental research on shear capacity of reinforced high-strength lightweight concrete beams. *Journal of Harbin Institute of Technology* 40 (4): 46–49. (in Chinese).

99 Ren, Z. and Wang, F. (2005). Analysis of structural design parameters of high strength lightweight aggregate concrete large span bridges. *Overseas Building Materials Technology* 26 (3): 105–108. (in Chinese).

100 Wang, F. (2003). *Research on High Performance Lightweight Aggregate Concrete (HPLC) and Its Application*. Wuhan: Wuhan University of Technology (in Chinese).

101 Wang, F., Zhou, B., Peng, Y. et al. (2005). Research on the elements distribution characteristics in the interfacial transition zone (ITZ) between lightweight aggregate and cement paste. *Journal of Wuhan University of Technology* 27 (3): 30–33. (in Chinese).

102 Hu, S., Wang, F., and Ding, Q. (2005). Interface structure between lightweight aggregate and cement paste. *Journal of the Chinese Ceramic Society* 33 (6): 713–717. (in Chinese).

103 Hu, S., Ding, Q., He, Y. et al. (2006, July 19). High-strength lightweight aggregate with surface reactivity and its preparation method. CN 200410012911. (in Chinese).

104 Ding, Q., Tian, Y., Wang, F. et al. (2005). Study on the influence of aggregate composition on the macro-performance of semi lightweight concrete. *Journal of Wuhan University of Technology* 27 (5): 37–39. (in Chinese).

105 Hu, S., Tian, Y., and Ding, Q. (2007). Influence of lightweight aggregates on the physical and mechanical properties of high-strength sub-lightweight concrete. *China Concrete and Cement Products* 2007 (6): 1–4. (in Chinese).

106 Wang, F., Ding, Q., Chen, Y. et al. (2003). Study on shrinkage of high strength lightweight aggregate concrete. *Journal of Building Materials* 6 (4): 431–435. (in Chinese).

107 Tian, Y., Ding, Q., Wang, F. et al. (2005). Study on early autogenous shrinkage of high strength lightweight aggregate concrete (HSLWAC). *Concrete* (2): 29–31. (in Chinese).

108 Ding, Q., Hu, S., Tian, Y. et al. (2010, February 24). Preparation method of high strength and high toughness lightweight aggregate concrete. CN 200710053634. (in Chinese).

109 Wang, T. (2004). *Research on Brittleness and Toughening Technology of High-strength Lightweight Aggregate Concrete*. Wuhan: Wuhan University of Technology (in Chinese).

110 Jiang, C., Wang, T., Ding, Q. et al. (2004). Influence of polymer addition on performance and mechanical properties of lightweight aggregate concrete. *Wuhan University Journal of Natural Sciences* 9 (3): 348–352.

111 Tian, Y., Shi, S., Jia, K. et al. (2015). Mechanical and dynamic properties of high strength concrete modified with lightweight aggregates presaturated polymer emulsion. *Construction and Building Materials* 93: 1151–1156.

112 Ding, Q., Wang, T., Zhang, F. et al. (2004). Research on mechanics of mixed fiber reinforced aggregate concrete. *Journal of Wuhan University of Technology* 26 (1): 42–45. (in Chinese).

113 Ding, Q., Tian, Y., Wang, F. et al. (2005). Autogenous shrinkage of high strength lightweight aggregate concrete. *Journal of Wuhan University of Technology-Mater. Sci. Ed.* 20 (4): 123–125.

114 Liu, M., Liu, Q., Ding, Q. et al. (2006). Local bearing analysis of prestressed lightweight aggregate concrete hollow slab. *Journal of HUST (Urban Science Edition)* 23 (2): 13–16. (in Chinese).

115 Liu, M., Liu, Q., Ding, Q. et al. (2006). Finite element analysis model on prestressed LWAC hollow slabs. *Journal of Wuhan University of Technology* 28 (4): 56–59. (in Chinese).

116 Liu, M., Ding, Q., Gao, Z. et al. (2009, October 07). A continuous rigid bridge structure with varying concrete density gradient. CN 200710052952. (in Chinese).

117 Hu, S., Ding, Q., Wang, F. et al. (2006, November 29). A method of preventing cracking at the anchorage end of prestressed lightweight concrete constructions. CN 200510018123. (in Chinese).

118 Ding, Q., Hu, S., Wang, F. et al. (2008, February 06). A test method for homogeneity of lightweight aggregate concrete. CN 200710053100. (in Chinese).

119 Wang, F., Hu, S., Ding, Q. et al. (2008, March 19). A kind of light aggregate concrete pumping agent. CN 200710053099. (in Chinese).

120 Hu, S., Ding, Q., Wang, F. et al. (2008, March 05). A preparation method of pumping lightweight aggregate concrete. CN 200710053101. (in Chinese).

121 Wang, F., Zhou, Y., Wang, J. et al. (2008). Mix ratio design and pumping technology of super-ought aggregate concrete. *Construction Technology* 37 (9): 109–111. (in Chinese).

7

Novel Functional Concrete Technologies

7.1 Recyclable Cement Concrete

7.1.1 Overview

With the large-scale construction of infrastructure and the expiration of the replacement of a large number of old concrete buildings, a large number of cement and concrete structures have been abandoned, occupying limited urban space and under increasing pressure for disposal. On the other hand, as the key raw material of cement and concrete, the huge resources, energy consumption, and environmental load brought by cement production have been increasingly emphasized, and the raw materials for cement production as well as natural sand and gravel and other natural resources are becoming increasingly scarce. The recycling of cement and concrete is an important task in the field of building materials.

At present, for the study of regenerated concrete, the main technical idea is to make regenerated concrete into recycled aggregate and hardened slurry by crushing and separation [1, 2] (Figure 7.1). The recycled aggregate is used for the preparation of concrete, and the hardened cement paste can be disposed of at a suitable high temperature to prepare regenerated cementitious materials due to a certain degree of rehydration activity [3, 4]. Although this method can utilize waste concrete in a whole-component resourceful way, the crushing and sorting process consumes large amounts of energy, and the recycled aggregates have many intrinsic defects after crushing, with high water absorption, which affects the quality of regenerated concrete [5]. Recycled cementitious materials have low activity compared with ordinary cement [6] and are often used as an admixture to replace cement, and the replacement rate is generally not more than 20%, so the utilization rate of resource utilization is relatively low [7, 8].

Aiming at the above problems, the authors propose a new design and processing technology for recyclable cement concrete, that is, designing concrete based on the chemical composition of cement composition, so that its composition is similar to the cement raw material. Consequently, at the end of its service life, the concrete can be crushed and directly used as raw material for calcining

High-Tech Concrete Materials: Design, Preparation, and Applications, First Edition.
Fazhou Wang and Shuguang Hu.

Figure 7.1 Separation of waste concrete components. *Source:* With permission [4] / Journal of the Chinese Ceramic Society.

cement clinker, with little to no need for corrective raw materials. This approach achieves "Infinite" recycling within a closed-loop lifecycle, turning waste concrete to cement.

7.1.2 Design Concepts for Recyclable Cement Concrete

In the design of recyclable cement concrete, it is not only necessary to consider the strength, durability, and other indexes of the existing concrete to meet the requirements of the actual project but also to consider the recycling performance of the concrete after disposal. Therefore, its design needs to integrate the principles of both concrete proportioning and cement raw material formulation, and investigate the constraints of the relationship among the water–cement ratio, cementitious material composition and dosage, cement raw material rate value, the raw material rate value. The proportioning design is conducted by selecting the aggregates or admixtures with required chemical components to satisfy the target rate values for cement production. At the same time, it is also necessary to take into account the influence of the introduced admixtures on the service performance of cement concrete. After the recyclable cement concrete meet the engineering design requirements and complete the service process, a part of them can be reused as the main raw material for cement production, with the rest crushed into aggregate. Both the recycled aggregate and cement are then used to produce the next cycle of recyclable cement concrete. Alternatively, all the recyclable concrete can be utilized to produce the regenerated cement, which in turn be used for the subsequent cycle of recyclable cement concrete [9].

According to the $CaO–SiO_2–Al_2O_3$ ternary system phase diagram, conventional Portland cement clinker typically requires $CaO/SiO_2 = 2.5–3.5$ in the raw material. However, chemical analysis of common waste cement concrete shows that its $CaO/SiO_2 < 2$ (with variations depending on concrete strength grade and raw materials). This indicates that when waste concrete is crushed and finely ground as raw cement materials, the siliceous component is high while the calcium component is low. Moreover, the content of Al_2O_3 and Fe_2O_3 in most waste concrete is low. Therefore, when preparing recyclable cement and concrete, it is necessary to incorporate additional calcium, aluminum, and iron corrective raw materials.

7.1.3 Properties of Recyclable Cement Concrete

7.1.3.1 Mechanical and Impermeability Properties

The chemical composition of each raw material is shown in Table 7.1. The mix proportions of conventional C40 concrete and recyclable concrete (#1 and #2 in the following figures) are shown in Table 7.2. The mechanical and impermeability properties of the two types of concrete were compared, as shown in Figure 7.2. It can be seen that the early strength of the recyclable concrete is low, with the 3 d compressive strength reduced by 35.14% compared to the blank group. However, it has good potential for strength development in the later stage, with the 90 d compressive strength reduced by only 13.99% compared to the blank group. The main reason is that the steel slag and fly ash with low hydration activity in the early stage replaced the cement with a high activity of 40 wt%, which retarded early strength development. As cement hydration proceeded, the late reaction activity of fly ash and steel slag was activated, contributing to the later strength gain. In terms of capillary water absorption, the water absorption of recyclable concrete was lower than that of conventional concrete, mainly due to the addition of stone powder in the manufactured sand, which improves the densification of the concrete. Additionally, the physical filling and late hydration reaction of steel slag and fly ash further refine the pore structure, thereby improving impermeability.

7.1.3.2 Thermal Behavior During Heating

Figure 7.3 shows the thermal analysis curves of TG (a) and DSC (b) from room temperature to 1450 °C using waste recyclable concrete (1#) and pure reagent for

Table 7.1 Chemical composition of raw materials (%).

	SiO_2	Al_2O_3	Fe_2O_3	CaO	MgO	SO_3	Na_2O	K_2O	Loss
Cement	23.089	6.496	2.907	60.494	2.075	2.367	0.155	0.725	1.190
Coarse aggregate	3.217	0.123	0.047	52.13	2.661	0.109	0	0.016	40.970
River sand	69.570	13.314	4.088	3.082	1.133	0.034	4.055	2.925	0.770
Machine-made sand	3.412	0.407	0.366	53.615	1.711	0.120	0	0.074	40.150
Slag	17.884	4.382	21.749	42.855	3.137	0.564	0.079	0.216	4.768
Fly ash	54.699	30.093	4.326	3.436	0.569	0.357	0.500	1.029	3.210

Table 7.2 Mixing ratios of plain concrete and recyclable concrete (kg/m^3).

Type	Cement	Slag	Fly ash	Coarse aggregate	River sand	Machine-made sand	Water
Ordinary concrete	477	0	0	1125	579	0	215
Recyclable concrete	287	114	76	1125	250	329	215

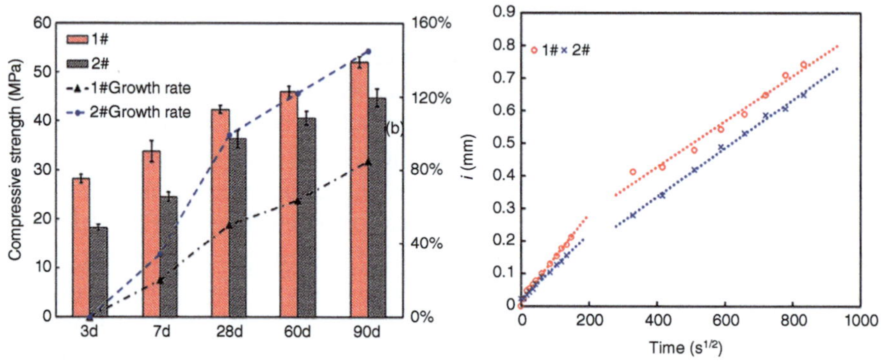

Figure 7.2 Basic properties of recyclable concrete.

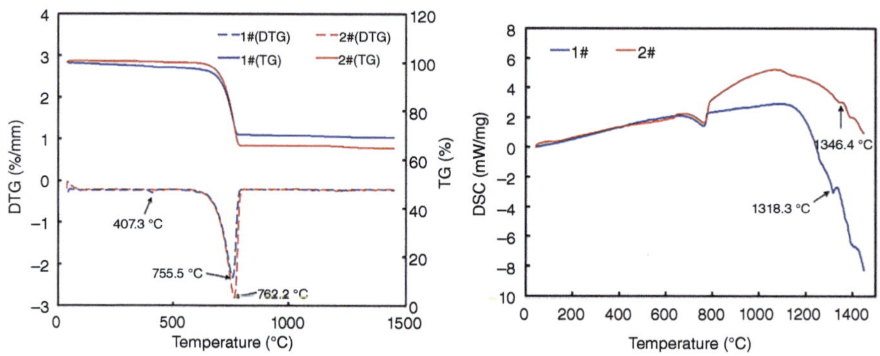

Figure 7.3 Thermal analysis curves for different cement raw materials.

chemical analysis (2#) with the same rate value as cement raw material. Compared with the chemical analysis of pure reagent as cement raw material, the thermal characteristics of the waste recyclable concrete as cement raw material mainly have the following changes:

(1) A small endothermic peak is observed at 407.3 °C, attributed to the decomposition of $Ca(OH)_2$, a hydration product of cement present in the waste concrete.
(2) Although the endothermic peak for calcium carbonate decomposition occurs at a similar temperature , its peak area is reduced, and the corresponding mass decrease in the TG curve is lower. This is because a portion of the CaO in the waste concrete exists in the hydration products of cement (C-S-H gel and $Ca(OH)_2$), which do not undergo a decomposition reaction.
(3) The DSC curves of the waste concrete as cement raw material showed an obvious heat endothermic peak around 1318.3 °C, indicating the formation of the liquid phase. This temperature was 28 °C lower than that of the control, which indicated that trace elements and impurities in the waste concrete function as both mineralizers and fluxes. In addition, the DSC curves of the two samples showed a significant difference after the decomposition reaction of calcium carbonate, which needs further investigation.

Table 7.3 *f*-CaO content of each group of cement clinker.

Calcine (°C)	1300	1350	1400	1450
f-CaO (wt%)	4.51	0.71	0.62	0.56

7.1.3.3 Flammability

The free calcium oxide content in clinkers calcined at different temperatures (1300, 1350, 1400, and 1450 °C) was measured to evaluate the quality differences of the recycled cement clinker, as shown in Table 7.3. When the calcination temperature is increased to 1350 °C and above, the *f*-CaO content of cement clinker decreases sharply, and all of them are lower than 1.5%, which follows the requirements of GB/T21372-2008 "Portland cement Clinker." This is because 1350 °C is higher than the temperature at which C_3A, C_4AF, and other components melt to form the liquid phase. Furthermore, this temperature is 100 °C lower than the maximum temperature required in conventional cement production, thereby significantly reducing energy consumption.

7.1.3.4 Mechanical Properties of Recyclable Cement

Table 7.4 shows the mix design of recyclable cement prepared using waste recyclable concrete as cement raw meal. Waste concrete component accounts for 97.8% of the entire mass of the raw material, demonstrating good recyclability. Figure 7.4 presents the compressive strengths of the recycled cement slurries (denoted as K1, K2, K3, K4, respectively) obtained by calcining the waste recyclable concrete at different temperatures (1300, 1350, 1400, 1450 °C). The results indicate that the 3 d strengths of all recycled cement groups are similar to that of the control group (K0). However, their 28 d strengths are reduced by about 11%, and the strength of the recycled cement calcined at 1350 °C is the highest.

7.1.3.5 Trends in Recyclable Cement Concrete

The design idea of recyclable cement concrete is of great importance in the current situation of increasing scarcity of resources and energy and tight supply of raw materials in the building materials industry. The work presented and investigated in this section provides a technically feasible approach to full-lifecycle recyclable

Table 7.4 Formulation of cement raw material and calcination temperature.

Sample	Raw material (g)					Calcination temperature (°C)
	Recyclable concrete	$CaCO_3$	SiO_2	Fe_2O_3	Al_2O_3	
K0	0	395.28	76.42	10.48	17.82	1450
K1	488.91	8.76	1.07	1.27	0	1300
K2	488.91	8.76	1.07	1.27	0	1350
K3	488.91	8.76	1.07	1.27	0	1400
K4	488.91	8.76	1.07	1.27	0	1450

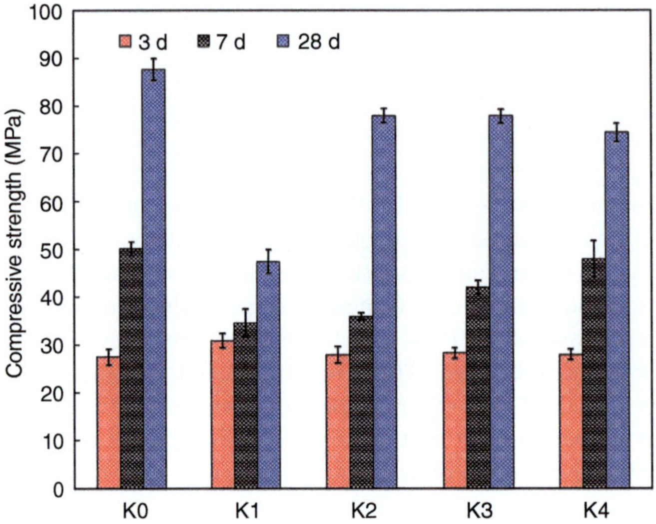

Figure 7.4 Effect of different calcination temperatures on the compressive strength of recycled cement net paste.

cement concrete, but more work is required for the application of this technology. It should be noted that modern concrete components are becoming more and more complex, and the introduction of a large amount of industrial waste has changed the composition of cement raw materials from the traditional Ca-Si-Al-Fe quaternary system to a multiphase component containing a variety of impurities, whose multiphase complexity will inevitably change the calcination characteristics of cement clinker and clinker properties during the recycling process. In addition, the continuous enrichment of impurity components in concrete and the change of rate value under multiple cycles will inevitably have a greater impact on the formation process of the cement clinker mineral phase, and the performance of cement clinker will become more difficult to stabilize and control.

7.2 Resin Aggregate Concrete

7.2.1 Overview

The authors' team developed a novel concrete by utilizing superabsorbent polymer (SAP) particles as a replacement for sand and gravel aggregates in conventional concrete – resin aggregate concrete [10]. This concrete is a new type of concrete with load-bearing capacity originates from a dense hydration layer structure that formed by modulating the high water absorption and release behavior of spherical SAP particles. The team studied the structure and performance law of resin aggregate concrete in-depth. We used the large-size spherical SAPs as the in-situ pore-making template [11], constructing regular spherical pore inside the concrete, and endowing the concrete with multifunctional properties such as lightweight, high-strength, thermal insulation, and acoustic noise absorption [12]. These findings establish a foundation for advancing the practical application of resin aggregate concrete.

7.2.2 Design Principle for Resin Aggregate Concrete

The core component of resin aggregate concrete is spherical SAP particles; however, spherical SAP particles have a small density and smooth surface, and their poor stability inside the concrete mix is the difficulty in the preparation process of this concrete. Since the solid particles (e.g., sand, stone, cement, spherical SAP, etc.) in the concrete mix do not have the same density and relative motion speed, relative motion occurs between the solid particles, resulting in the inhomogeneity and discontinuity of the concrete mix. Here, we take the spherical SAP particles in SAP resin aggregate concrete as the research object; when the spherical SAP is in the cement paste mix, it will be subjected to gravity, viscous resistance, and buoyancy at the same time. Referring to the equations of motion of lightweight aggregate in concrete studied by Fazhou Wang [13], the equations of motion of spherical SAP can be similarly expressed as shown in Eqs. (7.1)–(7.3):

$$\frac{4}{3}\pi r^3 \rho \frac{\mathrm{d}v}{\mathrm{d}t} = \frac{4}{3}\pi r^3 \rho g - 6\pi r \eta v - \frac{4}{3}\pi r^3 \rho_c g \tag{7.1}$$

where ρ is the density of spherical SAP, kg/m^3; ρ_c is the density of cement paste, kg/m^3; r is the radius of spherical SAP particles, m; η is the viscosity coefficient of the cement slurry, Pa·s; and v is the velocity of movement of spherical SAP particles, m/s.

Organizing the above equation gives

$$\frac{\mathrm{d}v}{\mathrm{d}t} + \frac{9\eta}{2\rho r^2} v = g\left(1 - \frac{\rho_c}{\rho}\right) \tag{7.2}$$

This is obtained by integrating the above equation and assuming the initial conditions $t = 0$ and $v = 0$:

$$v = \frac{2r^2 g(\rho - \rho_c)}{9\eta}\left(1 - \mathrm{e}^{-\frac{9\eta}{2\rho r^2}t}\right) \tag{7.3}$$

Let v be the final velocity of the spherical SAP particles, then we have

$$v = \frac{2r^2 g(\rho - \rho_c)}{9\eta} \tag{7.4}$$

Since the density of spherical SAP (about 1.04 g/cm^3 or so) is smaller than the density of cement paste, it can be seen from Eq. (7.4) that the speed of movement of spherical SAP is negative, i.e., it has a tendency to float (as shown in Figure 7.5), and

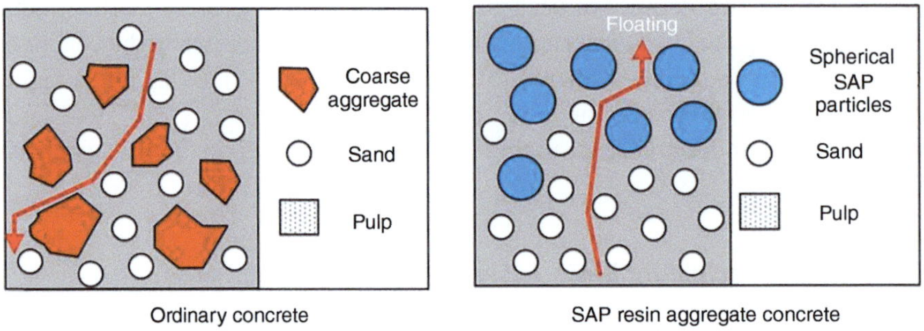

Figure 7.5 Motion tendency of spherical SAP particles inside concrete.

the greater the uplift speed is, the more pronounced is the tendency of the concrete delamination and segregation of SAP resin aggregate. From Eq. (7.4), it can be seen that the uplift velocity of spherical SAP is proportional to the square r^2 of the spherical SAP particle size and the density difference $\rho - \rho_c$ and inversely proportional to the viscosity coefficient η of the cement paste. As a result, the following design principles are proposed:

(1) Reduce the maximum particle size of spherical SAP. The maximum particle size of spherical SAP should be controlled to maintain the overall concrete performance. Excessive large SAP particles result in a high uplift velocity and the formation of oversized voids, which can adversely affect the concrete strength.
(2) Increasing the viscosity of the mix. The incorporation of viscosity-increasing chemical admixtures can increase the viscosity of the cement paste. Mineral admixtures such as silica fume could also increase viscosity due to their high specific surface area while simultaneously improving concrete matrix compactness and late strength development.
(3) Reduce the density difference between spherical SAP particles and cement paste. On the premise of not significantly reducing the strength of concrete, through appropriate air-entraining, it can help increase the air content of the cement paste, and thus reduce the overall density of the cement paste, thereby reducing the floating problem of spherical SAP particles. In addition, proper air entrainment can improve the fluidity of the mix and the frost resistance of the concrete.
(4) Addition of fibers. The addition of fibers can form a three-dimensional framework structure within the concrete mix, which can restrain the movement of SAP to move through the mix.
(5) Accelerated structure formation. The formation of fast-hardening composite cementitious material by combining fast-hardening cement such as sulfoaluminate with ordinary Portland cement can help accelerate the formation of cementite structure, which is also beneficial for the rapid stabilization of spherical SAP particles in the mix.

7.2.3 Characterization of Resin Aggregate Concrete

7.2.3.1 Workability

The main feature of resin aggregate concrete is its high content spherical SAP resin aggregate. These aggregates have good water retention properties, a regular spherical appearance, and a smooth surface, allowing them to play a lubricating role inside the concrete mix, thereby providing high fluidity. Concrete with good working performance requires a good fluidity (general slump in more than 200 mm) during the mixing, transportation, and pouring, no water seepage, no segregation, and self-leveling behavior. For instance, at a 40% resin aggregate substitution rate, the resin aggregate concrete slump can reach more than 240 mm, confirming good working performance. Figure 7.6 shows the state of the resin aggregate concrete mix.

Since the spherical SAP particles have the "lubrication effect," the concrete has good fluidity, the slump is 240 mm, the slump extension is 500 mm, the T_{500} is 4s, and the hysteresis rate measured by the L-meter is $H_2/H_1 = 0.45$. The hysteresis rate reflects the filling capacity of the concrete and the gap passing capacity; according to the Design and Construction Guidelines for Self-Compacting Concrete, to prepare the self-compacting concrete with good filling capacity, the hysteresis rate H_2/H_1 should be more than 0.8. It can be seen that although the prepared resin aggregate concrete has good flowability, its filling performance still needs to be improved, and this is mainly because the high flowability and filling ability of the self-compacting concrete mainly depend on its gravity as the power source. Due to the small density of spherical SAP particles used in resin aggregate concrete (about $1.04\,g/cm^3$) and the light weight of the mix, the power it can generate is limited. Nonetheless, the resin-aggregate concrete has good workability, high flowability, is lightweight, and is ease of work.

7.2.3.2 Physical and Mechanical Properties

Figure 7.7 shows the effect of different spherical SAP particle dosages on the spherical porosity and density of resin aggregate concrete. Overall, the higher the amount of spherical SAP, the higher the spherical porosity inside the concrete

Figure 7.6 Working condition of resin-aggregate concrete.

and the lower the density. When the spherical SAP substitution rate (volume rate of substituted aggregate) reaches 40%, the density of concrete can be reduced to below 1900 kg/m^3. If ceramic sand (apparent density 1680 kg/m^3) is used as fine aggregate, its density can be further reduced to below 1600 kg/m^3 and as low as 640 kg/m^3.

To investigate the effect of spherical SAP particles on the later strength of concrete, the specimens were sealed with cling film and stored after 28 d of standard curing until 1 year. Figure 7.8 shows the effect of different spherical SAP dosages on the strength development of concrete. When the spherical SAP substitution rate (volume fraction, the same as below) increases from 0% to 100%, the 28 d compressive strength of concrete decreases from 68 to 4 MPa, in which the concrete specimen C3 with 40% of spherical SAP substitution rate can reach more than 40 MPa compressive strength after 1 year. Therefore, concrete materials with different strength grades can be designed and prepared by controlling the admixture of spherical SAP particles so that this kind of resin aggregate concrete can meet different working requirements such as structural load-bearing or external maintenance.

Figure 7.9 shows the effect of the spherical SAP substitution rate on the elastic modulus of concrete. The SAP admixture has a significant effect on the modulus of elasticity of concrete, showing a decreasing trend similar to that of the compressive strength [13]; the modulus of elasticity of the concrete decreases from 31 to 2.5 GPa when the spherical SAP substitution rate increases from 0% to 100%. The principle of decreasing modulus of elasticity with the increase in the spherical SAP

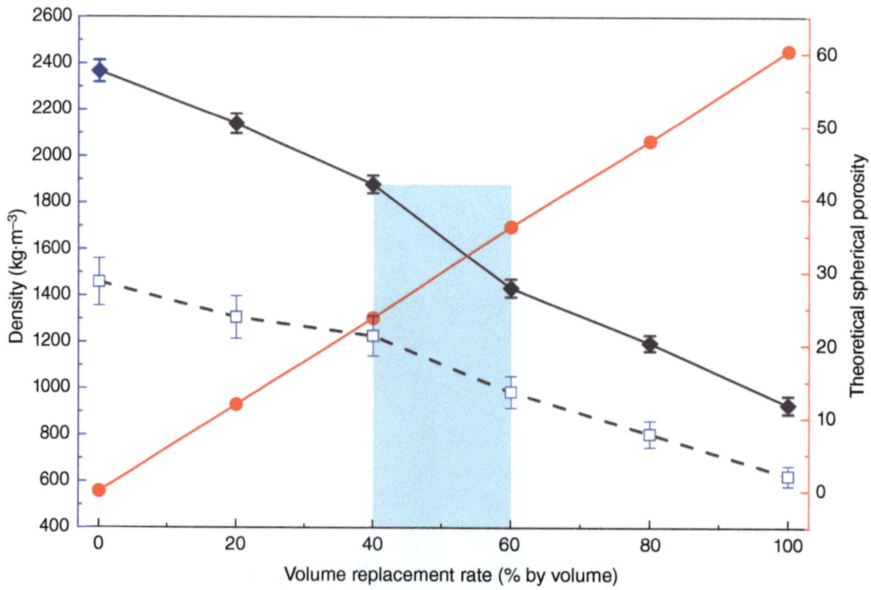

Figure 7.7 Effect of different spherical SAP particle volume substitution rates on the spherical porosity and density of test pieces.

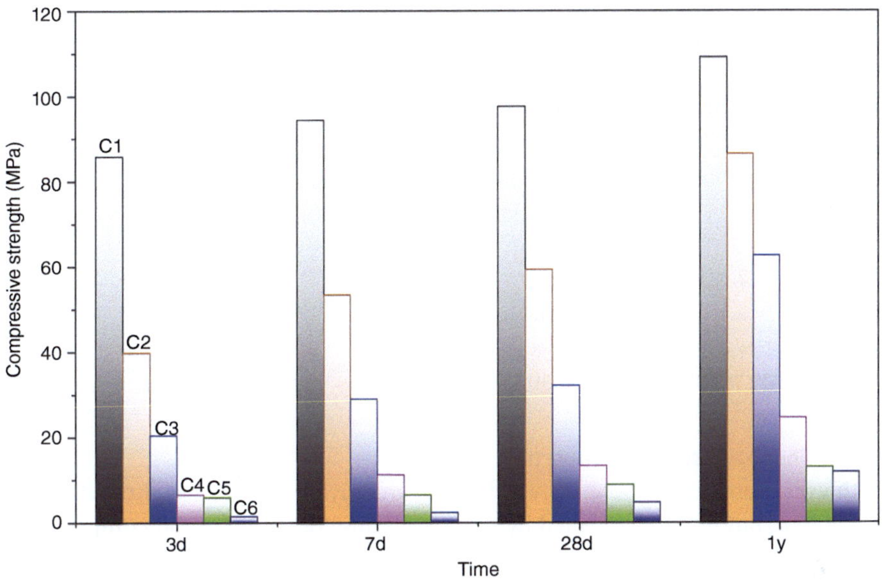

Figure 7.8 Compressive strength of different series of resin aggregate concrete.

substitution rate in Figure 7.9 is similar to that of replacing gravel aggregates with lightweight aggregates: The increase in spherical SAP substitution rate decreases the content per unit volume within the concrete. In general, the lower the aggregate dosage, the lower the modulus of elasticity of concrete. In addition, the increase in porosity within the concrete contributes to the reduction in its modulus of elasticity.

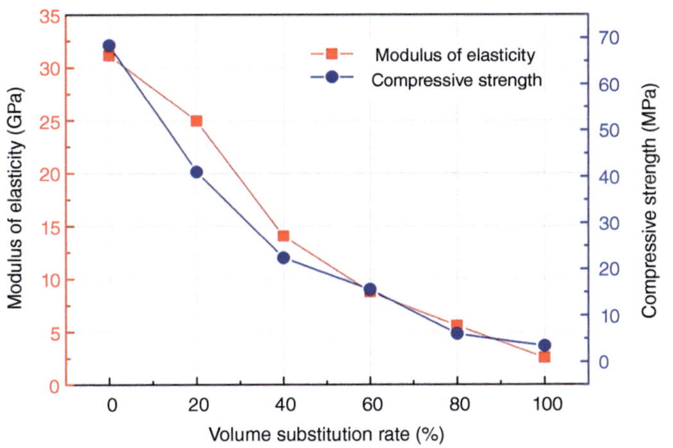

Figure 7.9 Spherical pore structure and elastic modulus of resin aggregate concrete.

Table 7.5 Thermal conductivity of plain concrete prepared with different aggregates.

Coarse type	Thermal conductivity (W/(m·K))
Quartz	3.5
Dolomite	3.2
Limestone	2.6–3.3
Granite	2.6–2.7
Rhyolite	2.2
Basalt	1.9–2.2

7.2.3.3 Thermal Insulation Properties

The thermal conductivity of concrete depends mainly on its bulk weight and moisture content, but it is also affected by the distribution of voids, a solid composition such as sand and gravel aggregates, and other aggregates. Table 7.5 shows the typical thermal conductivity of plain concrete prepared from different aggregates.

Figure 7.10 shows the effect of the spherical SAP substitution rate on the thermal conductivity of concrete. From the figure, it can be seen that with the increase of the substitution rate of spherical SAP, the thermal conductivity of concrete specimens is significantly reduced, and its thermal insulation performance is improved. The measured thermal conductivity of the blank control group without spherical SAP is 2.626 W/(m·K), and when the spherical SAP completely replaces the aggregate,

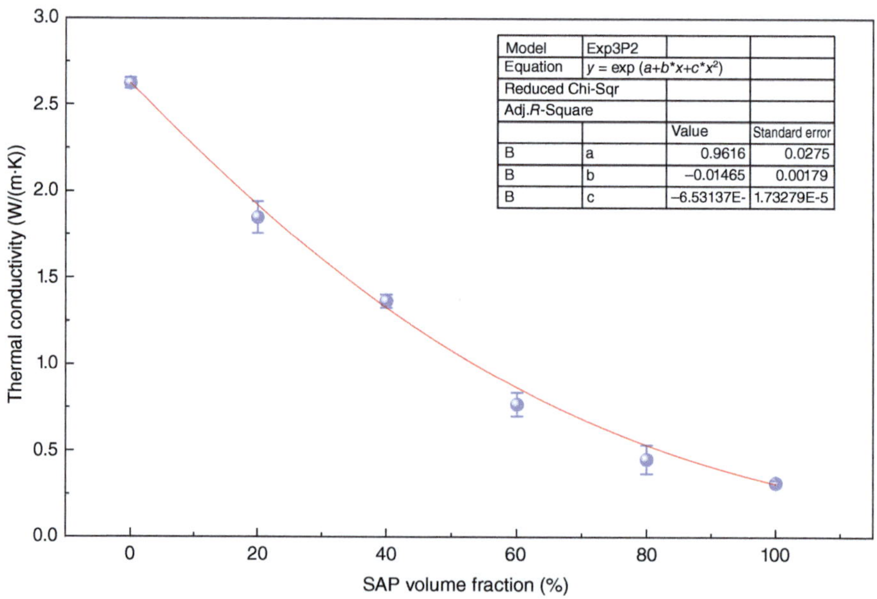

Figure 7.10 Spherical pore structure and thermal conductivity of resin aggregate concrete.

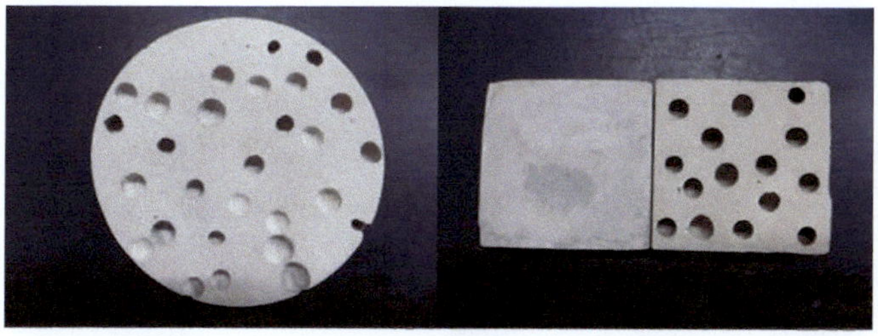

Figure 7.11 Spherical pore section of resin aggregate concrete.

the thermal conductivity of concrete is reduced to 0.316 W/(m·K). The reason why spherical SAP can significantly improve the thermal insulation performance of concrete is that it forms a large number of closed, independent spherical pores inside the concrete (see Figure 7.11), which both reduces the content of aggregates and increases the spherical porosity inside the concrete. In comparison, the thermal conductivity of aerated concrete is usually in the range of 0.08 to 0.25 W/(m·K), but the strength is lower, and the thermal conductivity of clay-terra cotta concrete with a dry density of 1200–1600 kg/m^3 is 0.53–0.84 W/(m·K). It can be seen that the rational design of utilizing spherical SAP can make the thermal insulation performance of concrete exceed that of lightweight aggregate concrete of the same density class, which has good application value.

7.2.3.4 Sound Absorption and Noise Reduction Performance

The contact noise of tires and pavement is the main source of traffic noise, and the noise frequency of pavement and tires is mainly concentrated at 600–1250 Hz. The sound absorption coefficients and noise reduction coefficients of the concrete with different spherical SAP admixtures are shown in Figure 7.12, in which the

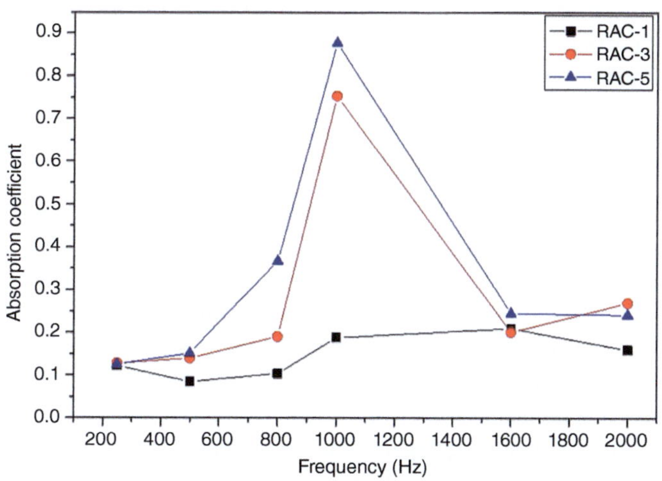

Figure 7.12 Spherical pore structure and sound absorption coefficient of resin aggregate concrete.

Figure 7.13 Conceptual framework of CO_2-driven 3D printing concrete.

RAC-1, 3, and 5 are the resin aggregate concrete with the substitution rate of 0%, 40%, and 80%, respectively. Compared with the control group without spherical SAP, the concrete specimens with spherical SAP showed peak sound absorption at 1000 Hz, and the peak value reached about 0.9, which is in the frequency of traffic noise from 600 to 1250 Hz, indicating that it has a good absorption effect on the main noise sources in traffic noise. According to the literature [14], the sound absorption coefficient of ordinary concrete pavement is 0.09, that of dense-graded asphalt pavement is 0.22, and that of low-noise asphalt concrete pavement is 0.43. It can be seen that the admixture of spherical SAP changes the internal pore structure of the concrete so that its sound absorption and noise reduction performance exceed that of the dense-graded asphalt pavement, and it exhibits a good potential for sound absorption and noise reduction. Considering from the perspective of pavement, noise reduction mainly involves sound absorption and vibration damping. Spherical SAP can significantly improve the sound-absorbing performance of concrete materials, the main reason is that spherical SAP produces more spherical pores inside and on the surface of concrete, which causes sound waves to undergo multiple reflections and energy dissipation at the pore walls. In addition, spherical SAP is possesses damping characteristics similar to rubber particles, which enables the concrete absorb tire vibration and impact at the same time thereby providing an effective noise-reduction function.

7.2.4 Development Trends of Resin Aggregate Concrete

Resin aggregate concrete not only provides a new approach for lightweight and multifunctional concrete but also, more importantly, a technical idea for the design and regulation of macroscopic pore structures within concrete. It offers the possibility of "computational designability" for concrete properties. Future research efforts should focus on two main areas. On the one hand, it is necessary to refine the molecular structure of SAPs to enhance their gel strength and mitigate fragmentation during the mechanical mixing of concrete. On the other hand, research on particle grading of resin aggregate, the relationship between the micromechanical characteristics of the arch-shell structure formed after the release of water from resin aggregate, and the macro-mechanical properties of concrete should be further explored. This will aid in developing methods to construct specific high-strength arched shell structures, enabling the preparation of lightweight high-strength concretes with higher specific strength or functional concrete for specific functional requirements.

7.3 CO_2-driven 3D Printing Concrete

7.3.1 Overview of CO_2-driven 3D Printing Concrete

3D printing technology, a cornerstone of intelligent manufacturing, enables the fabrication of physical objects through computer-aided design and layer-by-layer material deposition under automated control. As one of the hallmarks of the Fourth Industrial Revolution [15], this technology has been strategically prioritized in the "Made in China 2025" initiative. In recent years, 3D printing has been increasingly adopted in concrete manufacturing to address the industry's persistent challenges of low productivity and substantial environmental impact. This technological integration shows promising potential in promoting green manufacturing and sustainable development in the building materials sector [16].

Contemporary cement-based 3D printing construction materials predominantly utilize liquid-mediated systems, where material structuration and strength development are achieved through chemical reactions between cement and either water or alkaline activators (e.g., NaOH and Na_2SiO_3 solutions) [17–20]. In these water-hardened or alkali-activated systems, the liquid phase serves a dual function: regulating material flowability during extrusion while simultaneously initiating the chemical reactions essential for strength development and structural integrity. However, this inherent dual role creates fundamental challenges, manifesting as trade-offs between extrudability and buildability, as well as between open time and construction efficiency.

To address these challenges, Zhichao Liu and colleagues developed an innovative gas-driven cement-based material system for 3D printing applications [17, 18]. As illustrated in Figure 7.13, this system employs highly carbonation-reactive calcium silicate minerals (e.g., γ-C_2S) or industrial byproducts (e.g., steel slag powder) as the binder component. The printing material is initially processed into a rheologically controlled slurry through water mixing and subsequently printed layer-by-layer. The printed structure rapidly develops high strength and durability through carbonation reaction with CO_2 gas, forming a matrix predominantly composed of calcium carbonate. This CO_2-driven printing material system offers several distinct advantages.

(1) Decoupled Printability and Buildability: The cementitious components exhibit minimal hydration reactivity, ensuring stable rheological properties during preparation and pumping. Rapid strength development and interlayer adhesion are achieved through targeted CO_2 carbonation post-printing, effectively resolving the traditional trade-offs inherent in liquid-driven systems.

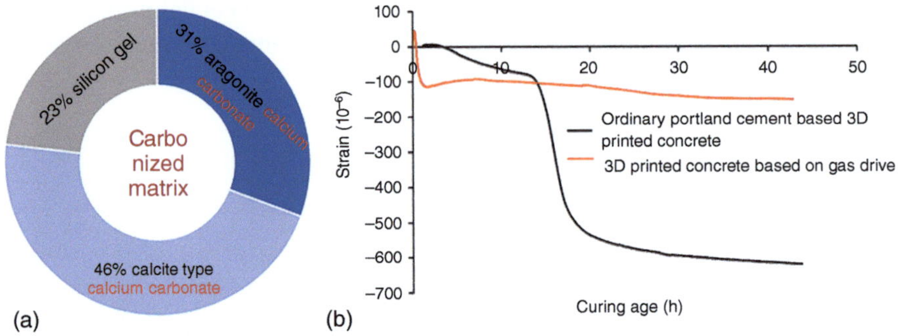

Figure 7.14 (a) Matrix composition and (b) volume stability of CO_2-driven 3D printing concrete.

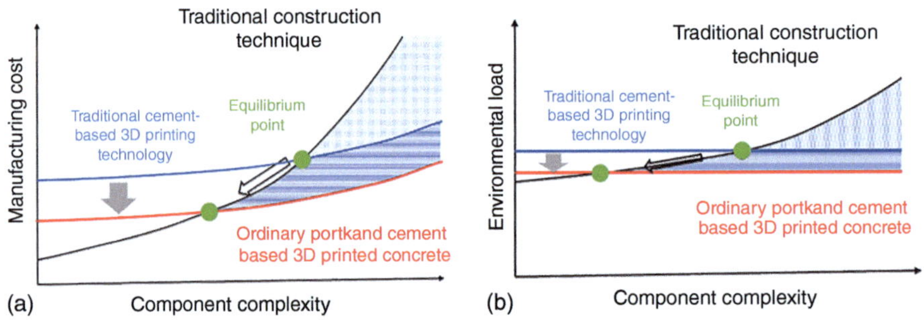

Figure 7.15 Impact of CO_2-driven 3D printing concrete based on (a) manufacturing cost and (b) environmental loads.

(2) Enhanced Durability and Environmental Resilience: The primary binding phase consists of chemically stable calcium carbonate crystals (Figure 7.14(a)). Superior volume stability and crack resistance during carbonation (Figure 7.14(b)) contribute to improved structural durability.

(3) Sustainability and Cost-Effectiveness: The system can incorporate up to 95% industrial byproducts in the binding materials, leveraging their high carbonation reactivity [19–22]. CO_2 sequestration is achieved through its role as a reaction-driving medium. As shown in Figure 7.15 (adapted from Wangler et al. [16]), this approach significantly reduces material costs and environmental impact, advancing sustainable manufacturing in 3D printing construction.

7.3.2 Preparation of CO_2-driven 3D Printing Concrete

The CO_2-driven 3D printing concrete comprises three primary material categories: (1) carbonatable binders, including γ-C_2S, steel slag powder (SS), and ordinary Portland cement; (2) carbonation-enhancing agents, consisting of amorphous silica (SF) and chitin-based additives; and (3) rheological modifiers, specifically polyvinyl alcohol (PVA), sodium carboxymethylcellulose (CMC), and superplasticizer (SP). Table 7.6 presents the optimized mix proportions of this CO_2-driven 3D printing

Table 7.6 Mix proportion of CO_2-driven 3D printing concrete.

SS (g)	Cement (g)	SF (g)	PVA (%)	CMC (%)	SP (%)	W/B ratio
179.4	44.8	28.5	0.2	2	2	0.28

Figure 7.16 Actual printing performance of CO_2-driven 3D printing concrete.

concrete, wherein the dosages of rheological modifiers (PVA, CMC, and SP) are expressed as mass percentages relative to the combined content of carbonatable binders and carbonation-enhancing agents.

The initial mixing procedure of the CO_2-driven 3D printing concrete follows conventional concrete mixing protocols. Subsequently, the fresh mixture is extruded through a 3D printing system, followed by CO_2 curing to achieve structural hardening. Prior to CO_2 exposure, the printed structure undergoes a controlled drying process to achieve an optimal water loss ratio of 60–70%, which is crucial for efficient carbonation. The CO_2 curing process is conducted under mild conditions, and the printed components are placed in a carbonation chamber filled with pure CO_2 (100% concentration) at 0.3 MPa pressure and room temperature for 8–12 hours. The physical appearance of the CO_2-cured printed components is illustrated in Figure 7.16.

7.3.3 Performance of CO_2-driven 3D Printing Concrete

The performance criteria for 3D printing concrete encompass two critical aspects: extrudability and buildability. Extrudability refers to the material's capacity for continuous and uninterrupted extrusion through the printing nozzle without blockage or flow disruption. Buildability, on the other hand, describes the printed structure's ability to maintain geometric stability under the progressively increasing load of subsequent layers throughout the printing.

7.3.3.1 Printing Continuity

The extrudability assessment of 3D printing concrete is quantified by evaluating printing continuity over a 3600 mm filament length, where printing interruptions are recorded as a performance metric. Table 7.7 presents the mix proportions and

Table 7.7 Continuity test ratios and results for CO_2-driven 3D printing concrete.

				Number of printing interruptions		
Ratio number	**W/B ratio**	**PVA (%)**	**CMC (%)**	**Test 1**	**Test 2**	**Test 3**
1	0.14	0.1	0.5	3	3	2
2	0.17	0.1	1	1	2	2
3	0.24	0.1	2	6	9	8
4	0.15	0.2	0.5	1	2	2
5	0.17	0.2	1	2	3	2
6	0.24	0.2	2	4	2	8
7	0.17	0.3	0.5	3	2	6
8	0.18	0.3	1	2	1	4
9	0.25	0.3	2	6	4	5
10	0.28	0.2	2	0	0	0

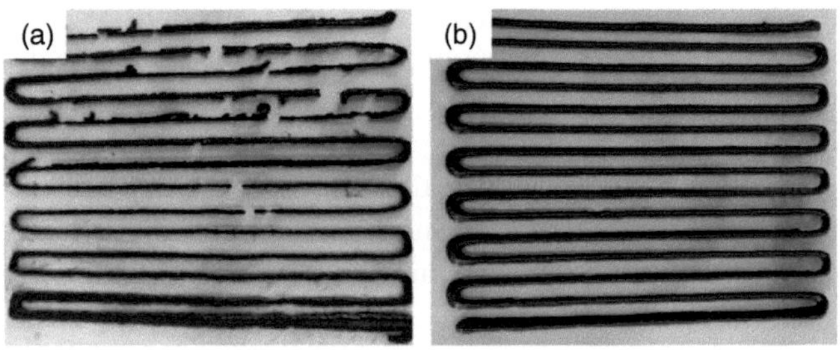

Figure 7.17 Continuity test results of CO_2-driven 3D printing concrete.

corresponding continuity test results. The experimental findings reveal that increasing sodium carboxymethylcellulose (CMC) content progressively deteriorates printing continuity, with a 2% dosage (Mix 3) resulting in significant print interruptions, as illustrated in Figure 7.17(a). In contrast, variations in polyvinyl alcohol (PVA) content demonstrated minimal impact on printing continuity, suggesting negligible influence on mixture rheology. Figure 7.17(b) depicts the successful continuous extrusion achieved with Mix 10, demonstrating uninterrupted filament deposition throughout the test duration.

7.3.3.2 Printing Buildability

In 3D concrete printing, layer-by-layer construction requires each layer, particularly the bottom layer, to maintain stability under multiple forces: its own self-weight, the cumulative weight of subsequent layers, and the interlayer extrusion pressures. The test methodology evaluates the buildability of CO_2-driven 3D printing concrete using a cylindrical specimen (height: 48 mm, outer diameter: 15 mm, inner diameter: 12.6 mm). The assessment focuses on two key

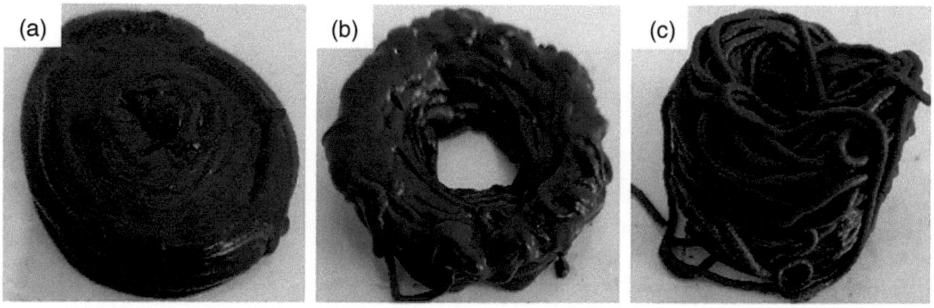

Figure 7.18 Printing stackability t of CO_2-driven 3D printing concrete ((a), (b), and (c) correspond to ratios 1, 5, and 9 in Figure 7.8, respectively).

parameters: print completion degree and vertical stacking accuracy. These metrics are crucial since accumulated deformation errors during layered printing can compromise dimensional accuracy and potentially lead to structural failure. The experimental program utilized the mix compositions detailed in Table 7.7. The initial mix designs demonstrate inadequate buildability, resulting in significant structural collapse, as illustrated in Figure 7.18. The primary mechanism of failure is traced to excessive fluidity: the bottom layers experience flow deformation during initial printing, which destabilizes subsequent layers and leads to progressive deformation, ultimately causing structural collapse. Increasing the sodium CMC content has a positive effect on buildability by enhancing plastic viscosity, increasing yield stress, and improving layer stacking stability. Based on these findings, an optimized mix design is developed with fixed contents of CMC at 0.2% and PVA at 2%. The study then proceeds to investigate various dosages of polycarboxylic acid superplasticizer, with the specific proportions detailed in Table 7.8, to balance extrudability with buildability performance. As shown in Figure 7.19, reducing the superplasticizer content increases the water demand while significantly improving the buildability. At 3% superplasticizer dosage, the bottom layers exhibit satisfactory filamentary stacking, but as printing height increases, cumulative flow deformation causes the upper structure to collapse, preventing complete printing. When the superplasticizer content is reduced to 2%, printing performance improves substantially, enabling defect-free completion of the model that meets requirements. However, further reduction to 1% superplasticizer results in excessive loss of fluidity, making the mixture too stiff for extrusion through the 3D printer.

Table 7.8 Superplasticizer dosages for stackability tests.

Ratio number	PVA (%)	CMC (%)	SP (%)	W/B ratio
1	0.2	2	1	0.35
2	0.2	2	2	0.28
3	0.2	2	3	0.26

(1) Unable to print (2) Complete print (3) Top collapse

Figure 7.19 Stackability test results.

Table 7.9 Stacking height loss values.

Number	Actual measurement height (mm)				Mean value (mm)	Height loss (%)
1	47.81	47.94	48.05	47.78	47.90	0.21
2	46.93	46.83	46.77	46.87	46.85	2.40
3	47.50	47.71	46.81	47.01	47.26	1.54

Based on the optimized mix proportions identified in previous tests, further characterization of buildability is conducted by analyzing stacking height loss to assess printing accuracy ranges, with results presented in Table 7.9. The height measurements reveal consistent values across multiple points for all three thin-walled cylinder specimens, indicating uniform extrusion throughout the printing process without significant stacking errors. When compared to the target model dimensions, the printed specimens show minimal overall height loss ranging from 0 to 2.5%. The low deviation values demonstrate that the developed 3D printing mix proportions meet the required printing specifications for dimensional accuracy and stability.

7.3.3.3 Mechanical Property

In this section, the investigation focuses on how functional admixtures affect the mechanical properties of 3D-printed concrete. The experimental mix designs are detailed in Table 7.10, where admixture quantities are expressed as mass percentages of the total mix, with a consistent water–cement ratio of 0.20. Figure 7.20(a) illustrates that the addition of sodium carboxymethylcellulose (CMC) causes a progressive decline in compressive strength, with a significant 63.3% reduction compared to the control mix when the CMC content reaches 2%. This strength reduction is attributed to two primary mechanisms of CMC: its film-forming and water-retention properties. These properties impede strength development by both preventing carbon dioxide penetration through the capillary structure and limiting carbon dioxide dissolution in water, thereby reducing the extent of carbonation in the sample. Figure 7.20(b) shows that adding polyvinyl alcohol (PVA) at 0.1% content leads to a 7.9% strength increase compared to the control mix. However, further increases in PVA content result in declining strength. This behavior can be explained by PVA's film-forming properties in aqueous solutions. After drying and pretreatment, the

Table 7.10 CO_2-driven 3D printing concrete compressive strength test ratios.

	PVA (%)	CMC (%)	SP (%)
A1	0.2	0	4
A2	0.2	0.5	4
A3	0.2	1	4
A4	0.2	2	4
M1	0	1	4
M2	0.1	1	4
M3	0.2	1	4
M4	0.3	1	4

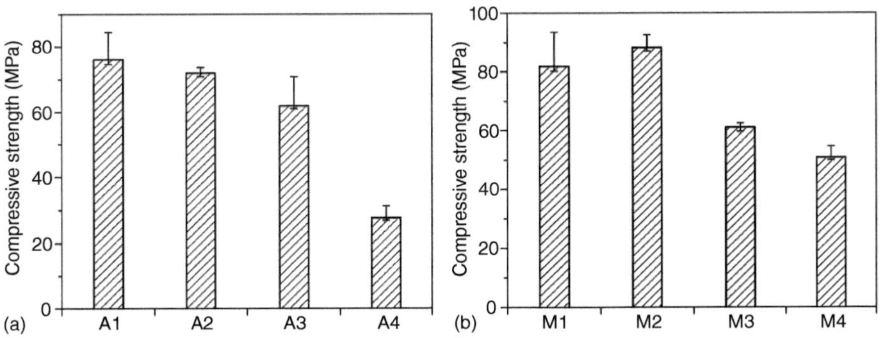

Figure 7.20 (a) Effect of CMC dosage on compressive strength and (b) effect of PVA dosage on compressive strength.

PVA film initially enhances strength by filling internal capillary structures. However, at higher dosages, the PVA film becomes a barrier to carbon dioxide penetration, reducing the extent of carbonation and consequently decreasing strength development.

Figure 7.21 illustrates the inherent anisotropy in direct-writing 3D-printed specimens resulting from the layer-by-layer printing process. This anisotropic behavior arises from the differences between the XY-axis printing direction (within each layer) and the Z-axis printing direction (across layers), leading to a nonhomogeneous internal structure. To quantify this mechanical strength anisotropy in printed specimens, a characterization method is proposed based on the eccentricity formula, as expressed in Eq. (7.5).

$$Q = \frac{2(R_Z - R_{XY})}{R_Z + R_{XY}} \times 100\% \tag{7.5}$$

where Q is the anisotropy parameter, %; R_Z is the compressive strength of the printed sample in the Z-axis direction, MPa; and R_{XY} is the compressive strength of the printed sample in the XY-axis direction, MPa.

Figure 7.21 Difference in stacking of 3D-printed specimens in the (a) *XY*-axis printing direction and the (b) *Z*-axis printing direction.

Table 7.11 Mechanical strength anisotropy test results.

Direction	Strength test result (MPa)			Average value (MPa)	Anisotropy parameter (%)
	Test 1	Test 2	Test 3		
XY-axis	19.11	14.79	18.95	17.62	75.08
Z-axis	40.61	35.89	39.91	38.80	

As shown in Table 7.11, the compressive strength of the printed samples in the *Z*-axis direction is higher than that in the *XY*-axis direction, and the discrepancy index is larger, reaching 75.08%. For 3D printed products, the time interval during the layer-by-layer printing process can affect the accumulation connection between layers. This part is a weak point of the overall structure and is prone to damage under force. When testing the compressive strength in the *Z*-axis direction, the load direction is perpendicular to the interlayer interface, and the interface is mainly subjected to pressure and is not easy to be damaged; while when testing in the *XY*-axis direction, the load is parallel to the interlayer interface, and the tensile stresses generated by the transverse expansion are also parallel to the interface, and the microcracks generated by the force are more likely to be generated in the interface and to develop along the direction of the interface, which results in the compressive strength in the *XY*-axis direction being significantly smaller than that in the *Z*-axis direction.

7.3.4 Trends in 3D-printed Concrete

Currently, CO_2-driven 3D-printed concrete faces stability challenges in printed structure construction. The key to controlling extrudability and buildability lies in two main properties: the shear-thinning rheological properties of freshly mixed slurry and the thixotropic properties that enable rapid, stable structure formation after layer printing. The plastic formation of printed layers before carbonation stabilization is especially critical for maintaining geometry. While conventional methods use passive regulation through coagulants, these struggle to balance competing rheological and plasticity requirements. The authors' previous research demonstrated good electromagnetic sensitivity of nano-Fe_3O_4 particles [23]. Future research could explore doping printed materials with stimulus-responsive media that form ordered structural arrangements under stimulus signals like magnetic fields, enabling active regulation of material rheological properties.

References

1 He, Y. and Hu, S. (2008, July 02). The invention relates to a method for separating waste concrete components. CN 200610019242.0. (in Chinese).

2 Lu, L., He, Y., and Hu, S. (2008). Structural characteristics of dehydrated phase of hardened cement paste and its rehydrating ability. *Journal of the Chinese Ceramic Society* 36 (10): 1343–1347. (in Chinese).

3 Lu, L. (2008). Experimental study on component separation of waste concrete. *Journal of Building Materials* 11 (6): 721–725. (in Chinese).

4 Hu, S. and He, Y. (2007). Preparation of regenerated binding materials using waste concrete. *Journal of the Chinese Ceramic Society* 35 (5): 693. (in Chinese).

5 He, Y. (2006). *Research on Recycling of Cement-Based Solid Waste*. Wuhan: Wuhan University of Technology (in Chinese).

6 Hu, S. and He, Y. (2006). The effect of regenerated binding materials on the hydration system of portland cement. *Journal of Wuhan University of Technology* 28 (10): 4–7. (in Chinese).

7 Wu, J., Ding, Q., and He, Y. (2006). Make high reactivity cement admixture by waste concrete. Cement. *Engineering* (4): 13–15. (in Chinese).

8 Hua, T. (2012). *Modification and Properties of Recycled Cementitious materials*. Wuhan: Wuhan University of Technology (in Chinese).

9 Zhu, M., Yang, C., and Hu, S. (2020, March 31). A recyclable cement concrete and its preparation method. CN 201911112504.1. (in Chinese).

10 Wang, F. and Yang, J. (2013, October 02). The invention relates to water-absorbing expanded resin aggregate concrete and a preparation method thereof. CN 201310276977.1. (in Chinese).

11 Wang, F. and Yang, J. (2013, September 25). A design and control method of concrete hole structure based on internal mold. CN 201310276868.X. (in Chinese).

12 Wang, F., Yang, J., and Cheng, H. (2015). Preliminary exploration on pore structural/functional integration for concrete. *Journal of Building Materials* 18 (4): 608–613+639. (in Chinese).

13 Hu, S. and Wang, F. (2006). *Light Aggregate Concrete*. Beijing: Chemical Industry Press (in Chinese).

14 Ma, B., Wei, D., and Li, X. (2009). Primary investigation on sound-absorption character of low-noise asphalt pavement. *New Building Materials* 36 (08): 18–20. (in Chinese).

15 De Schutter, G., Lesage, K., Mechtcherine, V. et al. (2018). Vision of 3D printing with concrete—technical, economic and environmental potentials. *Cement and Concrete Research* 112: 25–36.

16 Wangler, T., Roussel, N., Bos, F.P. et al. (2019). Digital concrete: a review. *Cement and Concrete Research* 123: 105780.

17 Wang, F., Zhong, K., Liu, Z. et al. (2020, June 12). A gas-driven 3D printing material and its preparation method and application. CN 202010066118.X. (in Chinese).

18 Zhong, K. (2019). *Research on 3D Printing Materials and Technology Based on Carbonization Hardening*. Wuhan: Wuhan University of Technology (in Chinese).

19 Liu, Z., Zeng, H., Wang, F. et al. (2020 February 21). A preparation method of high-performance carbonized reinforced concrete. CN 201911212719.0. (in Chinese).

20 Mu, Y., Liu, Z., Wang, F. et al. (2018). Carbonation characteristics of γ-dicalcium silicate for low-carbon building material. *Construction and Building Materials* 177: 322–331.

21 Mu, Y., Liu, Z., and Wang, F. (2019). Comparative study on the carbonation-activated calcium silicates as sustainable binders: reactivity, mechanical performance, and microstructure. *ACS Sustainable Chemistry & Engineering* 7 (7): 7058–7070.

22 Mu, Y. (2019). *Carbonation-induced Hardening Mechanism of Calcium Silicate Minerals and the Performance-enhancing Strategies of the Carbonated Materials*. Wuhan: Wuhan University of Technology (in Chinese).

23 He, Y., Lu, L., Sun, K. et al. (2018). Electromagnetic wave absorbing cement-based composite using nano-Fe_3O_4 magnetic fluid as absorber. *Cement and Concrete Composites* 92: 1–6.

Index

High-Tech Concrete Materials: Design, Preparation, and Applications, First Edition.
Fazhou Wang and Shuguang Hu.